END PAPER—1. Eastern mountains; the Great Smoky Mountains, Tennessee and North Carolina. After King and Stupka, 1950 (drawn by P. B. King).

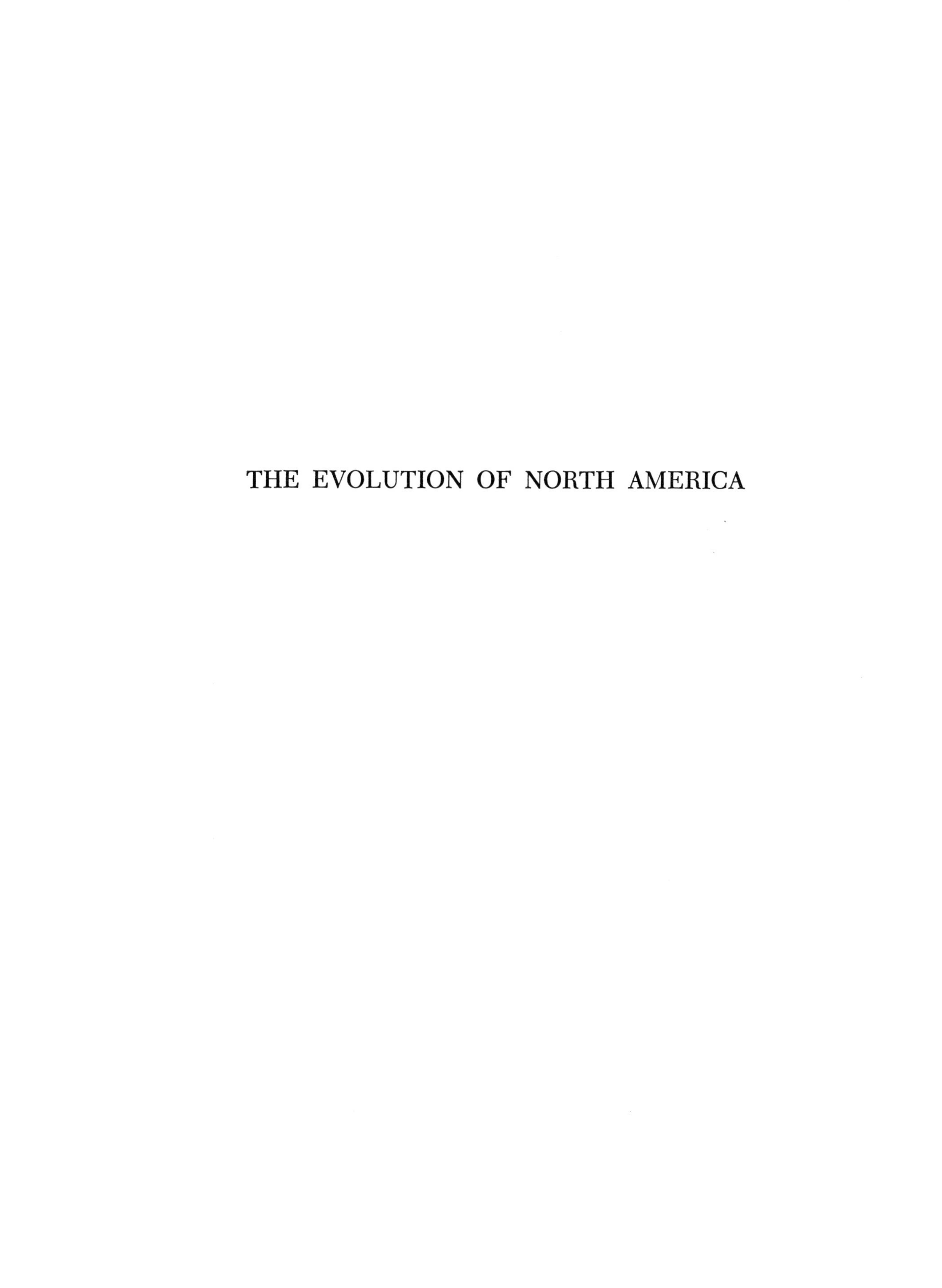

THE EVOLUTION OF NORTH AMERICA

By the same author

◆

THE TECTONICS OF MIDDLE NORTH AMERICA

THE EVOLUTION OF NORTH AMERICA

By PHILIP B. KING

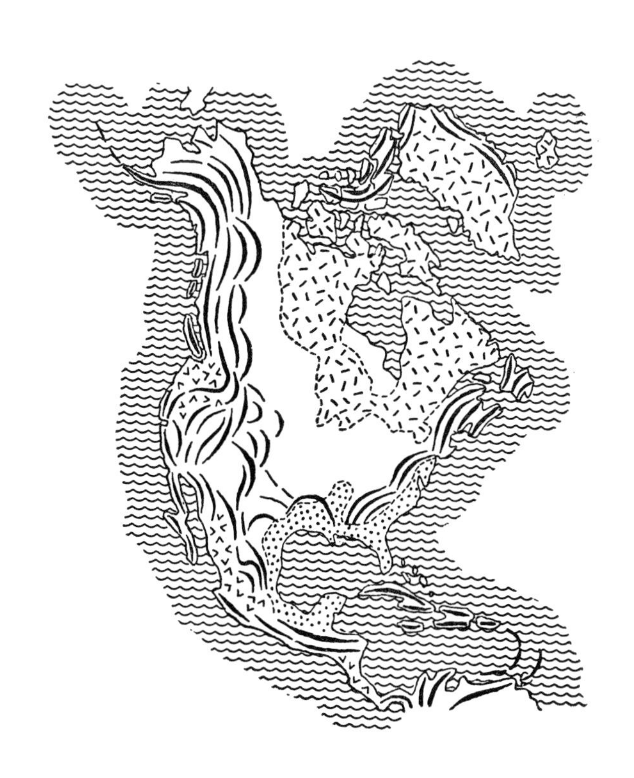

PRINCETON, NEW JERSEY
PRINCETON UNIVERSITY PRESS
1959

The Library of Congress catalog entry for this book appears at the end of the text.

Third Printing, 1961

Dr. Philip B. King has been a geologist of the U.S. Geological Survey since 1930. He has written many journal articles, professional papers, bulletins, and guidebooks. He has compiled and edited several regional maps as well as the first *Tectonic Map of the United States* (1944), and he is currently compiling a Geologic Map of the United States and revising and expanding his previous book, *The Tectonics of Middle North America* (Princeton University Press, 1950).

▪▪▪

Printed in the United States of America
by Princeton University Press, Princeton, New Jersey

PREFACE

For many years I have been interested in the regional geology of North America, and its implications in an understanding of the broader architecture of the earth—partly as an outgrowth of duties for the U.S. Geological Survey, partly as an avocation. A product of this interest has been my "Tectonics of Middle North America," a work in progress, some of which has been published, some not yet completed.

The present book is likewise a regional study of North America, but with a different method and viewpoint. It is essentially the revised text of a one-semester series of lectures, given in 1955 and 1956 while I was visiting professor of geology at the University of California, Los Angeles. Limitations in the amount of material which could be covered in twenty-eight lectures and in what could be comprehended by an undergraduate audience have imposed restrictions of treatment which impart to the book many of its peculiarities and imperfections, and possibly some of its virtues.

The objective of the work, as the title implies, is to set forth the story of North America and its changing aspect through time, back to its earliest beginnings. Instead of presenting this story chronologically, as in historical geology, or discussing the processes which brought it about topically, as in physical geology, treatment is regional. My regional treatment, however, is neither encyclopedic nor an end in itself. Rather, selected regions are treated which illustrate the principles of continental evolution, or significant stages in that evolution. The regions are not the only ones which could be selected for the purpose; others might do as well. They are those with which I am particularly familiar, and are also those mostly from the middle part of the continent (United States and southern Canada), for which the most detailed data are available.

In using the book the reader should therefore be forewarned:

He will probably miss discussion of some favorite area, and he will seek in vain for much information on Greenland, Alaska, Central America, or other out-of-the-way places. But it is thought that principles established in the regions discussed will provide tools for rationalization of the rest.

Also, although the book is based on a college course, the work is not a textbook in the usual sense. A textbook should be more comprehensive and impartial in its coverage both of regions and of theory. Instead, the book sets forth the observations, contemplations, and speculations of one geologist as he looks upon the continent in which he lives. It is avowedly "slanted" in directions of his current thinking, and contains willful prejudices and outrageous hypotheses, some of which may not stand the test of time. These prejudices and hypotheses would certainly corrupt the tender minds of undergraduate students and uninitiated laymen as no textbook should do. But they are honest expressions of opinion and it is hoped that they will be as interesting to read as they have been to formulate.

Philip B. King

Los Altos, California
January 1958

ACKNOWLEDGMENTS

During preparation of this manuscript, one or more of its successive versions have been read by friends and colleagues, including Edgar H. Bailey of the U.S. Geological Survey, Walter H. Bucher of Columbia University, Arthur F. Buddington of Princeton University, Erling Dorf of Princeton University, Warren B. Hamilton of the U.S. Geological Survey, Chester R. Longwell of the U.S. Geological Survey, John C. Maxwell of Princeton University, Peter Misch of the University of Washington, Clemens A. Nelson of the University of California, Los Angeles, John Rodgers of Yale University, John E. Sanders of Yale University, Franklin B. Van Houten of Princeton University, Robert E. Wallace of the U.S. Geological Survey, and Alfred O. Woodford of Pomona College. I herewith express my gratitude to these geologists for the time which they have so generously given, for their encouragement in what has proved to be a long undertaking, and for their many helpful criticisms and suggestions. While I have carefully considered the latter in revising the manuscript, it is hardly to be expected that the critics will agree fully with all the results.

No list of acknowledgments would be complete that did not mention as well the help given by the sixty-eight graduate students in my seminar on structural geology at the University of California, Los Angeles, during the years 1954 to 1956. Together we explored many geological and geophysical problems and the publications thereon, which are the theoretical foundations of this work. The enthusiasm and interest of these young geologists has aided materially in preparation of the work, and to them, one and all, I offer my gratitude and appreciation. Finally, I wish to express my personal pleasure at the drawings which introduce the different chapters, that were prepared for this book by Trudy King.

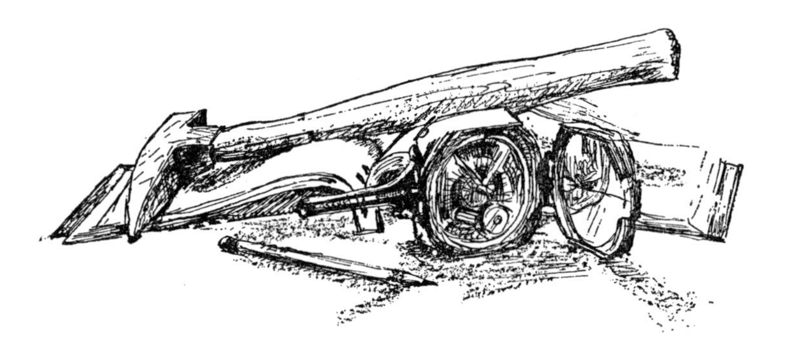

REFERENCE MATERIAL

MAPS

The story of North America rests on various kinds of maps. Some maps are included in this book to illustrate special features, but larger reference maps cannot be, even though they would aid in comprehension of the text. Large reference maps have been compiled and published under auspices of various agencies; if possible, the reader should seek them out and refer to them (see list below).

Geographic maps. North America is represented, first of all, by geographic maps. The most familiar are those which show political and governmental subdivisions, but these give little idea of the fundamental nature of the continent. Better than these are maps which show physical configuration by layer tints, contour lines, or relief shading; in recent years, also, several excellent relief models of the middle part of the continent have become available. The maps used by the author during his compilations were the sheets of the American Geographical Society's "Map of the Americas."

But although physical maps or models give information on the height and steepness of the land, its true character and makeup are difficult to infer; a clearer idea is given by "landform maps" or "physiographic diagrams," in which the distinctive forms of the land are shown in a semi-pictorial manner. Especially recommended is Erwin Raisz's "Landform map of the United States."

All geographic maps represent the appearance of North America at a moment in time—merely one "frame" of the ceaselessly changing and evolving aspect of part of the earth's surface. To understand past history of North America and how it evolved, recourse must be had to other maps.

Geologic maps. Geologic maps show distribution of different rocks which lie at the surface, from which the various land forms have been created. On these maps rocks are partly differentiated as to kind—whether igneous, sedimentary, or metamorphic, and the subdivisions thereof—and partly by ages during which they were formed. Most geologic maps published in the United States follow a conventional scheme of colors. Igneous rocks are shown in vivid tints of red, purple, green, etc., without particular regard for age. Sedimentary rocks, and to a large extent the metamorphic, are shown in a sequence that follows a prismatic scale. Precambrian rocks appear in greys and browns. Paleozoic rocks are shown in reds, purples, and blues—red for Cambrian and Ordovician, purple for Silurian and Devonian, and blue for Mississippian, Pennsylvanian, and Permian. Mesozoic rocks are shown in green—blue-green for Triassic and Jurassic, and olive-green for Cretaceous. Cenozoic rocks—the Tertiary and Quaternary—appear in shades of yellow or orange.

Geologic maps are available on various scales—the map of North America on a scale of 1:5,000,000, that of the United States on 1:2,500,000, and maps of the states on 1:500,000 (or ten times the scale of the geologic map of North America). Maps on still larger scales cover smaller areas in greater detail, showing not only distribution of the rocks by color patterns, but their geometry by structure symbols; such maps are the basic data from which the smaller-scaled ones have been assembled.

From geologic maps the skilled eye can infer many things about the makeup and history of an area—belted outcrops which indicate a homoclinal dip in one direction, concentric outcrops around a dome or basin, truncated outcrops which indicate an unconformity and superposition of unlike structures. Geologic maps, nevertheless, show only surface features, and much of the story is still concealed. To explain the hidden parts of the story, other forms of representation have been tried.

Tectonic maps. For example, tectonic maps are intended to show more clearly the structural makeup of the region. A small-scale example, a much generalized tectonic map of North America, is included with this book (Plate I); larger-scaled, more detailed examples are the tectonic maps of the United States and Canada.

On tectonic maps, folds and faults are represented by special symbols or, where rock configuration is more open, by means of structure contours. Many significant features are brought out which cannot be shown on the usual geologic maps, such as those that have been buried beneath younger deposits or broken up by later deformation. Some areas of special tectonic significance are shown by colors or patterns, such as the surface extent of basement and plutonic rocks of various ages. But because of special requirements of tectonic maps, and as this sort of cartography is newer and less tried, these colors do not follow the conventional scheme of ordinary geologic maps outlined earlier.

OTHER MAPS. Various additional kinds of maps have been used to extract geological history from the rocks. Of these perhaps the most familiar are *paleogeographic maps*, examples of which adorn most textbooks of historical geology. Paleogeographic maps of small areas may give a fairly accurate record of ancient geography, but those of the continent or its larger parts involve so much speculation and inference that they record little more than the opinions of their makers.

In more favor among professional geologists as aids to interpretation are *paleogeologic maps* that show ancient areal geology, *isopach maps* that show rock thicknesses, *lithofacies maps* showing distribution of different kinds of sedimentary deposits, and others, which will be explained more fully in later pages of this book. These are less speculative than paleogeographic maps, as they can be assembled from actual data of outcrops and drill records. The most comprehensive use of these cartographic methods to date is by the Paleotectonic Map project of the U.S. Geological Survey. Results of this project for one of the geological systems, the Jurassic, have appeared at the time of this writing, and a report on the Triassic is in preparation.

The following maps are recommended:

GEOGRAPHIC MAPS:

American Geographical Society, *Map of the Americas* (1:5,000,000): "Alaska, northern Canada, and Greenland" (1948); "United States, southern Canada, and Newfoundland" (1948); "Mexico, Central America, and the West Indies" (1942).

Raisz, Erwin (1939) *Landform map of the United States* (1:4,625,000): Institute of Geographical Exploration, Harvard University.

GEOLOGIC MAPS:

Stose, G. W. (1946) *Geologic map of North America* (1:5,000,000): Geol. Soc. America. (A revised version edited by E. N. Goddard is nearly ready for publication).

Stose, G. W., and Ljungstedt, O. A. (1933) *Geologic map of the United States* (1:2,500,000): U.S. Geological Survey. (A revised version is planned by the U.S. Geological Survey).

Geological Survey of Canada (1945) *Geological map of the Dominion of Canada* (1:3,801,600).

Sanchez Mejorada, Santiago Hernandez (1956) *Carta geologica de la Republica Mexicana* (1:2,000,000): Petróleos Mexicanos, for XX Int. Geol. Cong.

Roberts, R. J., and Irving, E. M. (1957) Geologic map of Central America (1:1,000,000): *U.S. Geol. Survey Bull. 1034*, pl. 1.

TECTONIC MAPS:

Committee on Tectonics, National Research Council (1944) *Tectonic map of the United States* (1:2,500,000): Am. Assoc. Petrol. Geol. (A revised version under the editorship of G. V. Cohee is in progress).

Geological Association of Canada (1950) *Tectonic map of Canada* (1:3,801,600): Geol. Soc. America.

OTHER MAPS:

McKee, E. D., and others (1956) Paleotectonic maps, Jurassic system: *U.S. Geol. Survey Misc. Inves. Map I-175*, 6 p., 9 pl.

PUBLICATIONS ON TECTONICS

For the reader who wishes to explore further the questions discussed in this book, the following suggestions are made:

Reading on general facts and principles of physical and historical geology is desirable. A number of excellent textbooks on these subjects is available, no one of which requires specific mention.

The larger architecture or tectonics of the earth is a more specialized subject, yet even here the number of publications is extensive and has appeared in many languages over a span of many years. Below are listed some of those published in English, especially in recent years. Papers relating more specifically to details are listed later under the appropriate chapters.

Bucher, W. H., 1933, *The Deformation of the Earth's Crust*: Princeton Univ. Press, 518 p.
(As this was published more than twenty years ago, some of its broader conclusions have been outmoded, even by later publications of its author. Nevertheless, its approach to tectonic analysis and its review of specific subjects retain a freshness and

vitality which will be an inspiration for many years to come.)

Umbgrove, J. H. F., 1947, *The Pulse of the Earth*: Martinus Nijhoff, The Hague, 358 p.
(Treats many of the same problems as the book by Bucher and some others, but from a different viewpoint and background, and with the benefit of data that became available in the intervening fourteen years.)

Kay, Marshall, 1951, North American geosynclines: *Geol. Soc. America Mem. 48*, 143 p.
(A comprehensive review of the sedimentary basins of North America and their relation to tectonic processes. Sound in principle, but burdened by an opaque style and a complex terminology. Further reference will be made to this work later in the book.)

Wilson, J. Tuzo, 1954, Development and structure of the crust, *in The Earth as a Planet*: chap. 4, p. 138-214, Univ. Chicago Press.
(A more condensed statement of earth tectonics than those of Bucher or Umbgrove, which sets forth some bold syntheses of classification and interpretation. Interesting reading, even in those parts which do not altogether inspire belief.)

Poldervaart, Arie, ed., 1955, Crust of the earth (a symposium): *Geol. Soc. America Spec. Paper 62*, 762 p.
(Forty-four papers by many geologists, geophysicists, geochemists, and paleontologists on a wide range of subjects that bear on problems of the earth's crust. Many of the papers are highly specialized and no attempt has been made to coordinate them. This in itself affords a good cross-section of current thought on the subject.)

de Sitter, L. U., 1956 *Structural Geology*: 552 p., McGraw-Hill Book Co., New York.
(An advanced text of structural geology. The first half deals largely with details of rock deformation and its problems; the last half (p. 325-518) takes up larger tectonic problems after the manner of Bucher and Umbgrove, and supplements them usefully.)

REFERENCE LISTS

The short lists of references at the ends of the chapters are intended to serve as a guide, should the reader desire to delve further into any subject. Some give basic factual data, others are elaborations of theories propounded in the text, and others take up alternative theories. The lists are neither exhaustive, nor are they intended fully to document the discussions in a particular chapter. They represent reading in which the author himself has found pleasure, instruction, or stimulus.

CONTENTS

LIST OF ILLUSTRATIONS

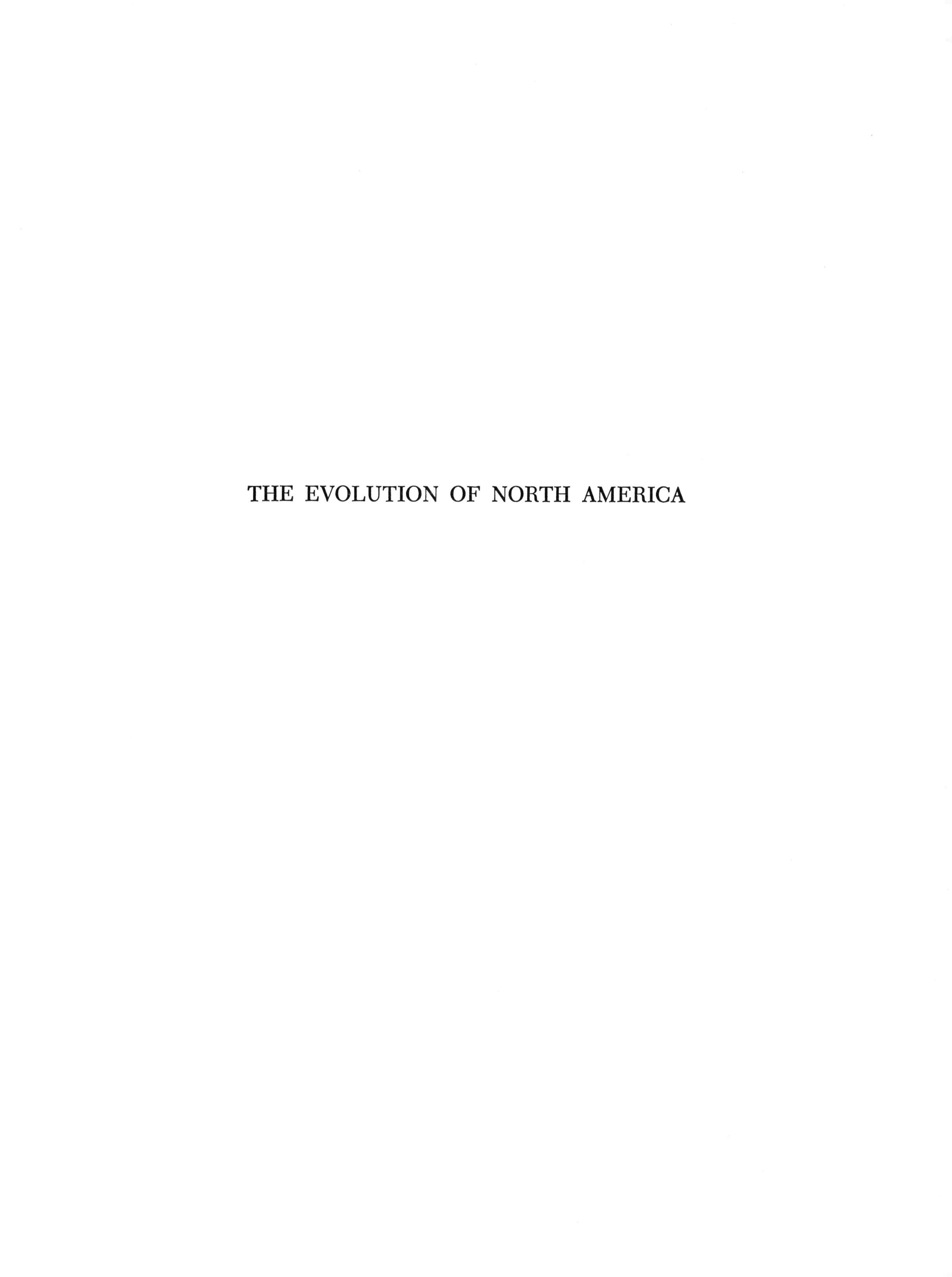

THE EVOLUTION OF NORTH AMERICA

CHAPTER I

THE NATURAL HISTORY OF CONTINENTS

1. LAYERS OF THE EARTH'S CRUST

RELIEF OF OCEANS AND CONTINENTS. To us land dwellers it always comes as something of a shock to realize that the normal surface of the earth is not land, but water. Seventy percent of the earth's surface is covered by water, mainly ocean; only thirty percent is dry land, mainly continents.

These differences are more than an accidental covering of parts of the surface by water, as there are certain remarkable features of the relief of this surface. The following figures have been calculated:

Depths below sea level

Thousands of meters:	−6	5—6	4—5	3—4	2—3	1—2	0—1
Percent area:	1.0	16.5	23.3	13.9	4.7	2.9	8.5

Heights above sea level

Thousands of meters:	0—1	1—2	2—3	3 plus
Percent area:	21.3	4.7	2.0	1.2

From this may be derived the following principle: *The frequency curve of elevations on the earth shows two pronounced maxima, corresponding to the ocean floors and to the continental platforms.*

Referring to our table, note that twenty-three percent of the surface of the earth lies at a level of 4,000 to 5,000 meters (about 12,000 to 15,000 feet) below the surface of the ocean, and twenty-one percent lies between the surface of the ocean and 1,000 meters (about 3,000 feet) above it. Standing out on these dominant surfaces are the mountain ranges, some of which project nearly to 30,000 feet above sea level, and the ocean deeps (that are mountains in reverse), which descend to as much as 35,000 feet below sea level. The continents stand as lofty platforms above the ocean floor, on the whole with very slight relief on their tops. Also significant in the table is the 8.5 percent of area between sea level and 1,000 meters (about 3,000 feet) below sea level. Much of this area lies at depths of less than 200 meters (600 feet) below sea level and forms the *continental shelves.* Although submerged, the shelves are actually parts of the continental structure; in other words, excess of water in the ocean basins drowns the edges of the continental platforms so that, on many coasts, shallow bottoms extend miles out to sea before breaking off into the ocean depths.

COMPOSITION OF CONTINENTS. Why does the earth's surface have this pattern?

Extensive geological and geophysical studies of the continents demonstrate that they consist everywhere, at relatively shallow depths, of granite-like rocks. These include true granites and such plutonic allies as granodiorites, quartz monzonites, and syenites, as well as most of the metamorphic schists and gneisses. It is true that these rocks lie at the surface over only small parts of the continents, much wider areas being covered by sediments. But the latter are a relatively thin blanket, less than a mile to a little more than ten miles thick at most, the granitic layer being much thicker. Besides, the sediments have been derived largely from granite-like rocks and more or less reflect the composition of their source; some sediments, such as certain sandstones, closely approach that composition.

Granites and their allies, we know, are composed dominantly of quartz and feldspar, hence are sometimes called acidic rocks. In chemical terms, their minerals are built principally of oxides of silicon and aluminum. For this granitic or continental type of earth's crust the name *sial* has been coined by combining the abbreviations *Si* and *Al*, representing the dominant elements silicon and aluminum. Sial has an average density of 2.7. Compare this with the average density of the earth as a whole, which is 5.5. Clearly, the stuff of continents is a skin of lighter crust lying on much denser interior material.

COMPOSITION OF CRUST BENEATH THE OCEANS. The crust beneath the oceans is evidently quite different from that beneath the continents. For example, study of earthquake waves indicates that these travel at different rates through the crust of the continents and the crust beneath the oceans, those crossing the oceans advancing more rapidly because of greater density of the sub-oceanic rocks. Moreover, volcanoes in the oceans, such as those which have built up the Hawaiian Islands, erupt basalt rather than lavas of granitic composition. Basalt is a dark dense rock containing such minerals as pyroxene and olivine, and is dominantly composed of oxides of silicon, magnesium, and iron. There are many reasons for believing that much of the sub-oceanic crust has a composition like that of basalt.

For this sub-oceanic type of crust the name *sima* has been coined, as with the word sial, by combining the abbreviations *Si* and *Ma* for its two dominant elements, silicon and magnesium. Sima has an average density of 3.0, and is therefore denser than sial.

Study of the earthquake waves leads us to believe that the bottoms of the continental blocks also become more dense and probably approach the composition of oceanic sima, although there is no sharp boundary between this lower part and the thicker, overlying sialic part. Like cakes of ice floating on water, the continents are broad plates, about 23 miles thick, whose tops rise not only above the surface of the sima layer on the ocean floor, but whose bases extend deeper. A cross-section of the edge of a continent may therefore be indicated diagrammatically as in Fig. 1.

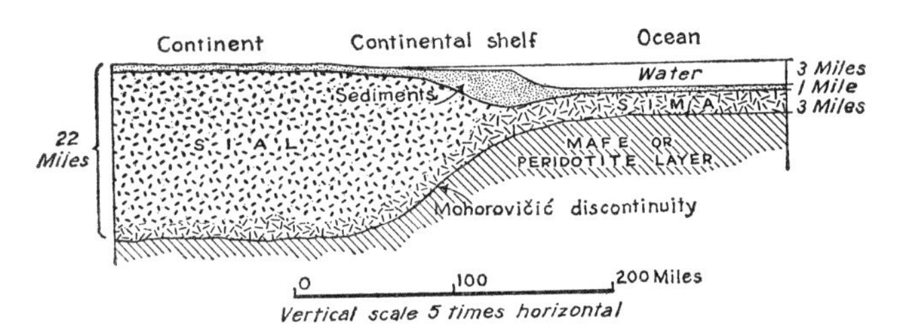

FIG. 1. Diagrammatic section across the edge of a continent into an adjacent ocean basin showing crustal layers which underlie them. After J. T. Wilson, 1954; and Ewing and Press, 1955.

DEEPER LAYERS OF THE CRUST. Although sima is denser than sial, it is still not as dense as the average of the whole earth; still heavier, more compact material must lie beneath it. Moreover, study of earthquake waves indicates that both the sima of the ocean floors, and the sial and sima of the continental plates have well-defined bases; they are separated from heavier underlying material by a fairly abrupt discontinuity lying at a depth of about four miles beneath the ocean floor, and at a depth averaging about 22 miles beneath the continental surface. This was first recognized about sixty years ago by a Yugoslav seismologist with the rather difficult name Mohorovičić, hence has become known to geophysicists as the *Mohorovičić discontinuity*—for short the "M-discontinuity" and irreverently "The Moho." It is the base of the crust of the earth as we know it, and lies beneath continents and oceans alike, as shown by the lower line of our figure (Fig. 1).

Rocks beneath the discontinuity lie far below our limits of observation, and although we can deduce their density from seismic properties, it is obviously difficult to translate this physical fact into terms of mineral and chemical composition. Nevertheless, certain rocks that form rare exposures at the surface—the peridotites—are thought to have been injected upward through the entire simatic and sialic crust, and to have been derived with little change from the layer beneath the discontinuity. Peridotites are even denser than basalt and its allies, with oxides of magnesium and iron dominant, those of silicon and aluminum relatively subordinate. With some hesitancy we can refer to the rocks beneath the M-discontinuity as the *periodotite layer* or (to continue our synthetic terminology) as the *mafe*, by combining the abbreviations *Ma* and *Fe* for magnesium and iron.

Although some notion as to the character of rocks and layers still deeper in the earth is afforded by seismic evidence, we have probed deeply enough for our purpose. Let us now return to the surface features.

2. SURFACE FEATURES OF CONTINENTS

GEOLOGICAL CLASSIFICATION OF SURFACE FEATURES. Our geographies of grade-school days taught us that the surface of the land is divisible into *plains, plateaus,* and *mountains.* This is commendably objective, as it classifies the features purely by form, without becoming involved in theories of their origin.

But our interest here is in origin and evolution, so that it is desirable to convert this classification into something geological. Thus, plateaus are in pretty much the same geological category as plains, the plateaus merely having been lifted higher so that they are more dissected. Moreover, there are some lowland areas, as in the Piedmont province of the southeastern United States, which share the rock structure of adjacent mountain systems and are merely worn down parts thereof; geologically they are more properly included with the mountains. Our classification then becomes:

LOWLANDS

Shields

Areas of exposed ancient rocks; former mountain

belts that were long ago stabilized and worn down.

Interior lowlands

Ancient rocks, covered by later sediments which remain nearly flat-lying.

Coastal plains

Areas of young sediments along coasts, with low seaward dips; outer edges commonly submerged to form the continental shelves.

MOUNTAINS (including parts now worn down into low country)

Folded mountains

Made up dominantly of deformed sedimentary rocks.

Complex mountains

Made up of crystalline rocks, i.e. of plutonics and metamorphics.

Block mountains

Terranes broken by faults into blocks of various shapes and sizes which have been raised, lowered, and tilted.

Volcanic mountains

Built by piling up of ejected material, rather than by deformation.

And various other types

Classification of mountains by origin appears to be more complex than for lowlands, yet actually is less fundamental; the varieties listed above and various others commonly occur together in disturbed regions. In the western mountains of the United States all the varieties are present, mixed together or superposed. The disturbed regions from which mountains are produced may be called *orogenic belts* or *deformed belts.* They are products of crustal unrest, hence during the time of their formation they were *mobile belts.* These belts are by far the most interesting parts of the continents, and we will have much to say about them later.

SYMMETRY OF NORTH AMERICA. With respect to its surface features, North America is almost ideally symmetrical (Plate 1). In the north-central part, mainly in Canada, is the *Canadian Shield,* made up of ancient (that is, Precambrian) rocks, mostly granite but including various kinds of old folded sediments and lavas. The Shield has been worn down into a low rolling surface which passes southward and southwestward beneath sedimentary rocks of the *Interior Lowlands.*

Nearly encircling the Shield and Interior Lowlands are various mountain systems, formed at different times since the Precambrian. On the southeast is the *Appalachian system,* with a partly buried extension across the Mississippi River that closes in the Lowlands on the south. On the west is the *Cordilleran system,* extending the entire length of the continent along the Pacific. In the far north, where geological exploration is just beginning, we know that there are other systems of folded mountains (the *Innuitian* and *East Greenland* of Plate 1) along the oceanward sides of the Arctic Islands and Greenland.

Finally, along the edges of the continent, especially on the southeast, are *Coastal Plains* that were built after the mountains were deformed. They are made up of Mesozoic and Cenozoic sediments laid over the worn down edges of the mountain structures.

From North America it would seem that we could make a generalization about the plan of continental structure. From the central core of the continent, proceeding outward, we would find:

SHIELD—INTERIOR LOWLANDS—MOUNTAINS—COASTAL PLAINS—CONTINENTAL SHELVES—OCEAN BASINS, which would appear diagrammatically as in Fig. 2.

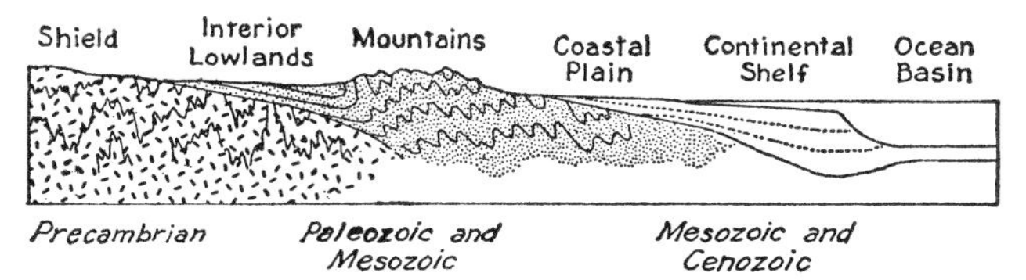

FIG. 2. Section from the center of a continent to its margin showing ideal arrangement of lowlands and mountain belts.

COMPARISON OF NORTH AMERICA WITH OTHER CONTINENTS. Turning to the other continents, though, we find it difficult to apply this generalization.

Asia is perhaps most symmetrical, with a shield and interior lowlands in Siberia and great festoons of mountains surrounding them on the south and southeast, in part submerged along the edges of the Pacific Ocean.

Europe is somewhat like Asia, but curiously separated from the latter by the Ural Mountains which run directly north across the continental block. Europe has a shield area in Scandinavia with plains to the south and east, then mountains near the Mediterranean culminating in the Alps. But Europe is cut off along its western and northwestern coasts where the shield breaks away and the mountain ranges run out into the Atlantic.

Still more anomalous are the continents in the Southern Hemisphere:

Africa is nearly all shield and plains with only fragments of mountains at the extreme north and south.

South America has only mountains on the west and a shield in Brazil on the east.

Australia has a shield and plains on the west and mountains in the east.

From what we know of *Antarctica*, it appears to be lopsided somewhat like South America and Australia.

This rapid survey of the continents of the world suggests that, however beautiful the symmetry of North America may be, it is not necessarily "normal" for all continents. Or, if North America is "normal," most of the other continents are "abnormal." The remainder of this book will deal almost entirely with North America, and certain broad generalizations will emerge; it is well to bear in mind that, before these can be applied to any theory of world structure, they will require checking from facts obtained in other continents.

3. CONTINENTS IN TIME

We have so far examined the continents in terms of their present substructure and surface features. Let us now examine them in terms of another dimension, that of time—geologic time, that is, which reaches back to the earth's beginning.

PERMANENCE OF CONTINENTS AND OCEANS. One of the great debates in geology has been on the permanence of the continental platforms and ocean basins. Have oceans and continents ever changed places? Has our continent of North America ever been part of the ocean, and are there vanished continents beneath ocean waters, such as the lost Atlantis, supposedly sunk to great depths in the middle of the Atlantic?

The reader may recall that much of the record of historical geology from Cambrian time onward has been that "the seas came in and the seas went out"; that almost every part of the continent of North America has at one time or another been submerged beneath the seas—parts many times in successive brief transgressions, parts for very long periods. But this really has little bearing on the gross history of the continent. There is scant evidence that deep sea marine deposits were laid on it, except perhaps near its edges. Most of the seas of the geological record were shallow; the term *epicontinental seas* has been used for them. They resembled the present-day seas which submerge the edge of the continent to form the continental shelves, and Hudson Bay, a great shallow downwarp in which the seas have been admitted to the heart of the continent.

It would appear, therefore, that the continents have been continental for a long time in geologic past. Moreover, the radically different compositions of continental and oceanic crusts suggest that the ocean basins have been at least as persistent as the continental platforms. It seems unlikely that any former continental areas have foundered into ocean basins, for it would be difficult by any known geological processes to convert continental sial into oceanic sima.

Long persistence of continents and ocean basins has led many geologists to assume that they are features inherited from the earth's beginning—the continents are presumed to be a scum or slag of light crust which floated to the surface of the still molten, mainly heavier planet. We shall see presently that this idea may require modification.

THEORY OF CONTINENTAL DRIFT. Together with the question of permanence of the continents and ocean basins, there is another: if the continents are permanent features, have they always remained permanently in their present positions? We have said that the light, sialic continental plates "float" on their substratum like cakes of ice on water. Is this analogy more real than we have supposed, and can the continents possibly have been able to "drift" like floes of ice? This is the theory of *continental drift*, frequently called the *Wegener theory* after the German meteorologist, Alfred Wegener, who cast it into its most complete form.

The idea is outrageous, of course, yet perhaps no more so than others put forward in the past to explain anomalies of nature which have subsequently been proved correct. And there are anomalies of continental pattern and continental history for which there is still no ready explanation.

The asymmetrical appearance of many of the continents, especially in the Southern Hemisphere, suggests that they may be fragments of formerly more extensive continental areas, of which there is now no trace in the intervening ocean basins. There are, as well, strange resemblances between rocks, structures, fossils, and geologic history in the opposing continental fragments, especially between continents on the two sides of the Atlantic Ocean, as though there had been former land connections between them. Perhaps, then, the present continents were all once part of a continuous sialic plate which broke into pieces with time, the parts sliding away from each other across a simatic substratum. No additional land would then be required to provide a transoceanic connection; the land indicated beyond the edge of a modern continent would be the continent on the opposite side of the ocean.

Is the Wegener theory a unifying principle which does for the understanding of earth structure what Darwin's theory of evolution did for the understanding of the plant and animal kingdoms? As when Darwin's theory was first proposed, formidable objections were raised against Wegener's. With Darwin's theory the objections have been accounted for and have melted away so that the theory is now generally accepted as fact. But with Wegener's the objections have persisted and few geologists now accept the theory in its original form. What we know of strengths of different crustal layers and forces that would be required to move the continents seem against the idea of drift. Nevertheless, problems of transoceanic resemblances and connections remain as outstanding riddles of geology for which no satisfactory alternative explanation has been devised. Some part of the Wegener theory may yet find a place in a still unborn master explanation of world structure.

If we were dealing in this book with one of the continents in the Southern Hemisphere we would have to say much more about questions of former land connections and possibilities of drift. North America, fortunately, is enough of a unit in itself that we need not explore the matter much further. Our discussion will proceed in other directions.

SEARCH FOR THE EARTH'S ORIGINAL CRUST. A corollary of the belief in permanence of the continents is another long-held tenet of geology—existence of an original granitic crust beneath the continents. That is, if the continental platforms were differentiated as a slag on the surface of the globe when its rocky crust first took form, the foundations of the continents have always been granitic, and this foundation was the source, directly or indirectly, of most of the younger continental rocks—the sedimentary by erosion, the plutonic and volcanic by fusion. Moreover, if such an original granitic foundation existed, its surface was that on which the first sedimentary rocks were laid.

Much labor has been expended by geologists studying the earth's oldest rocks in a search for some trace of this original granitic crust. All such work has shown, however, that the earth's oldest granites intrude still older rocks which were once surface lavas and sediments. Any original granitic crust on which the latter might have been laid has never been discovered.

Indeed, when the oldest identifiable rocks have been found they turned out to be lavas and their sedimentary derivatives, largely basaltic in composition, hence simatic rather than sialic. For example, some of the earliest ages that have been determined by radiometric methods (see chapter II, Section 3), amounting to about two and a half billion years, are from rocks of northern Ontario and eastern Manitoba. These are plutonic rocks that intrude basaltic lavas, indicating that the latter were still more ancient. Both the igneous rocks and lavas are very early Precambrian in age, five times as old as all time since the beginning of the Cambrian, or the beginning of well-recorded geologic history.

THEORY OF CONTINENTAL ACCRETION. That the earliest identifiable rocks of the continent are basaltic leads to the intriguing possibility that much of the original crust of the present continents was not sial, but sima. This implies that the continental plates evolved through time by *accretion*—that sialic material was manufactured by various earth processes and was built outward from nuclei over the original simatic layer.

Let us speculate on how this might have taken place, but with awareness of our presumption at probing so deeply into the distant past.

If the primitive crust of the earth were all simatic material, its surface would have been much more monotonous than the present one, and most of it would have been covered by ocean waters. But even a nearly homogeneous crust would possess some irregularities. The primitive crust was probably under stress like the present crust, and in places was forced up along lines of buckling and fracture. Basaltic lavas were piled higher in some places than others, and maintained their height for brief periods or over small areas. Also, knots and patches of the crust were perhaps more sialic than the rest, and being less dense, projected above their surroundings. Probably none of these protuberances amounted to very much, but the waters of the primitive ocean had not attained the volume of the present ocean so that even modest irregularities may well have stood above the waters as dry land.

Exposed protuberances were subject to weathering and erosion, part of the product going into solution in the sea, part being deposited as sediments along the edges of the protuberances. Much of the lime, iron, and magnesia of the original sima was thus sorted out and lost by solution, concentrating as sediments what were relatively minor constituents in the source rocks—the silica and alumina.

Part of the sediments laid down along the edges of the primitive protuberances probably accumulated in the first *geosynclines*. We will have much more to say about geosynclines later (see chapter IV, Section 4); suffice it to say here that geosynclines were primarily areas of sedimentation, and especially areas where sediments accumulated to greater thicknesses than usual. Such an excess could occur only where a place was made for the sediments by subsidence of the crust, and this could only happen by failure along a zone of weakness; from their beginnings geosynclines were thus mobile belts. Commonly, by a continuation of crustal mobility, the geosynclinal sediments became deformed, and eventually projected as mountain ranges.

Now, it is a noteworthy fact that all granites and allied plutonic rocks occur in deformed belts, or belts which had a history as mobile belts and geosynclines immediately prior to emplacement of the plutonic rocks. On the tectonic map of North America (Plate I) note the great granitic masses in the Appalachians and Cordillera in areas which had been geosynclines in Paleozoic and Mesozoic time. It is true, of course, that granites and their allies occur in other places—under the flat-lying Paleozoic and Mesozoic rocks of Kansas, for example. But we believe that such places were geosynclines and mobile belts at a far earlier period, before flat-lying sediments were laid over them.

Be that as it may, there seems to be some causal relation between geosynclines, mobile belts, and granites. Thick sedimentation took place during initial stages, accompanied by downwarping of the crust. Downwarping was accentuated by downfolding during deformation, the two contriving to bring the geosynclinal sediments to such depths that they were subject to the internal heat of the earth, and therefore susceptible to transformation into granite. At first the transformed material was merely a static mass of semiliquid rock, but at later stages parts of it became mobile and were able to rise into the overlying deformed and metamorphosed geosynclinal material.

Let us suppose, then, that our primitive geosyncline, filled with sediments more sialic than their source rocks, became deformed and raised into a mountain range, and that the most altered and heated parts were transformed into granite which penetrated widely the higher geosynclinal material. By these processes the former geosyncline became *consolidated* into a sialic crust which was added as land to the initial protuberance.

This cycle from geosyncline to consolidated mountain belt took place many times, and in the early stages perhaps rather rapidly. Most of the Paleozoic and later geosynclines were hundreds of miles broad and thousands long, and required hundreds of millions of years for consolidation. But in primitive times the crust was probably thinner and weaker, so that individual geosynclines were smaller and had a shorter life. Although some of them were haphazardly placed, many seem to have formed next outside a preceding geosyncline, so that more sialic crust was added to the growing continent with each consolidation.

Continental accretion during recorded geologic time. So much is fancy; whether it has any relation to actual happenings in the remote past, we can hardly say. From now on the record is plainer.

A frequently cited illustration of continental growth is on the southeast side of the Canadian Shield, in Quebec, the Maritime Provinces, and New England. In the Labrador Peninsula on the northwest, near the center of the shield, are very old rocks, probably as old as the two and a half billion years mentioned earlier. These are followed farther southeast by the Grenville mountain belt, also part of the shield and thoroughly metamorphosed, granitized, and worn down; its deformation took place about one billion years ago, or twice as long ago as the beginning of the Cambrian. Still farther southeast, across the St. Lawrence River, is the Appalachian mountain belt, deformed in mid-Paleozoic time, or 200 to 300 million years ago. Since this deformation, more sediments have been accumulating on the continental shelf farther southeast. No new mountain belt has formed there, and as we cannot look with certainty into the future there is no way to determine whether one ever will.

For this side of North America we can supplement our earlier cross-section from the center of the shield to the continental margin (Fig. 2) by showing the time relations of the different parts, and thus illustrate the progressive outward growth of continental structure (Fig. 3).

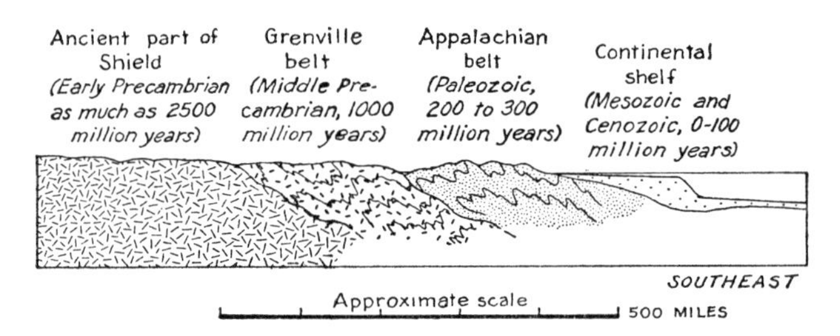

Fig. 3. Generalized section southeastward from Labrador Peninsula to Atlantic Ocean, to show age relations of the Canadian Shield, the Grenville and Appalachian belts, and the Continental shelf.

Do not infer, however, that the continent is visibly expanding before our eyes. In the example cited note that successive accretions were spread over two and a half billion years—most of the span of known earth history. If the process is in operation, it must be inconceivably slow.

Moreover, much of the continental area as we know it probably was consolidated by middle Precambrian time. Radiometric age determinations on Precambrian rocks exposed outside the shield indicate that these were deformed and granitized during or before the Grenville mountain making just referred to. Edges of these consolidated Precambrian rocks extend out well beneath the Paleozoic and Mesozoic geosynclines of of the Appalachians and the Cordillera; ages of about a billion years for the basement rocks are reported as far west as the Grand Canyon and as far east as the Blue Ridge.

Any additions to the continent in Paleozoic and later time were along its present edges. We shall see later that there is some indication that the oceanward sides of the Appalachian and Cordilleran geosynclines were built over a simatic rather than a sialic foundation. These parts were originally ocean floor, but during the last half billion years of earth history were consolidated into mountain belts and added to the continental crust. There is, moreover, indication that parts of the coastal plain deposits are similarly being built over the sima; if so, processes of continental accretion still continue.

So far, we have touched briefly on many fundamental questions of geology. We will elaborate on them further in the succeeding chapters during our discussion of the different regions of North America.

REFERENCES

1. *Layers of the earth's crust*:
Ewing, Maurice, and Press, Frank, 1955, Geophysical contrasts between continents and ocean basins: *Geol. Soc. America Spec. Paper* 62, p. 1-6.
Gilluly, James, 1955, Geologic contrasts between continents and ocean basins: *Geol. Soc. America Spec. Paper* 62, p. 7-18.

2. *Surface features of continents*:
Cady, W. M., 1950, Classification of geotectonic elements: *Am. Geophys. Union Trans.*, v. 31, p. 780-785.

3. *Continents in time*:
Bucher, W. H., 1950, Megatectonics and geophysics: *Am. Geophys. Union Trans.*, v. 31, p. 495-507.
Du Toit, A. L., 1937, *Our Wandering Continents*: Oliver & Boyd, Edinburgh, 366 p.
Mayr, Ernst, ed., 1952, The problem of land connections across South Atlantic, with special reference to the Mesozoic: *Am. Mus. Nat. Hist. Bull.*, v. 99, p. 85-258 (with papers by fifteen authors).
Wilson, J. Tuzo, 1949, The origin of continents and Precambrian History: *Roy. Soc. Canada Trans.*, v. 43, ser. 3, sec. 4, p. 157-185.

CHAPTER II

THE CANADIAN SHIELD AND ITS ANCIENT ROCKS

1. CENTRAL STABLE REGION

Geological analysis of the continent of North America can best begin with its central part, lying within the encircling Paleozoic and younger mountain chains. In contrast to the latter, with their long records of crustal mobility, this part of the continent has been stable since the beginning of Cambrian time—a *Central Stable Region* whose subsequent deformation has seldom been greater than gentle movements upward or downward, or mild warping and flexing.

Basement of this Central Stable Region consists of Precambrian rocks. They are a prime control of its stability, as they are now strong, rigid, consolidated, and capable of resisting deformation of the crust. This consolidation was long in creation, however, as the rocks themselves bear the imprint of an earlier mobility—of formation of successive orogenic belts not unlike younger belts along the present continental margins. In the Interior Lowlands, or outer part of the Central Stable Region, Precambrian rocks are covered by a progressively thickening wedge of Paleozoic and younger sediments, nearly flat-lying or gently dipping. In the Canadian Shield, or nuclear area, the basement emerges and forms most of the surface. It is to the latter area that we shall first turn.

2. SURFACE FEATURES OF THE SHIELD

Location and extent. The Canadian Shield has also sometimes been called the "Laurentian Shield." This designation stems from the southeastern margin, where the shield surface projects along the St. Lawrence River in the Laurentian Upland, but the shield as a whole forms nearly half of the Dominion of Canada, and hence the widely used name *Canadian Shield* is more appropriate.

The shield makes up a sizeable fraction of the northeastern part of North America (Plate I) and has an area of more than two million square miles. It forms the whole Labrador Peninsula, much of the provinces of Quebec and Ontario, the northeastern parts of Manitoba and Saskatchewan, much of the Northwest Territories, and part of the Arctic Islands. Points of the shield extend into the United States in the Lake Superior region and the Adirondack Mountains of New York State. The island of Greenland also seems to be an outlier of the shield, although it is separated by water from the remainder and its surface is largely covered by ice cap. On the far side of Greenland and the Arctic Islands are younger belts of folds like those which elsewhere border the stable region, but as yet we know little about them.

Shields defined. The geologic term "shield" derives from military shields of older times, which had various forms, some roughly heart-shaped, all with a convex curvature. Lying flat, one of these might resemble a geological shield. That is, it should have a hard, convex surface sloping away from the center in all directions and possibly coming to a pointed end in some direction.

Topography. The Canadian Shield has many of the features of an ideal shield although the analogy is not perfect; most of its departures from ideal form result from damage it has suffered since Cambrian time.

It is most like an ideal shield along its south, southeast, and southwest borders. Here, the old surface of the Precambrian rocks rises gently inward away from the sedimentary rocks which cover it in the surrounding Interior Lowlands. In part, the latter form outward-sloping cuestas whose scarps face toward the shield. Lower ground between the scarps and the rising shield is drowned by a remarkable chain of lakes; of these, the Great Lakes between the United States and Canada are the largest and most familiar, but the chain also includes Lake Winnipeg, Lake Athabaska, Great Slave Lake, and Great Bear Lake farther northwest in Canada. Basins of these lakes were excavated in weak rocks along the border of the shield late in geologic time by glacial erosion.

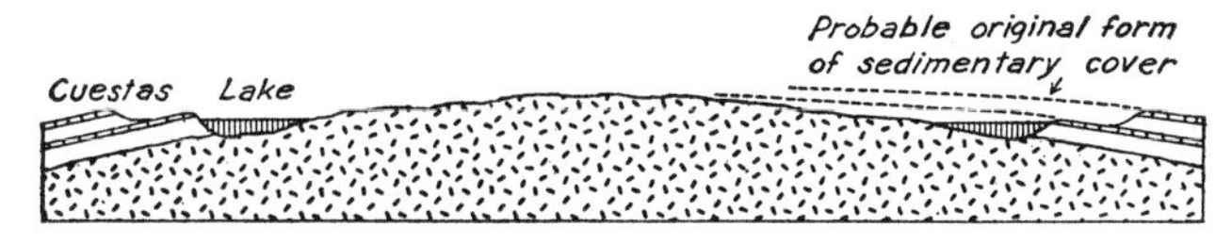

FIG. 4. Idealized section across a shield, to show origin of characteristic features in southern part of Canadian Shield.

Inland from the southern border of the shield the surface of the Precambrian rocks rises into low hills and rough highlands which seldom project more than 1,500 feet above sea level. Much of the southern part of the surface is forested but trees play out northward, and beyond is barren land and tundra. Surface of the Precambrian has undergone erosion intermittently since Cambrian time or even earlier. Parts may have remained bare ever since this remote period, although other parts, especially toward the edges, were covered from time to time by later sediments, most traces of which have now been eroded.

Latest vigorous erosion was during the Pleistocene period when glacial ice cleared off the accumulated soil making the land rocky and sterile, and so disordered the drainage that the surface is dotted with a myriad of lakes. Sterility of the land is compensated for by the mineral wealth of the ancient rocks that has been revealed; the rocks contain valuable deposits of *iron, nickel, cobalt, gold, uranium,* and many other minerals and metals. Mining is thus one of the great industries of Canada.

ACCIDENTS THAT HAVE MODIFIED THE SHIELD. Away from the southern border to the north and northeast the Canadian Shield departs more from the ideal form because of various accidents that have befallen it since the beginning of Cambrian time.

(a) Most recent damage resulted from Pleistocene glaciation, some of whose effects have just been noted. During the Pleistocene the shield was covered by great continental ice caps: one in Labrador, one in the Northwest Territories, and another in Greenland. The first two melted away 11,000 years or so ago, the third remains.

Glacial erosion beneath the ice caps and deposition of morainic material along their edges disordered the drainage, producing many lakes—the larger ones at the edge, the smaller ones in the interior. It also cleared away the surface material leaving wide areas of bare rock, and hence robbed the shield of its soil but exposed its mineral deposits.

Weight of the ice also overloaded the crust, so that in the center of the great ice caps its surface was depressed many hundreds of feet. Although much of the ice has melted, rebound of the crust has been slow, so that parts of the shield are still submerged and much of its surface is still rising.

(b) One of the most conspicuous depressed areas is that of *Hudson Bay*, squarely in the center of the shield—a broad shallow sea that fills a downwarp in the continental surface. Although the modern downwarp may be due mostly to glacial loading, the area seems to have been one of earlier crustal weakness. On the geologic map note the Paleozoic rocks, capped in part by Mesozoic, which cover an extensive area along the southwest shore; note also the Paleozoic rocks of Southampton Island at the north end. These emerged areas of younger rocks probably mark the edges of a shallow basin that extends under much of the bay, and which was filled from time to time by sediments of encroaching seas.

(c) Even greater modifications of the simple form of the shield have taken place to the east and northeast. Labrador faces the St. Lawrence Estuary on the southeast and the Atlantic Ocean on the northeast with bold, forbidding coasts toward which streams flow in canyons as much as a thousand feet deep. In northernmost Labrador, in fact, peaks rise to heights of 5,000 feet above the sea, and have been called the *Torngat Mountains*. But while these may be mountains in a topographic sense, they are hardly "mountains" as we have defined the word geologically; rather, they are merely an uplifted and dissected edge of the Canadian Shield. Similar highly uplifted parts of the shield continue across Hudson Strait into *Baffin Land*, where some of the peaks attain heights as great as 9,000 feet.

It is difficult to escape the conclusion that this part of the Canadian Shield has been broken up by faulting late in geologic time, with the steep, straight coasts of Labrador and Baffin Land raised, and the floor of the seas along their edges downdropped.

(d) The western part of the island of *Greenland* east of Baffin Land consists of ancient rocks like those of the shield; it probably was also once part of the shield. Greenland is, indeed, nearly connected with the rest of the North American continental block at its north end, but farther south it is separated by Davis Strait and Baffin Bay, parts of whose floors lie under 4,000 to 6,000 feet of water. Like Baffin Land, Greenland faces the strait and bay along bold coasts. Separation of Greenland from the rest of the shield may have begun during Cretaceous time, when marine sediments were laid over what is now the western part of the island, but main dismemberment probably took place later, during the Tertiary. The reason for this separation is not clear. If we extended our idea of breakup of the northeastern part of the shield by faulting, we might assume that Davis Strait and Baffin Bay were underlain by great downdropped blocks of shield rocks.

CONTINENTAL BREAKDOWN VERSUS CONTINENTAL ACCRETION. Whatever the cause of separation of Greenland from the rest of the shield, it illustrates a form of continental activity the reverse of that considered in the previous chapter. In general, continents seem to have grown through time at the expense of the ocean basins, by an extension of the sialic plate over simatic crust, through formation of geosynclines, mountain belts, and granitic bodies peripheral to the central nucleus. These additions to the continental area, once consolidated, mostly resisted further changes by crustal deformation. But in some places and under certain conditions, the supposedly permanent continental plates later broke down, in this case by faulting, elsewhere by other forms of reactivation. We will consider other sorts of breakdown later, especially in the region of the Southern Rocky Mountains (Chapter VII, Section 2).

3. INTERPRETATION OF HISTORY OF PRECAMBRIAN TIME

So far, we have discussed the surface of the Canadian Shield without saying much about the rocks which compose it, except that they are ancient, strong, and rigid. We have indicated, though, that while they are strong and rigid now, they bear the imprint of an earlier mobility—of formation of successive orogenic belts, each consolidated in their turn, that were probably not greatly different from those formed in later times around the borders of the present continent. The shield is, then, not a monotonous expanse of indecipherable altered rocks, but is divisible into many parts, each of which has had its own eventful history.

PROPOSED SUBDIVISIONS OF PRECAMBRIAN TIME. In these varied rocks and events geologists have long sought for some story of general significance—for some means of subdividing rocks and time before the Cambrian into systems and periods comparable to those which have proved useful in the younger rocks higher in the geologic column.

The traditional idea, still widely expressed, is a division of Precambrian time into *Archeozoic* and *Proterozoic eras*, or (to use comparable terms) into *Archean* and *Algonkian systems*. Rocks of the first are supposed to be complex to the point of indecipherability, metamorphosed, and full of granite. Rocks of the second are supposed to be more gently tilted, little metamorphosed, and consisting of sedimentary and volcanic rocks not greatly different from those of later periods and systems.

These concepts form the main basis for classifications shown in the table below; they have been used by geologists of the U.S. Geological Survey and the Geological Survey of Canada, and apply to the Precambrian rocks of parts or all of the Canadian Shield (and by implication to Precambrian rocks in general).

CLASSIFICATIONS OF PRECAMBRIAN ROCKS OF THE CANADIAN SHIELD

U.S. Geological Survey (Leith, Lund, and Leith: U.S. Geol. Survey Prof. Paper 184, 1935)	*Geological Survey of Canada* (H. C. Cooke: Canada Geol. Survey Econ. Geol. ser. 1, 1947)
ALGONKIAN TYPE:	*PROTEROZOIC:*
Killarney granite	
Keweenawan	Keweenawan
Upper Huronian	Upper Huronian (Animikie)
Middle Huronian	Middle Huronian (Cobalt)
Lower Huronian	Lower Huronian (Bruce)
Algoman granite	Post-Archaean interval
	ARCHAEAN:
Knife Lake (Classification uncertain)	Timiskaming
ARCHEAN TYPE:	
Laurentian granite	
Keewatin	Keewatin in northwest and Grenville in southeast.

All such classifications involve certain assumptions which, until recently, ran through work done by geologists on the Precambrian of the Canadian Shield. Precambrian time was supposed to have been marked by a series of cycles, each consisting of: (1) sedimentation and volcanism, (2) folding and faulting, (3) intrusion of granite, (4) erosion, followed by a new cycle. It was assumed that each of these cycles and their parts took place all over the Canadian Shield at the same time. There were thus supposed to have been great, universal intrusions of Laurentian granite, later of Algoman granite, and finally of Killarney granite.

But we can well be suspicious of these traditional views of the Precambrian. This is not the way things happened in times since. Later orogenic belts have been long and narrow, and accumulation of thick sediments, their deformation, and their intrusion by granite during any epoch was confined to one or more of these narrow belts. Deformation in the various orogenic belts took place at different times, while other areas were not deformed greatly; thick deformed rock masses at one place are thus equivalent to quite different, thin, undeformed rocks elsewhere. Furthermore, on general grounds we may suspect that Precambrian time was far longer than geologic time since, so that many more such orogenic belts must have been formed successively in Precambrian time than later.

Not much could be added to this picture until recently, because ordinary methods of dating and correlation could not be used in the Precambrian rocks. In later rocks geologic history is based ultimately on *fossils*; significant fossils such as trilobites and brachiopods appear first in the Cambrian. Precambrian rocks are virtually without fossil remains. Some other means of dating the rocks must therefore be found before progress can be made in unraveling Precambrian history.

Radiometric dating. So far, the most valuable of these other means that has been found is *radiometric dating*. Principles on which this is based have been known for some time, but sufficient results to yield significant conclusions have only become available in recent years; even now, many more determinations are needed before we can approach a complete story. Nevertheless, what has been found up to now has revolutionized our thinking about Precambrian history.

As we will have occasion to mention dates obtained by radiometric means from time to time in this book, it is desirable to say something about this method.

Uranium and other chemical elements of high atomic weight have the heaviest atoms and the most complex aggregates of neutrons, protons, and other atomic parts. Being heavy and complex they tend to be unstable and break down by loss of various units that make up the atoms. Since World War II it has been possible to break them down artificially to produce atomic energy.

But in nature there is also a gradual breakdown of the heavier elements which goes on at a constant rate regardless of the environment, the temperature, the pressure, or other outside influences to which the material might be subjected. Uranium goes through various transformations, passing first into radium and giving off the gas helium. Eventually stability is attained in the metallic element lead, still heavy and complex, but much farther down the periodic scale. The lead derived from uranium is an isotope of the element with a slightly different weight from that of lead derived from other sources.

Now, the age of a rock that originally contained its full quota of uranium should be obtainable from the ratio between the uranium still left in it and the isotope of the lead formed from it. The greater the amount of this lead isotope, and the less the amount of original uranium, the older the rock should be.

These calculations yield a figure indicating how many years ago a rock formed, which has been termed its *absolute age* to distinguish it from the conventional age assignment to a named period or epoch in earth history. In terms of human history, it is as if we used a date of 300 A.D., rather than spoke of an event as occurring during the Later Roman Empire.

This method of "absolute" dating sounds simple enough, but there are many complications, some chemical, some geological, which still prevent the results from being absolute in fact.

The chemical complications we need not go deeply into here. As an example of possible difficulties, thorium as well as uranium goes through a series of transformations ending in lead, but at a rate different from that of uranium. If there were both uranium and thorium in a rock, a complex formula would have to be adopted to make allowance for unlike decomposition of two heavy elements rather than one alone.

Geologically one difficulty has been that not all rocks bear uranium, or enough uranium to yield accurate dates. Most of the uranium-bearing rocks that can be measured are igneous and plutonic, and especially the pegmatitic offshoots of igneous bodies which occur as veins or dikes; until recently most radiometric determinations have been made on pegmatites. Refinements in methods, however, have permitted the use of minerals in ordinary igneous rocks that contain small amounts of uranium, such as zircon, and have opened the way for making determinations on almost any plutonic or granitic rocks. Also, it has become possible to measure decomposition of the light element potassium; this leaves as a residue the gas argon which is well preserved in certain minerals—notably the micas which are common minerals in igneous rocks.

Even so, all these determinations are made on igneous rocks, which would not bear fossils in any event, and hence cannot be fitted directly into the conventional time scale. We can determine their relation to this time scale only in case the igneous rock cuts a sedimentary rock that can be dated by fossils or otherwise. Nevertheless, some progress has been made in determining the ages of sedimentary rocks themselves by means of various contained minerals, especially glauconite which is abundant in many sedimentary rocks from the Cambrian onward.

Many of the geological difficulties are thus being overcome, and such varied techniques are being devised each year, making use of unthought of elements and minerals, that the only limit in radiometric dating now seems to be the ingenuity of the geochemists themselves. Amidst this progress the geologist waits eagerly for enough determinations to give him convincing tie points to the geological story which he has deduced by other means.

The following table gives a widely accepted time scale, based primarily on radiometric determinations, that was compiled by the U.S. Geological Survey. This will be helpful for comparison with dates given in the present chapter and others later in the book, although it should be realized that the figures will be subject to modification as additional data accumulate.

Approximate absolute ages of divisions of geologic time as inferred from radiometric data.
(After table by U.S. Geological Survey compiled from reports of Committee on Measurement of Geologic Time, National Research Council, and other sources)

ERAS AND PERIODS	APPROXIMATE NUMBER OF MILLIONS OF YEARS AGO	APPROXIMATE LENGTH IN MILLIONS OF YEARS
Cenozoic		
Quaternary	0-1	1
Pliocene	1-12	11
Miocene	12-28	16
Oligocene	28-40	12
Eocene and Paleocene	40-60	20
Mesozoic		
Cretaceous	60-130	70
Jurassic	130-155	25
Triassic	155-185	30
Paleozoic		
Permian	185-210	25
Pennsylvanian	210-235	25
Mississippian	235-265	30
Devonian	265-320	55
Silurian	320-360	40
Ordovician	360-440	80
Cambrian	440-520	80
Precambrian	520-2,100 plus	1,600 plus

Comparison between Precambrian and later time. Radiometric methods and others suggest that Cambrian time began about 500 million (half a billion) years ago—it may have been somewhat earlier as suggested by the table, or somewhat later, but the round figure is easiest to remember as a standard of comparison.

This does not mean that Cambrian time came in with a bang—any more, say, than the Middle Ages of human history gave way overnight to the Renaissance. No trilobite (any more than some Medieval man) woke up one morning to the dawn of a new era and said to his neighbor, "Look, it is now 500 million years B.C., and today is the beginning of the Cambrian!" Actually, our determinations still do not permit us to know within wide limits the date of the beginning of the Cambrian. Besides, conditions which we associate with the Cambrian probably stole in slowly over an interval of millions of years.

Now, the oldest reliable radiometric dating yields a figure of about 2,500 million (two and a half billion) years. Still earlier dates have been reported but need further checking for their chemical reliability. The earth itself is believed to have originated four or five billion years ago, so that these oldest dated rocks are still nowhere near the earth's beginning.

But consider the figure of 2,500 million years. This means that the earliest rocks which we can date are five times older than all time since the beginning of the Cambrian, or all the time in which, up to now, we have been able to read geologic history accurately. Also, that Precambrian time, or the time we are now discussing, was at least four times longer than the Paleozoic, Mesozoic, and Cenozoic eras combined. Human history, by comparison, is a mere moment, and civilized human history, the last 7,000 years or so, an infinitesimal figure.

This can be put in the form of a diagram, as follows:

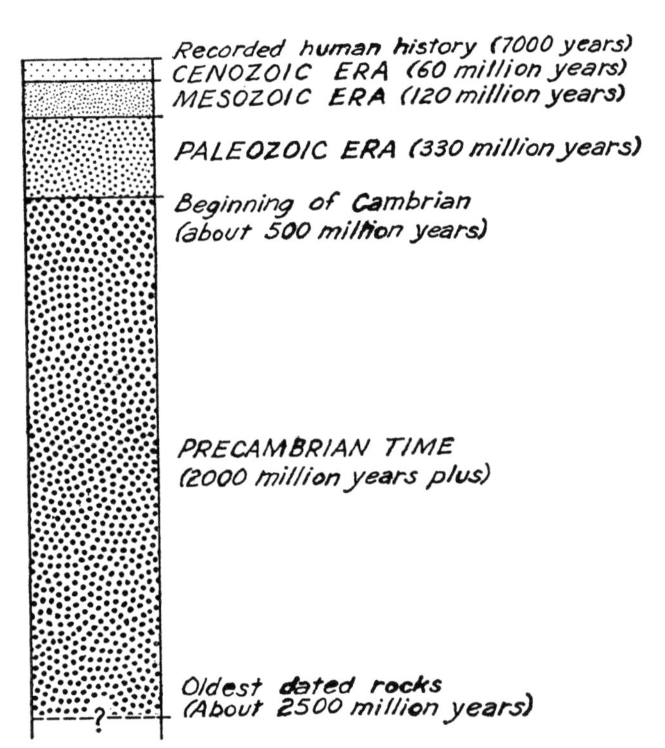

Fig. 5. Diagram showing relative proportion of different divisions of geologic time as known from radiometric determinations and other evidence.

Bearing of radiometric dating on Precambrian history. Consider what this means in terms of the Canadian Shield and the history of its rocks, which was the purpose of bringing up the subject of radiometric dating in the first place:

(a) The relatively few, simple divisions of the Precambrian commonly assumed must have slight relation to its true history.

(b) Instead of the few, universal cycles that have been inferred, there were probably many more cycles which were local in their effects.

(c) Cycles of one area that have been correlated with those of another may very well not be equivalent.

An outstanding example of the latter is the widely accepted correlation of the Laurentian granite. *Sir William Logan* in his pioneer work on the Precambrian rocks of the Canadian Shield in the 1840's and 1850's found extensive masses of granite in the Laurentian Highlands immediately northwest of the St. Lawrence River in the southeastern part of the shield, and therefore called them by that name. When *Andrew C. Lawson* began his work in the 1880's in the Lake of the Woods region more than a thousand miles to the northwest in the south-central part of the shield, he found equally extensive bodies of granite which he assumed were the same as those of the type area, hence called them Laurentian also. But radiometric determinations show that the true Laurentian has an age of about 1,100 million years, and the assumed Laurentian has an age of about 2,500 million years and is thus two and a half times as old. Lawson's assumption of their contemporaneity, though probably justifiable at the time, stands as one of the greatest miscorrelations in geological investigations.

(d) Finally, we can conclude that actual correlation of Precambrian events from place to place, in the shield or elsewhere, must await many more radiometric determinations than are now available.

4. PRECAMBRIAN ROCKS OF THE CANADIAN SHIELD

Kinds of Precambrian rocks. With this theoretical background in mind, let us see what the rocks themselves tell us of the time before the Cambrian in the Canadian Shield. We will speak here mainly of the Precambrian rocks in its southern part, where most information is available, but should keep in mind that this is probably only a fragment of a much larger and more complex picture.

To start with, it will be best to put aside any preconceived ideas of general classification and list merely the different kinds of Precambrian rocks. Various assumptions have been made as to relations of the kinds of rocks to each other and their respective ages, as noted below. Such assumptions are valid enough in local areas, but correlation of rocks and events over wider areas needs more verification from radiometric data.

Two principal kinds of Precambrian rocks occur in the shield:

(I) *Plutonic rocks*, which form perhaps the greatest surface area. They are mainly granites and granite gneisses, but in the Grenville area on the southeast they include quartz-poor syenites and the peculiar rock anorthosite. The plutonic rocks are never a primary granitic crust; at one place or another each body cuts through and invades some surficial rocks, although it may form the basement of others. (The plutonic rocks have been called variously *Laurentian, Algoman,* and *Killarnean.*)

(II) *Surficial rocks*, that is, sediments and lava flows of various kinds with associated shallow intrusives, the latter mainly diabase and gabbro. Although their areal extent is less than that of the plutonic rocks, they provide more clues as to the interpretation of Precambrian history. They may be divided into groups on the basis of kind, structure, metamorphism, and their relations to rocks above and below them (Fig. 6).

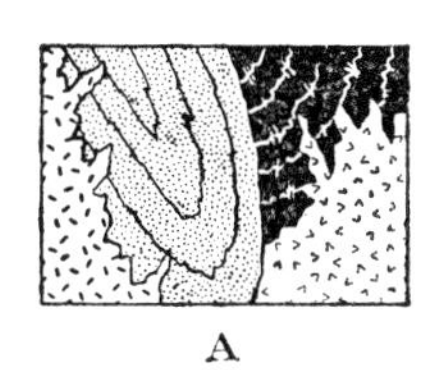
A

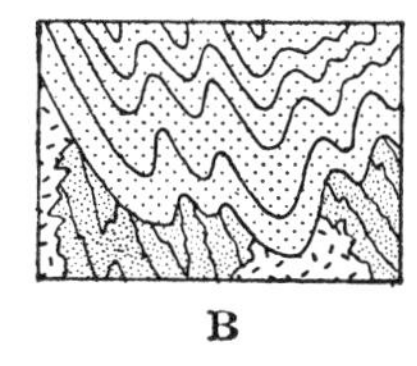
B

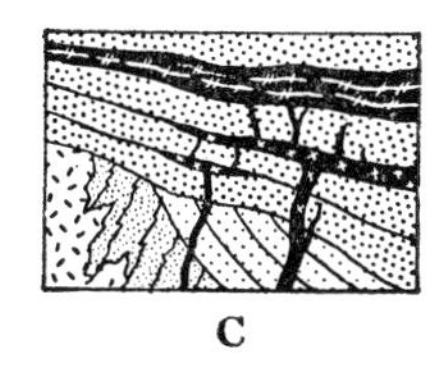
C

Fig. 6. Sketch sections showing characteristic structure of the three classes of surficial rocks in the Precambrian of the Canadian Shield which are described in the text.

(A) Complex, steeply tilted, generally heavily metamorphosed surficial rocks which form narrow belts in the dominant granitic terrane. In most of the shield they are poorly sorted clastic sediments, and lavas that are mainly basaltic. Large parts of both sediments and lavas accumulated under water and were probably marine; some may have been subaerial. (The sediments are commonly termed *Timiskaming, Knife Lake,* or *Coutchiching*; the lavas, *Keewatin.*) On the southeast the surficial rocks (called *Grenville* series) include much marble and quartzite.

(B) Less tilted, less metamorphosed, mostly openly folded sediments with fewer lavas. These sediments are better sorted than the preceding and include slates, quartzites, limestones, dolomites, and iron formations; most were subaqueous and probably marine. They were laid over a foundation of earlier granites, sediments, and lavas, and are infolded with them in widely spaced orogenic belts. (They are commonly called *Huronian* or *early Proterozoic.*)

(C) Gently tilted or flat-lying sediments and lavas, unaltered or little metamorphosed. The sediments are mostly coarse arkosic clastics, partly of subaerial and continental origin. They form irregular patches over other rocks of the shield and occupy the smallest areas of any. Whatever their true age, they are clearly the latest Precambrian rocks of their respective areas. (They are commonly termed *Animikie*, i.e. latest Huronian or *Keweenawan*, as well as *later Proterozoic.*)

PROVINCES OF CANADIAN SHIELD. Further treatment of Precambrian rocks and history can best proceed by dividing the Canadian Shield into provinces, mainly by differences between the foundation rocks. Each province forms a great block in which the foundation rocks differ in kind and structural pattern. The provinces are mostly separated by major faults, some of which are thrusts, others of which are transcurrent and have brought about lateral shifting between the blocks (Fig. 7).

In several of the provinces the surficial rocks in the foundation are basaltic lavas and their sedimentary derivatives; available radiometric dating suggests that such lavas and sediments are the oldest of any (i.e. more than 2,000 million years), and that they may approach closest to an original crust. The largest basalt-bearing province, the *Superior*, lies in the south-central part of the shield north of Lake Superior and south of Hudson Bay, across Manitoba, Ontario, and Quebec. A similar but smaller province, the *Yellowknife*, lies in the far northwest between Great Slave and Great Bear Lakes.

Between the Superior and Yellowknife provinces, in parts of Alberta, Saskatchewan, Manitoba, and Northwest Territories, is the *Churchill province*, distinguished from the other two by a different trend of its structures, as well as by greater amounts of sediments and lesser amounts of lavas in its surficial rocks. Radiometric data indicate that this province was consolidated somewhat later than the first two, perhaps about 1,900 million years ago. This part of the shield has long been stable, however, as the foundation rocks of the Churchill province are overlain near Lake Athabaska by young-looking, little-deformed sandstones cut by uranium-bearing veins that have yielded surprisingly ancient dates—nearly as old as the dates obtained from their basement.

Northwest of the Yellowknife province is the *Great Bear province*, resembling the Churchill province, whose consolidation was of the same age or a little later. Only a small part of it is exposed, and it passes beneath the Paleozoic cover on the west. Other provinces that have not yet been differentiated may occur in the far north and northwest.

Southeast of the Superior province along the southeast edge of the shield is the *Grenville province*, whose distinctive syenites, anorthosites, marbles, and quartzites we have already noted. Rocks like those of the Grenville province occur in the Adirondack Mountains of New York State, and also apparently border the northeast edge of the shield through Labrador and perhaps into Baffin Land. As mentioned earlier, rocks of the Grenville province were consolidated later than those of the provinces in the center of the shield; radiometric determinations on them yield ages of 800 to 1,100 million years. Note that the thorough deformation, granitization, and consolidation of this province took place at a time when equally ancient sandstones at Lake Athabaska in the Churchill province remained little disturbed.

ANCIENT ROCKS OF SUPERIOR PROVINCE. In much of the shield the geologic and tectonic maps show wide areas of pink color, representing granites and granite gneisses, crossed by narrower bands of other colors. In the Superior province the latter trend generally eastward, and consist of sediments and volcanics of the first class listed above, highly folded and steeply tilted. Some of the bands are poorly known, others are probably still undiscovered. But in northern Ontario and northwestern Quebec some of them contain gold ores, and great gold-mining camps are located upon them—Porcupine, Timmins, Kirkland Lake, Noranda, and others. Near these, the bands have been thoroughly explored and mapped.

The belts of sediments and volcanics are fragments of the earliest geosynclines known. All must be old; some are certainly very old, as at places in northeastern Manitoba and northwestern Ontario they are cut by pegmatitic offshoots of granitic rocks which have been dated at about 2,500 million years (the figure mentioned earlier); the surficial rocks cut by them are older.

The surficial rocks include great thicknesses of lavas, largely basaltic, but with rarer flows of more acidic or sialic composition; many of the flows were spread out under water, as they are marked by pillow structure. Associated sandstones are characteristically "dirty" or poorly sorted, and of the type called *graywacke*. Unlike the more familiar sandstones which are made up of well-worked grains, mainly quartz with a little cement, fragmental material of the graywackes is of all kinds, shapes, and sizes, and includes many rock fragments, some derived from the volcanics. The graywackes are products of rapid erosion of the source area, and transportation directly to the site of deposition without washing or reworking, where they were dumped into the water and settled helter-skelter to the bottom.

Graywackes younger than the Precambrian such as as those of the Franciscan group along the Pacific coast (see Chapter IX, Section 4), occur only in the earth's mobile belts, where mountains were uplifted so rapidly and troughs subsided so deeply that the sediments could not be modified by any sorting. In earliest Precambrian time such conditions appear to have been universal rather than special and local.

These earliest surficial rocks have been called by such names as Coutchiching, Keewatin, Timiskaming,

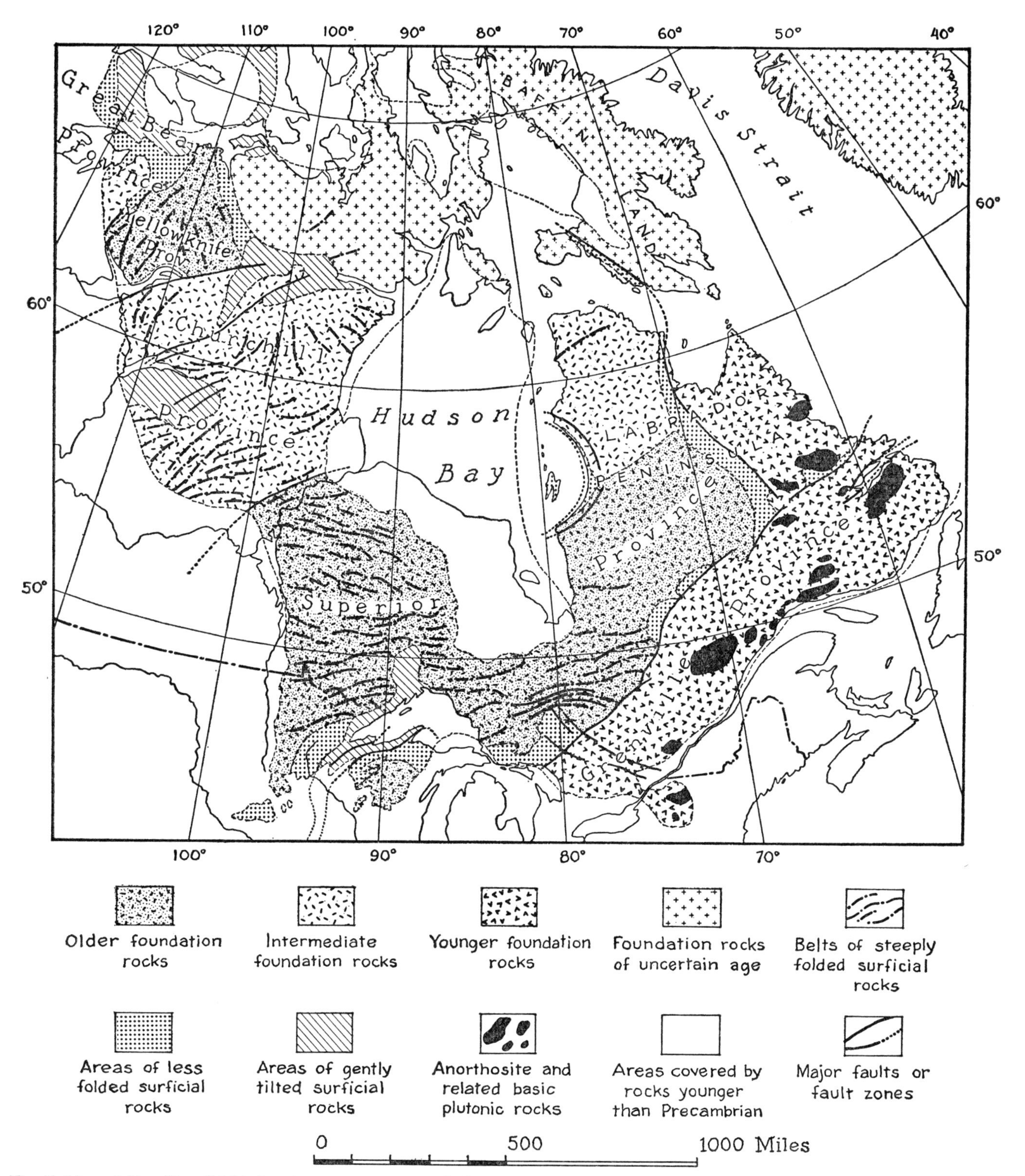

Fig. 7. Map of Canadian Shield showing provinces into which it is divisible and distribution of the different kinds of Precambrian rocks. Compiled from Tectonic Map of Canada, 1950; with additions from Gill, 1949; and J. T. Wilson, 1955

and Knife Lake—Keewatin for lavas, Coutchiching for sediments beneath lavas, Timiskaming and Knife Lake for sediments above lavas. But there is no assurance that rocks so called are of the same ages in the various belts. The lavas might occur at different levels from one belt to another, so that sediments beneath or above them may be low or high in the section, depending on locality. Probably rocks of the various belts formed at different times in the successive ancient geosynclines; some may have been earlier, some later than the determined date of 2,500 million years.

The pink areas of the maps are largely gneissic granite, but they may include belts of surficial rocks as yet undiscovered, as well as even more ancient sediments and volcanics, now so altered and granitized as to be nearly unrecognizable. Much of the granite seems to have welled up in geanticlines between the ancient geosynclines. All, so far as known, intrude earlier surficial rocks, but the probability of several ages of intrusion is suggested by unconformities by which later surficial rocks overlie earlier granites and invaded surficial rocks; some sediments invaded by granites contain cobbles of a granite of a still earlier cycle. The granites have been divided into the Laurentian and the Algoman, supposed to represent two great epochs of intrusion widely separated in time; undoubtedly the actual story was much more complex.

These surficial and plutonic rocks of the Superior province form one of the most ancient parts of the shield and of North America. A frequent fallacy which the geologist encounters in traveling about the country is the claim of local people that this mountain or that is "the oldest in North America." The geologist is puzzled as to what this is intended to mean. Does the mountain contain the oldest rocks? Or was the deformation of its rocks the oldest? Or is the present topographic mountain the oldest, and has it remained there through time immemorial? Such claims usually have little merit. But by any standards, the rocky country north of Lake Superior represents the *oldest known mountains of the continent*—old in rocks, old in deformation, old even in topography. Of course, this country is no longer truly mountainous in a topographic sense, having been worn down so many times that it is now merely hilly at most.

These ancient rocks are of the sort commonly called *Archean* or *Archeozoic*, and where later rocks are present they overlie them unconformably. A. C. Lawson, who worked and thought much on the rocks of the Canadian Shield, has written that "we look across the Eparchean unconformity at the ruins of an archaic world." This is a vivid, poetic picture, yet the skeptic wonders whether the break between older and younger rocks is as absolute as thus implied—whether the various observed unconformities are all of the same age, and whether there are not rocks in the shield which the geologist would be hard put to classify as between "Archean" and something later.

LATER PRECAMBRIAN ROCKS OF THE SUPERIOR PROVINCE (THE HURONIAN). Lying on the foundation of ancient granitic and surficial rocks of the Superior province are rocks of the second class listed above which cover smaller and more widely spaced belts. They were first observed years ago on the north shore of Lake Huron by *Sir William Logan*, pioneer geologist of Canada, hence have acquired the name *Huronian series*. Similar rocks elsewhere in the Superior province have also been so called; while these are all broadly of middle Precambrian age they are mostly so separated from the area of the original Huronian that precise correlation is not assured.

Rocks like those of the original Huronian reappear farther west—south of Lake Superior in Michigan and Wisconsin—but are separated from the original area by a 180-mile gap covered by Paleozoic rocks. Rocks apparently equivalent to those south of Lake Superior are exposed again northwest of the synclinal basin beneath the lake in Minnesota.

The Huronian sequence south of Lake Superior is broken into several parts by slight to moderate unconformities, so that it can be divided into a Lower, Middle, and Upper series. Deformation was thus progressive and the whole body has been steeply folded with a foundation of "Archean" rocks raised in the cores of the uplifts; the deformed belt south of Lake Superior has been termed the *Penokean Ranges.* Granitic dikes which cut the folds have been dated radiometrically at more than 1,400 million years; the Huronian itself is older. Northwest of Lake Superior only the Middle and Upper Huronian are present (the latter here sometimes called the Animikie). In this area, away from the Penokean deformation, the Huronian rocks are not folded, but are merely tilted southeastward at low angles toward the Lake Superior basin.

Parts of the Huronian south and northwest of Lake Superior much resemble sedimentary rocks of later ages; the Lower and Middle series consist of quartzite, dolomite, and slate. The thicker Upper series, however, is mainly slate, graywacke, and interbedded lava, not unlike deposits of the earlier Precambrian.

The only exceptional rocks, and those which give the Huronian its greatest economic interest, are the *iron formations*, the main sources of iron ore in North America. These were laid down as iron-bearing chert or siliceous shale which, when metamorphosed, became a rock called *taconite* containing perhaps ten to thirty percent iron. Through the ages, outcropping edges of the iron formations have been leached by weathering so that some of their constituents have been removed and the iron enriched, creating a high-grade ore that is nearly pure iron oxide.

The enriched belts of iron formation are the *iron ranges* of the Lake Superior region. Those to the northwest in Minnesota such as the Mesabi and Cuyuna Ranges are most prolific, as they occur in low-dipping beds. An elaborate economy has grown up around them—the great iron mines where ore is taken out of open cuts by power shovel, the railroads to Lake Superior where the ore is loaded into ships, ship transport down the Great Lakes to ports on Lake Erie where the ore is again carried by rail to steel mills in Pennsylvania and Ohio.

Vast though these iron deposits are, they are not inexhaustible. We hear now that the high-grade ore of even the great Mesabi Range will soon be mined out, and of how technological difficulties have been overcome so that mining can begin on the leaner taconite beneath.

One of the most interesting things about these middle Precambrian formations, however, is the recurrence of rock types—including the iron formations—in widely separated parts of the Superior province; as we have seen, this may or may not indicate similarity in age. Iron ore thus occurs in middle Precambrian rocks near the center of the Labrador Peninsula. It is now being developed for mining as a replacement for the dwindling supplies of the Lake Superior region; a railroad has been built from the deposits to a port on the Gulf of St. Lawrence; it is planned to ship the ore thence to steel mills in Pennsylvania by way of the St. Lawrence Waterway now under construction.

LATEST PRECAMBRIAN ROCKS OF THE SUPERIOR PROVINCE (THE KEWEENAWAN). The final chapter of Precambrian history in the Superior province is recorded by the *Keweenawan series*, a body of rocks of the third class listed above. The Keweenawan is more restricted in extent than any of the rocks heretofore considered, being exposed largely in the vicinity of Lake Superior; it also underlies much of the lake and probably extends a long distance farther southwest under the Paleozoic cover, as we shall see later Chapter III, Section 3).

Within this local area the Keweenawan was piled to a thickness of at least 50,000 feet (ten miles); much of the lower half is basaltic lava, the upper half is coarse red feldspathic sandstone derived from erosion of older parts of the shield roundabout. Both lavas and sediments probably formed in a terrestrial environment—that is, were laid down on land rather than under water. To make a place for this enormous body of surficial rocks, their floor subsided rapidly during accumulation, producing a basin that has much the same proportions as modern Lake Superior and its shores (Fig. 8); very likely neither lavas nor sediments were spread far beyond their present limits. After its accumulation the Keweenawan was little deformed or altered, and although its rocks have been broken by great faults, the form of the original basin is largely preserved.

While the flows were being poured out on the surface, the underlying part of the Keweenawan succession was filled by thin to thick sheets of basic intrusives. The largest of these is the *lopolith* of the Duluth gabbro, 140 miles long and nearly 50,000 feet thick, whose edge emerges along the northwestern shore of Lake Superior. In Ontario to the east a smaller lopolith of norite forms the saucer-like Sudbury basin. Radiometric determinations on the Duluth gabbro indicate that its age is about 1,100 millon years, which may be approximately the age of the whole series of basic intrusives.

The Keweenawan contains *copper deposits*, notable as having been the first to be mined in the United States. Copper occurs as flecks and masses of various sizes in the lavas and conglomerates of the series; the copper is *native* rather than combined in sulfides as in the great copper deposits of the west, discovered later. Being pure metal it can be extracted and used without smelting, and hence was even dug on a small scale by the Indians long before the coming of the white man. Copper artifacts are found in burial mounds and other Indian remains well over the Mississippi Valley, whither they had arrived by

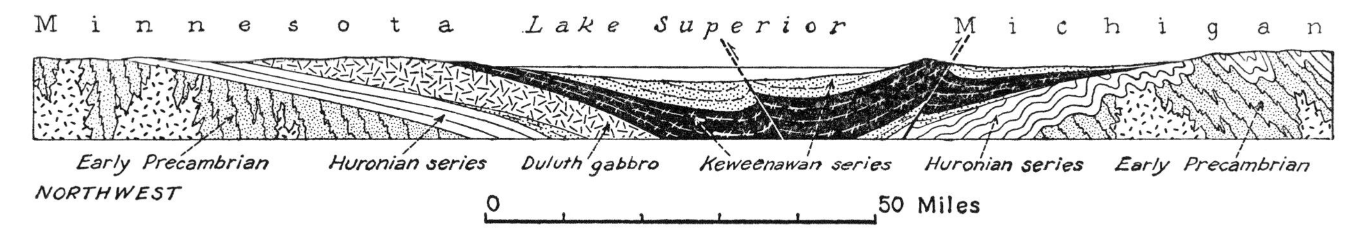

FIG. 8. Generalized section across Lake Superior basin to show structure and relations of Keweenawan and Huronian series. Compiled from Leith, Lund, and Leith, 1935.

trading from tribe to tribe. Besides the copper deposits of the Keweenawan the basal part of the Sudbury norite contains masses of *nickel-bearing sulfide* which are the principal source of that metal in North America.

The Keweenawan formed long after the Huronian of the same region had been laid down and deformed, differences in their ages being suggested by respective dates of more than 1,400 and of 1,100 million years for intrusives which cut the two series. In places the unconformity between the Huronian and Keweenawan is not very marked, but this is because of the variable deformation and metamorphism of the former—greatest on the southeast, less on the northwest.

The Keweenawan series is succeeded by thin, flat-lying, fossiliferous Upper Cambrian strata that rest on its tilted, faulted, and truncated structures. It is thus older than the Late Cambrian, but how much older? The freshness of the Keweenawan series—its lack of folding and metamorphism—has persuaded many geologists that it formed very near the beginning of Cambrian time, if not during Early Cambrian time itself. The recent determination by radiometric means of the age of the Duluth gabbro and related intrusives as about 1,100 million years indicates, however, that the whole series is very much older than anyone had expected, and that it formed long before the beginning of the Cambrian.

In the southern part of the shield, at least, breakdown of the crust to form the Lake Superior basin in which the Keweenawan accumulated was the final event before the region attained stability. Probably it is no accident that the basin formed over part of the earlier Huronian structures; the Keweenawan is *post-orogenic* to the Penokean Ranges of the latter. We shall encounter similar very thick, very local, basin-like, accumulations within younger orogenic belts. Nevertheless, it is somewhat odd that the belt of crustal disturbance thus created was never disturbed again; other belts which we will discuss later underwent an eventful subsequent history.

Rocks and structures of Grenville province. The Grenville province, mentioned several times earlier, borders the Superior province on the southeast, forming a belt 250 miles wide peripheral to the central part of the shield from Lake Huron northeastward along the St. Lawrence River through Ontario and Quebec into the Labrador Peninsula. Rocks like those of the Greenville province also occur, as noted, along the northeast side of the shield and in the Adirondack uplift of New York State.

The name Grenville originates in Grenville Township in southwestern Quebec; this is the type locality of the *Grenville series*, the most distinctive rock component of the province.

The Grenville series is a body of sediments quite different from the surficial rocks of the foundation of the Superior province. Limestones, now altered to marbles, occur in great masses. Some contain streaks of graphite derived from original carbonaceous material; small amounts of natural gas, a hydrocarbon, have been detected also; both suggest the existence of life of some sort during formation of the Grenville deposits. Besides, the limestones contain lenses of anhydrite and gypsum which seem to have been original deposits; if so, they are the first known evaporites, or precipitates from sea water of a kind that formed widely in Paleozoic and later times. Associated with the limestones are quartzites, or altered sandstones derived from cleanly washed sand, quite unlike the poorly washed graywackes of the Superior province. Interbedded with the altered limestones and sandstones are schists and gneisses, probably likewise of sedimentary origin; perhaps they were originally shales although they are now so altered that their initial character is obscure.

The steeply upturned, altered sedimentary rocks of the Grenville province must attain a great thickness. Geologists once believed that they might be as much as ten or twenty miles thick; later corrections for structural duplication reduce this to the still formidable figure of four miles, or 20,000 feet.

Grenville sediments must have originated under conditions quite different from those of the early surficial rocks of the Superior province. Instead of being derived from erosion of basaltic and other lands and deposited with little sorting in narrow troughs, Grenville sediments were probably laid down after much reworking in a broad geosyncline. Except for their extreme metamorphism they are much like the thick limestones, sandstones, and shales laid down in Paleozoic and Mesozoic times in parts of the Appalachian and Cordilleran geosynclines.

The Grenville series is intruded, replaced, and filled by great volumes of plutonic rocks—the *original Laurentian* of Logan—which, as stated, have yielded radiometric dates between 820 and 1,100 million years. The plutonic rocks are mainly concordant with the sediments, forming domes or sheets around which the sediments are wrapped.

Plutonic rocks of the province are as distinctive as the sediments. Their granites tend to be alkalic, that is, high in soda and potash; many are also low in silica and verge on syenites. But the most remarkable plutonic rocks of the province are the *anorthosites* which occur in great bodies (shown in black on Fig. 7). The most famous is that in the Adirondacks, but others as large or larger occur in Quebec and Labrador. Anorthosite is made up almost exclusively of plagioclase or soda-lime feldspar; if any other minerals are present they are basic silicates, the pyroxenes, and amphiboles. It is difficult to picture how anorthosite could have been derived directly from an igneous magma, and its origin has been much debated.

One of the most interesting features of the Grenville province is its contact with the Superior province on the northwest. The two meet along an abrupt line with which the Grenville rocks are parallel and into which the Superior rocks strike from the west (Fig. 7). Where the boundary has been studied carefully it is marked by one or more faults dipping southeast, by which rocks of the Grenville province have been thrust over those of the Superior province; the boundary is probably a fault zone through much or all of its course (it is labeled the *Grenville front or fault zone* on the Tectonic Map of Canada). After the Grenville belt was deformed and consolidated along the edge of the shield, it apparently was thrust inland toward the nucleus of the continent, just as the Appalachian and Cordilleran systems were thrust inland in later geologic time (Fig. 3).

The Grenville series is of uncertain age. Many geologists in the past have correlated it with the early surficial rocks of the Superior province—Keewatin, Timiskaming, and the rest—but the only basis is a similarity in complexity and metamorphism; otherwise they have little in common. All we know positively is that the Grenville series is cut by plutonic rocks as much as 1,100 million years old, and that the surficial rocks of the Superior province are cut by plutonic rocks several times older.

More likely the Grenville is of middle Precambrian age, in the same sense as the Huronian of the Superior province. Whether it is of the same age as the Huronian, older, or younger, does not matter greatly for our purpose; both terms—Grenville and Huronian—have been used so broadly of themselves that their rocks have a considerable time range from place to place. Structurally, anyway, both are closely related. As the reader will recall, the Huronian is steeply folded in the Penokean Ranges toward the southeast and is little deformed farther northwest. East of Lake Huron the Penokean structures extend along the Grenville "front" as if they were controlled by a thrust from the southeast. The Huronian rocks are thus comparable to rocks of the forelands of the later mountain belts—the Pennsylvanian of the Appalachians and the Cretaceous of the Rocky Mountains.

Precambrian rocks of the Grenville province are the first that show the characters of mobile and orogenic belts like those of the Paleozoic and later eras. Their sediments were deposited in a broad geosyncline, fringing the edge of the continent, and were later heavily deformed and injected by plutonic rocks. Toward the end of the orogenic cycle they were thrust inland toward the continental nucleus and the foreland basins of Huronian strata. By these processes the rocks were raised into a mountain belt which fringed the edge of the continent, and after consolidation were themselves added to the mass of the growing continent.

REFERENCES

Greenland and Arctic Islands

Fortier, K. O., McNair, A. H., and Thorsteinsson, R., 1954, Geology and petroleum possibilities in Canadian Arctic Islands: *Am. Assoc. Petrol. Geol. Bull.*, v. 38, p. 2075-2109.

Koch, Lauge, 1936, Uber den Bau Gronlands: *Geologische Rundschau*, v. 27, p. 9-30.

Teichert, Curt, 1939, Geology of Greenland, *in* Geology of North America, volume 1 (Ruedemann and Balk, eds.): *Geologie der Erde* (E. Krenkel, ed.), Berlin, p. 100-175.

Physical features of shield

Cooke, H. C., 1929, Studies of the physiography of the Canadian Shield, (I) Mature valleys of the Labrador Peninsula: *Roy. Soc. Canada Trans.*, ser. 3, v. 23, sec. 4, p. 91-120.

———, 1930, Studies of the physiography of the Canadian Shield, (II) Glacial depression and post-glacial uplift: *Roy. Soc. Canada Trans.*, ser. 3, v. 24, sec. 4, p. 51-87.

———, 1931, Studies of the physiography of the Canadian Shield, (III) The pre-Pleistocene physiographies as inferred from the geologic record: *Roy. Soc. Canada Trans.*, ser. 3, v. 25, sec. 4, p. 127-180.

Precambrian chronology

Collins, C. B., Farquhar, R. M., and Russell, R. D., 1954, Isotopic constitution of radiogenic leads and the measurement of geologic time: *Geol. Soc. America Bull.*, v. 65, p. 1-22.

Collins, C. B., and Freeman, J. R., 1951, Geological age determinations in the Canadian Shield: *Roy. Soc. Canada Trans.*, ser. 3, v. 45, sec. 4, p. 23-30.

Knopf, Adolph, 1957, Measuring geologic time: *Sci. Monthly*, v. 85, p. 225-236.

Lawson, A. C., 1914, A standard time scale for the pre-Cambrian rocks of North America: *12th Int. Geol. Cong. Compte Rendu*, p. 349-370.

Wilson, J. T., 1952, Some considerations regarding geochronology with special reference to Precambrian time: *Am. Geophys. Union Trans.*, v. 33, p. 195-203.

Precambrian rocks of the shield

Cooke, H. C., 1947, The Canadian Shield, *in* Geology and economic minerals of Canada: *Canada Geol. Survey, Econ. Geol.* ser. 1, p. 11-97.

Gill, J. E., 1949, Natural divisions of the Canadian Shield: *Roy. Soc. Canada Trans.*, ser. 3, v. 43, sec. 4, p. 61-69.

———, 1952, Mountain building in the Precambrian Canadian Shield: *18th Int. Geol. Cong. Rept.*, pt. 13, p. 97-104.

Gill, J. E., ed., 1957, The Proterozoic of Canada: *Royal Soc. Canada Spec. Publ. No. 2*, 191 p.

Thomsen, J. E., ed., 1956, The Grenville problem: *Royal Soc. Canada Spec. Publ. No. 1*, 119 p.

Precambrian rocks of the various provinces

Engel, A. E. J., and Engel, C. G., 1953, Grenville series in the northwest Adirondack Mountains, New York, (I) General features of the Grenville series: *Geol. Soc. America Bull.*, v. 64, p. 1015-1048.

Grout, F. F., Gruner, J. W., Schwartz, G. M., and Thiel, G. A., 1951, Precambrian stratigraphy of Minnesota: *Geol. Soc. America Bull.*, v. 62, p. 1017-1078.

James, H. L., 1958, Stratigraphy of pre-Keweenawan rocks in parts of northern Michigan: *U.S. Geol. Survey Prof. Paper 314 C*, p. 27-41.

Johnson, W. G. Q., 1954, Geology of the Temiskaming-Grenville contact southeast of Lake Temigami, northern Ontario: *Geol. Soc. America Bull.*, v. 65, p. 1047-1074.

Jolliffe, A. W., 1950, The northwestern part of the Canadian Shield: *18th Int. Geol. Cong. Rept.*, pt. 13, p. 141-149.

Leith, C. K., Lund, R. J., and Leith, Andrew, 1935, Precambrian rocks of the Lake Superior region: *U.S. Geol. Survey Prof. Paper 184*, 34 p., map.

Norman, G. W. H., 1940, Thrust faulting of Grenville gneisses northwestward against the Mistassini series of Mistassini Lake: *Jour. Geology*, v. 48, p. 512-525.

Pettijohn, F. J., 1943, Archean sedimentation: *Geol. Soc. America Bull.*, v. 54, p. 925-972.

Quirke, T. T., and Collins, W. H., 1930, The disappearance of the Huronian: *Canada Geol. Survey, Mem. 160*, 129 p.

Raasch, G. O., 1950, Current evaluation of the Cambrian-Keweenawan boundary: *Illinois Acad. Sci. Trans.*, v. 43, p. 137-150.

CHAPTER III

THE INTERIOR LOWLANDS, AND THE SCIENCE OF GENTLY-DIPPING STRATA

1. GEOGRAPHY

We pass now to that other part of the Central Stable Region, the Interior Lowlands, where the Precambrian basement is concealed beneath younger, little disturbed rocks—a region about as extensive as the Canadian Shield itself, lying mainly to the south, southwest, and west, forming the vast central part of the United States as well as a wide band in western Canada between the shield and the Cordillera (Plate 1).

In the United States the Interior Lowlands are the region drained by the Mississippi River and its tributaries—the Ohio, Missouri, Arkansas, and others which flow into the Gulf of Mexico. In Canada it is the region drained by the Nelson and Mackenzie Rivers and their tributaries which flow, respectively, into Hudson Bay and the Arctic Ocean.

The Interior Lowlands are largely plains country, mostly standing only a few hundred feet above sea level, widely masked by drift and morainal deposits, and leveled to the north by continental glaciers. Eastward toward the Appalachians and in the Ozarks of Missouri and adjacent states the surface rises into plateaus which have been intricately dissected. Westward the surface rises across the Great Plains also, but with little dissection, and attains altitudes of more than a mile above sea level along the foot of the Rocky Mountains.

2. GEOLOGICAL INVESTIGATIONS

The rather monotonous geologic features of the Interior Lowlands would seem of less interest than the complex rocks and structures of the Canadian Shield, the Appalachians, or the Cordilleras, nor can we, in this book, devote space to them commensurate with their surface area. But millions of our fellow citizens live in the Interior Lowlands and earn their livelihood from the land, either directly or indirectly —from the glacial drift and soils by agriculture and forestry, or from its rocks by exploitation of their fuels and mineral deposits. The lowlands have yielded a sizeable fraction of the nation's oil production; coal measures of Pennsylvanian age are preserved in many of its structural basins; the same basins and others farther west are mined for salt; veins in the upper Mississippi Valley, the Illinois-Kentucky district, and the Tri-State district (Missouri, Kansas, and Oklahoma) contain deposits of lead, zinc, fluorspar, and other valuable minerals.

Much patient labor has been expended to learn the geology of the Interior Lowlands, partly for the sake of research, partly for discovering and developing the mineral deposits; a body of special knowledge has thus accumulated which we might call *the science of gently-dipping strata.* The lowlands have, in fact, shared with the states of the eastern seaboard in the birth of American geology, and it is of interest here to sketch some of its beginnings.

The first real account and map of the geology of the United States—or any part of North America for that matter—was published in 1809 by *William Maclure*, a Scotch immigrant who had become a prosperous merchant in Philadelphia. Although Maclure traveled widely about the country east of the

Mississippi to assemble his information, today his results seem generalized and primitive.

For the next twenty years various other men interested in geology nibbled away at the subject in different areas. There were no true professional geologists in those days, only enthusiastic amateurs—men who had received some education in the professions of the ministry, law, or medicine, and who were curious about the surroundings in which they lived. For the most part their studies were a labor of love, but the various states began to realize the need for an appraisal of their resources as an aid to growth and development, and by the third decade of the century their legislatures were granting funds to geologists to make official surveys. Some of these surveys left an enduring mark on earth sciences, others are now forgotten; of them all, certainly the most notable was that of New York State.

In 1836 the State of New York set up a geological survey. It was organized along rather peculiar lines, the state being divided into four districts, each assigned to a geologist who operated more or less independently of the rest. The men and their assignments were:

W. W. Mather	Southeastern district
Ebenezer Emmons	Northeastern district
Lardner Vanuxem	Central district
Timothy A. Conrad	Western district

The first three districts, nearest the centers of population and with the most varied geology, were considered the choicest assignments. The western part of the state, which was rough pioneer country at the time and a region of gently dipping strata, was considered a poor fourth. At the end of the first year Conrad gave up the western district for other duties and it was assigned to a twenty-five-year old assistant of the survey, *James Hall.*

Out of this seemingly unpromising area Hall forged a career that was to occupy much of the remainder of his life. For here he found the Paleozoic strata to be unaltered and richly fossiliferous, sloping gently to the southwest, laid out in a series of gigantic steps which ascended from Ordovician on the northeast through Silurian and Devonian into Mississippian on the southwest. Here was found the stratigraphic key not only of the more complex and puzzling rocks elsewhere in the state, but of rocks in much of the adjacent part of the Interior Lowlands as well. For many years the western New York section thus became a standard of reference for the whole region. Hall's work on this section and his study of its innumerable fossils has given him an enduring place among the founders of American geology. That he was also perceptive to other geologic problems we will see presently when we discuss the subject of geosynclines (Chapter IV, Section 4).

It was Hall and his colleagues of the New York Survey who began the custom, now universal in American stratigraphy, of naming rock units *after geographic localities where they are typically exposed,* instead of using the earlier quaint, and often misleading, descriptive or mineralogical designations. "Calciferous sandrock" thus became Beekmantown, "Birdseye limestone" became Lowville, and "Corniferous limerock" became Onondaga.

The other geologists of the New York Survey labored valiantly on their supposedly choicer districts, but with less enduring results. They are little remembered today, except perhaps poor Emmons whose strange delusions still have a place in the annals of American geology. Some of the geology in the eastern part of the state has proved to be so difficult that there is little agreement on it even yet.

3. PRECAMBRIAN BASEMENT OF THE LOWLANDS

STRUCTURE OF THE BASEMENT. Before saying farewell to the Precambrian on which we have dwelt in the Canadian Shield, let us remember that the Interior Lowlands differ from the shield merely in possessing a sedimentary cover; that throughout the lowlands the sediments are underlain by Precambrian rocks. Of course we know much less about them here than in the shield as they are largely concealed, but some little patches do emerge—in the Sioux uplift of southeastern South Dakota, the Ozark uplift of Missouri, the Arbuckle and Wichita Mountains of Oklahoma, and the Llano uplift of Texas. At the western edge, also, Precambrian is exposed in the Black Hills of South Dakota and the Front Range of the Rocky Mountains in Wyoming and Colorado. Besides, Precambrian rocks have been reached in many of the deeper drill holes put down for oil and gas in the intervening areas, especially in the western half of the lowlands. Some notion of buried Precambrian structures has been obtained from geophysical data as well.

At this stage it is premature to speculate much on the nature of the buried Precambrian and it offers a promising field for research—piecing together outcrops, drill cores, and geophysical data, and establishing correlations by radiometric determinations. The older Precambrian in many parts of the region, as in the Llano uplift and Front Range, has yielded radiometric ages of around 1,000 million years, which suggest that the Precambrian in wide parts of this hidden region was deformed at about the time of the Grenville mountain making in the shield. It is known also that a belt of large positive gravity anomalies extends southwest from the Lake Superior basin and its Keweenawan rocks (Fig. 7), across Iowa and into northeastern Kansas, and probably marks a prolongation of the Lake Superior structure. Very likely the concealed Precambrian is divisible into provinces in the same manner as the Precambrian of the Canadian Shield, and consists of rocks of similar character and age.

SURFACE OF THE BASEMENT. The gentle surface of the Canadian Shield, cut on Precambrian rocks, passes beneath the Cambrian and younger rocks along its edges, and in the Interior Lowlands forms the floor on which these younger rocks were laid. Much of the planation of the Precambrian was thus accomplished before the Cambrian. In most places in the Interior Lowlands the surface had been reduced to a nearly level plain before sediments were laid over it; a few places retained considerable relief, as in parts of the Ozarks area of Missouri, so that knobs and peaks project into the cover of sedimentary rocks (Fig. 9 B).

Aside from this local relief the top of the Precambrian is broadly undulating, rising toward the crests of domes of the later rocks and descending deeply beneath their basins. These undulations result from later warping of the Precambrian; in fact, they

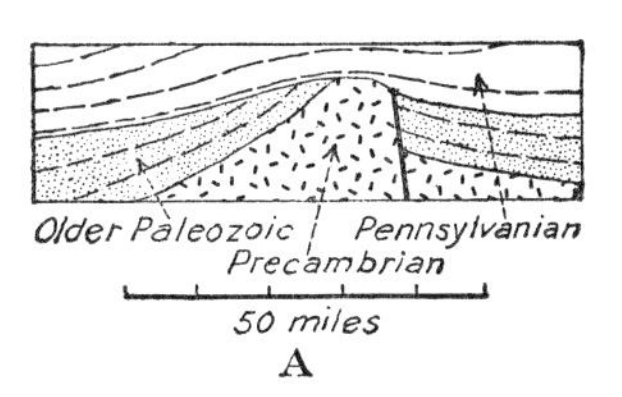

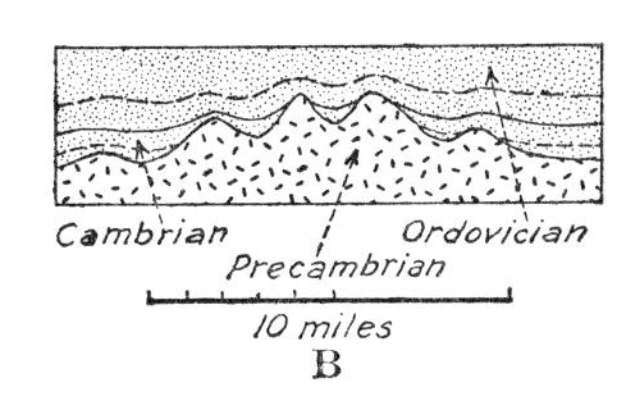

FIG. 9. Sections showing exceptional relief features of the Precambrian surface in the Interior Lowlands. (A) Nemaha uplift, or "granite ridge" of Kansas, produced by uplift and faulting of the surface in early Pennsylvanian time. (B) Buried hills in Ozark area, Missouri, which represent erosional topography of the surface prior to burial by Paleozoic rocks. Note difference in horizontal scales; vertical scales exaggerated. Adapted from Lee, 1956; and Bridge, 1930.

represent the total of all deformation that has affected the region since the beginning of the Cambrian.

In most places undulations of the top of the Precambrian are very gentle, but in some the surface is sharply bent or broken. Extending southward across Kansas into Oklahoma is a narrow strip where Precambrian projects to within a few hundreds or thousands of feet of the surface. This is the "granite ridge" of Kansas, or Nemaha uplift, not shown on the surface geologic maps and buried entirely by later Paleozoic rocks. The ridge is a block of Precambrian rocks that was raised in early Pennsylvanian time (Fig. 9 A).

4. SEDIMENTARY COVER OF THE LOWLANDS

Let us now consider the rocks of the Interior Lowlands which overlie the Precambrian floor. These, we find, are mainly sedimentary and of relatively small thickness—a few hundreds or thousands of feet thick in most places, but exceeding 10,000 in the deeper basins. Basal sedimentary deposits are Cambrian, or if not that, younger Paleozoic. Paleozoic rocks form the surface as far west as the hundredth meridian, beyond which they are covered by Cretaceous and Tertiary strata. A few lava caps occur along the western edge of the Great Plains; igneous intrusions are more widely scattered through the lowlands but are small, sparse, and inconsequential.

INITIAL, OR CAMBRIAN DEPOSITS. Events which took place during Cambrian time in the Interior Lowlands had an important influence on evolution of the continent as we know it, and are worth considering at some length.

Cambrian time appears to have been very long—we have earlier quoted an estimate of eighty million years for its duration, or longer than all time back to the beginning of the Tertiary. Where sedimentation was nearly continuous through the period, its deposits reach great thickness; in the Inyo Mountains of eastern California they amount to nearly 15,000 feet, or three miles (Chapter VIII, Section 2). But these thick Cambrian deposits are all outside the Interior Lowlands in areas which were geosynclines through long periods of Paleozoic and Mesozoic time and later were raised into mountain belts.

In considering these thick Cambrian deposits it is convenient to divide them into a *Lower, Middle,* and *Upper series*, each characterized by distinctive physical history and fossils. The terms Waucoban, Albertan, and Croixan have been used for the same subdivisions, but are not necessary for our purpose.

In contrast to these thick Cambrian deposits of the geosynclinal areas, those of the Interior Lowlands are a thousand feet thick at most, and in many places much thinner; with a few exceptions they are all Upper Cambrian. Typical Cambrian deposits of the lowlands are exposed along the edges of the shield as in Wisconsin and Minnesota, hence the alternative term "Croixan" for the Upper Cambrian, named for St. Croix River between the two states. They are mainly sandstones, probably derived from erosion of the Precambrian of the shield; the sandstones are marine—at many places they contain trilobites, brachiopods, and other sea-dwelling fossils.

In Cambrian time the sea thus covered first the geosynclines along the edges of the Central Stable Region, and through all Early and Middle Cambrian time the Central Stable Region (Interior Lowlands and Canadian Shield) was still land consisting of exposed Precambrian rocks. It was thus not until Late Cambrian time that marine waters began to spread over the central region and to cover large parts of the Interior Lowlands and even some of the shield. These relations of Lower, Middle, and Upper Cambrian to the geosynclines and Central Stable Region show that by the beginning of Cambrian time the latter had developed into a *continental platform* (Fig. 10).

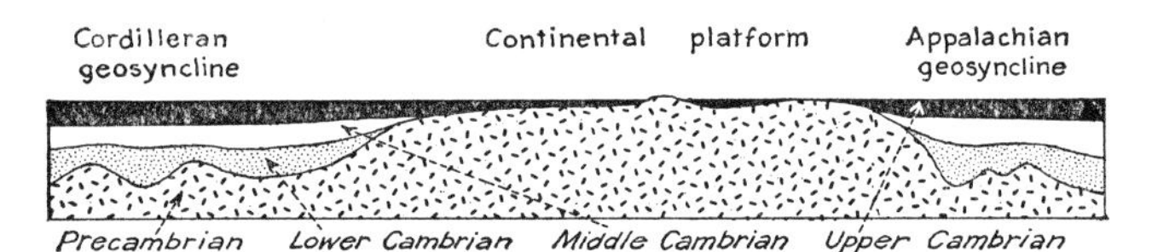

FIG. 10. Section from west to east across middle part of North America at end of Cambrian time, showing overlap of Lower, Middle, and Upper Cambrian series, and how the Central Stable Region acquired the character of a continental platform. Length of section about 2,500 miles; vertical scale grossly exaggerated. Adapted from Kay, 1951.

Abrupt thinning and wedging out of Lower and Middle Cambrian toward the edge of the Central Stable Region (continental platform) has been proved by occasional exposures and drill records, both along the front of the Appalachians on the southeast, and along the front of the Rocky Mountains in Canada on the west.

But there is an exceptional area along the southwestern side of the Central Stable Region. Note on the map that here the front of the Rocky Mountains extends southeast through Wyoming and south through Colorado into New Mexico. This part fails to obey the rule that younger mountain belts formed where troughs of greatest sedimentation existed. Where the Cambrian is turned up along the edges of the ranges it is thin Upper Cambrian, nearly like that in the Interior Lowlands of Wisconsin and Minnesota. The boundary between thin Cambrian and thick Cambrian lies well west of the mountain front near the western edge of the Colorado Plateau in central Utah.

In terms of continental history this means that the continental platform of Cambrian time extended westward beyond the edge of the present Interior Lowlands and included what are now the Southern Rocky Mountains and Colorado Plateau. It thus appears that the mountain structures of these provinces were created by dismemberment or reactivation of a corner of the continental platform after Cambrian time. We will return to this important inference later (Chapter VII, Sections 1 and 2).

CONTINENTAL BACKBONE. Another feature, partly the result of Cambrian overlap, partly a later development, is found in a belt extending southwest from Minnesota to New Mexico; here relatively young rocks generally lie on the Precambrian. This is not a structural feature in the usual sense, and it cannot be seen on any ordinary map; it is known mostly from comparison of stratigraphic sections shown either in surface outcrops or well records. It has been called the *continental arch* or *continental backbone.*

For example, behind El Paso, Texas rise the Franklin Mountains, an uplift which exposes a section that consists, above the Precambrian basement, of sedimentary rocks of Cambrian, Ordovician, Silurian, Devonian, Mississippian, Pennsylvanian, and Permian ages—in other words, representatives of all the Paleozoic systems (Fig. 11 A). But 250 miles to the north near Albuquerque, New Mexico rise the Sandia Mountains, consisting of Precambrian rocks nearly to their crests, overlain unconformably on their tops by Pennsylvanian rocks (Fig. 11 B). All the intervening Paleozoic systems are missing; in ranges between El Paso and Albuquerque it is found that the earlier systems gradually thin and disappear northward. In exposures within a hundred miles of Albuquerque even the Pennsylvanian is missing, so that Permian rocks lie directly on the Precambrian.

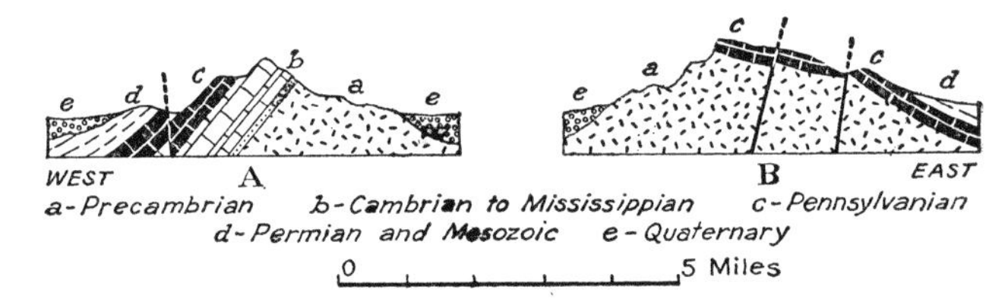

FIG. 11. Sections of Franklin Mountains (A) and Sandia Mountains (B) near El Paso, Texas and Albuquerque, New Mexico, respectively, showing difference in sequence of Paleozoic formations above Precambrian basement rocks, to illustrate stratigraphic relations on south margin of continental backbone. Sections are about 250 miles apart. After Richardson, 1909; and Darton, 1928.

This illustrates conditions near the southwest end of the continental backbone and how they are determined from surface outcrops. Outcrops and drill records indicate similar conditions elsewhere along this feature to the northeast. Cambrian or younger Paleozoic rocks were never deposited over parts of the belt, so that there was a tendency for it to persist as land through successive transgressions of the seas. Parts of the belt are marked by later uplifts along the same trend, so that, if any rocks were deposited on it earlier in the Paleozoic, they have been removed later by erosion.

OVERLAP FEATURES. By the gradual spreading of Cambrian and later seas, their deposits *overlapped* the Stable Region (Fig. 12 A). Each younger layer extended farther inland than the one before, deposits farthest inland being sandy near-shore deposits; these became progressively younger the farther inland the seas advanced. Sand bodies along the edges of the advancing seas are more porous than sediments farther out, and form *wedge belts of porosity* in which fluids such as oil or water can accumulate (Fig. 12 D). Some of the overlapping deposits thus form stratigraphic traps for oil, and have resulted in oil fields more extensive than those in the more familiar anticlinal traps, although they are more difficult to locate by ordinary means of exploration.

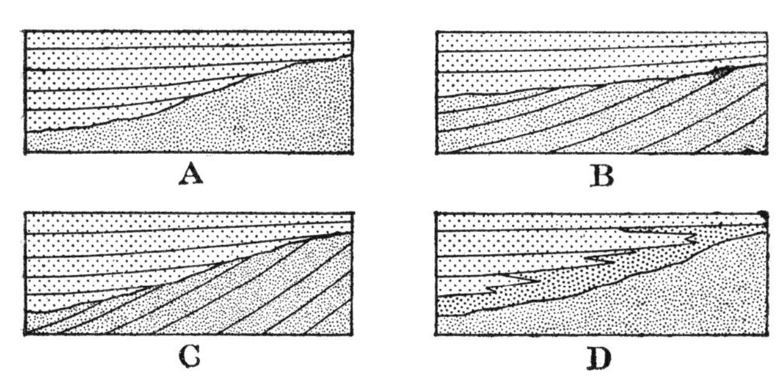

FIG. 12. Sketch sections illustrating: (A) Overlap. (B) Overstep. (C) Combination of overlap and overstep (the most common relation). (D) Wedge belts of porosity.

After Cambrian time successive advancing Paleozoic seas spread their deposits over earlier strata that had been somewhat tilted and eroded. Toward the positive or domal areas these deposits were thus laid across the truncated edges of successively older strata which they *overstep*, lying with small hiatus on rocks little earlier on the flanks, and with great hiatus on much older rocks on the crests (Fig. 12 B).

LATER PALEOZOIC DEPOSITS. Paleozoic deposits of the Interior Lowlands were products of shallow, ephemeral, constantly shifting seas, whose inferred outlines produce a most intricate pattern on the paleogeographic maps contained in textbooks of historical geology.

The lower half of the Paleozoic deposits, of Ordovician to Mississippian age, was of marine origin and includes much limestone, but has subordinate layers of sandstone and shale. The only conspicuous clastic layer is the St. Peter sandstone of Middle Ordovician age, as much as 300 feet thick and lying on an eroded surface of earlier carbonates which covers an area of more than 700,000 square miles in the north-central states. Much of the upper half of the Paleozoic deposits was non-marine and continental, including coal-bearing Pennsylvanian east of the Mississippi River and Permian red beds west of it.

Several exceptional but significant sorts of deposits are worth noting: Some of the marine deposits contain *reefs*, which are mound-like or wall-like masses of limestone or dolomite, built up by lime-secreting corals, algae, sponges, bryozoans, and other organisms, so that they projected above the sea floor on which contemporaneous deposits were accumulating. Reefs are common in the middle Silurian (Niagaran series) through a wide area in the northeastern states, and are mound-like or atoll-like bodies (Fig. 13 A). Devonian reefs occur in Alberta and Pennsylvanian reefs in central Texas, but the greatest reefs of all are those in the Permian of west Texas—barrier reefs or great walls built around the edges of subsiding basins (Fig. 13 B). We will discuss them in more detail later in the chapter (Chapter III, Section 6).

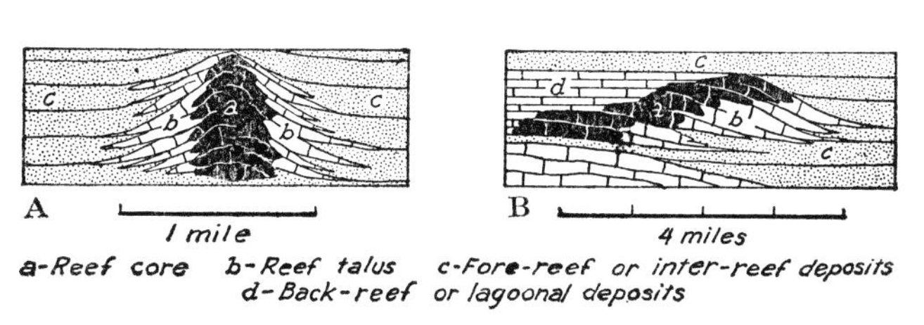

FIG. 13. Sketch sections showing limestone reef structures of Interior Lowlands: (A) Mound type, characteristic of Silurian of Great Lakes region. (B) Barrier type, characteristic of Permian of west Texas region. Note differences in horizontal scales; vertical scales somewhat exaggerated. Adapted from Cumings and Shrock, 1928; and King, 1948.

Reef structures arouse increasing interest because some of them, or deposits associated with them, are reservoirs for oil—a form of stratigraphic trap unlike the one previously mentioned, but equally difficult to locate by ordinary means of exploration.

Other exceptional deposits are the *evaporites* or deposits of gypsum, anhydrite, rock salt, and various potash minerals formed from sea water by precipitation of its dissolved constituents during evaporation. As with reefs, these are not confined to any one area or any particular part of the column. In general they occur at the farthest edge of any sea where barriers of some sort caused part of the sea to be cut off from free access to the ocean. Thickest evaporites are known only from drilling, as their outcropping edges dissolve and weather back so that they are concealed by younger deposits. Evaporites of the Permian are well known and extend from Texas to Kansas. Thick evaporites also occur in the upper Silurian (Salina series) in the northeastern states, notably in New York and Michigan. Others in the Devonian and Mississippian have more recently been discovered by drilling in the northwest from Montana into Alberta.

MESOZOIC AND TERTIARY DEPOSITS. We will say little here about the Mesozoic and Tertiary rocks of the Interior Lowlands, as it is more appropriate to consider them in connection with the Gulf Coastal Plain and the Cordillera (Chapters V and VII).

In the eastern half of the lowlands they occur only in the Mississippi Embayment, a sag extending into the lowlands along the Mississippi River, which allowed Coastal Plain deposits to encroach inland as far as southern Illinois. East of the Mississippi Embayment no Cretaceous or Tertiary deposits are present in the lowlands, and it is doubtful whether any part of them was submerged or covered during these later times.

Cretaceous deposits are extensive in the western half of the lowlands, however, forming a wide tract along the front of the Rocky Mountains from Texas into Canada. They are products of a sea that encroached inland from the Gulf of Mexico and spread eastward from the Cordilleran geosyncline. In places the Cretaceous is succeeded by continental or lacustrine early Tertiary deposits, and the whole is covered by a thin blanket of later Tertiary sands and gravels which forms the surface of the Great Plains.

5. STRUCTURES OF THE SEDIMENTARY COVER

DOMES AND BASINS. So much for the rocks of the Interior Lowlands—now, what of the structures in this area? In single outcrops the strata appear flat-lying, or so gently inclined that their dip is difficult to detect with the eye. Actually, they slope at very low angles in different directions for long distances, as one will find when he views all the outcrops of a larger area. Pattern of their structures comes into focus on small-scale geologic maps such as those of the United States or of North America.

In general, one would expect the dip of the strata to be away from the Canadian Shield and toward the bordering mountain belts, with sedimentary cover thinnest near the former and thickest close to the latter. Broadly, this is true. Cambrian and Ordovician rocks are exposed around most of the edges of the shield, whereas Pennsylvanian and Permian rocks lie along the Appalachians on the southeast, and Cretaceous and Tertiary rocks along the Cordillera on the west.

But the outward dip of the strata from shield to mountain belts is complicated by a series of domes and basins, well displayed in the map pattern (Plate 1).

To the north near the shield are the *Wisconsin* and *Adirondack domes* in which Precambrian basement rocks emerge due to stripping of the sedimentary cover which once arched over their crests. Farther south the large *Ozark dome* of Missouri and adjacent states and the small *Llano uplift* of central Texas bring smaller areas of Precambrian to the surface, surrounded by outcrops of Cambrian and early Ordovician rocks. East of the Mississippi are the *Cincinnati* and *Nashville domes* in which neither Precambrian nor Cambrian are exposed, but which reveal wide areas of Ordovician rocks on their crests.

The basins include the nearly circular *Michigan basin* which occupies the lower peninsula of that state and preserves Pennsylvanian rocks in the center (Fig. 16 F), closer to the edge of the shield and farther north than one would anticipate. The *Illinois basin* to the south also retains a wide area of Mississippian and Pennsylvanian rocks in the center.

FORELAND BASINS. So far we have not spoken of the larger and more extensive basins which follow the edges of the mountain belts. Because of their special relations to the latter, areas in which they occur are called *forelands*. Relation of foreland basins to mountain belts is twofold:

(1) During earlier stages of growth, when initial mountain ridges were being uplifted and eroded, clastic sediments were spread away from these over the foreland, tapering inland like wedges (Fig. 14 B).

(2) During the final stages of mountain building the forelands themselves were mildly folded and were downwarped into basins parallel to the mountain front (Fig. 14 C).

On the map the most conspicuous foreland basin on the southeast is the *Allegheny synclinorium* which

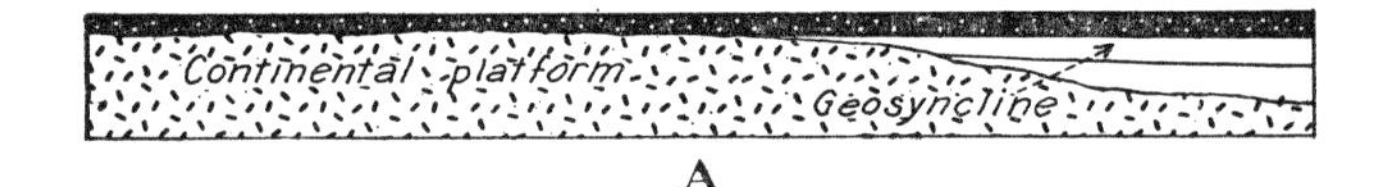

A

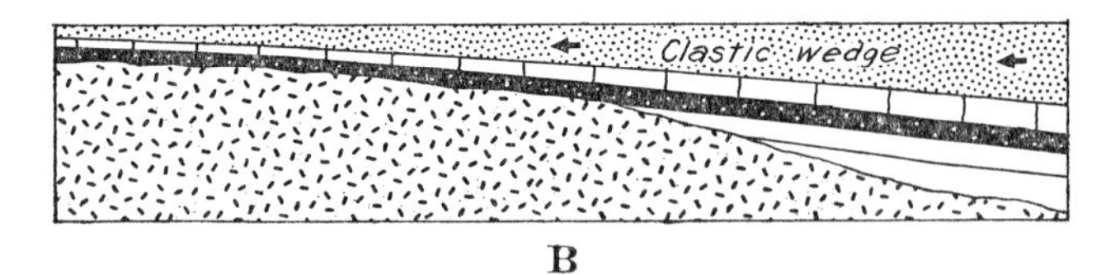

B

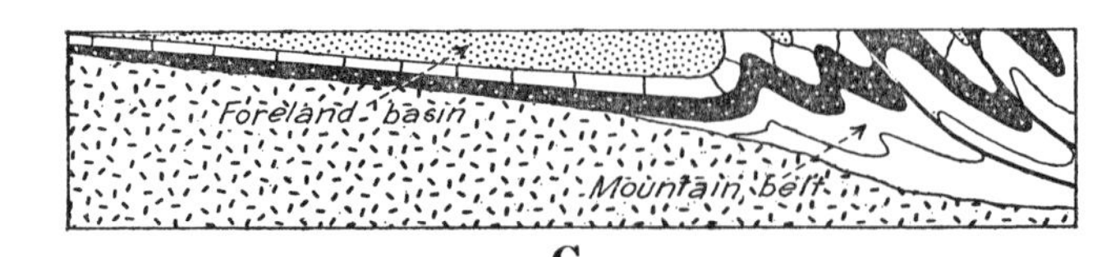

C

FIG. 14. Sequential sections showing development of a foreland area: (A) At end of Cambrian time, showing overlap from geosyncline toward continental platform. (B) After initial phase of mountain growth and deposition of clastic wedges derived from erosion of early mountain ridges, to right of end of section (arrows indicate direction of transport of sediments). (C) At close of mountain making, when a foreland basin developed in front of mountain belt. Length of sections about 300 miles; vertical scale greatly exaggerated.

extends along the front of the Appalachians from New York to Kentucky and preserves Pennsylvanian and early Permian continental beds and coal measures in its trough (Fig. 15 A). The most conspicuous foreland basins on the west are the *Williston basin* of North Dakota and adjacent states and the *Alberta basin* to the northwest in Canada. Other basins occur farther south along the edges of the Interior Lowlands, but they are more overlapped by younger deposits and are less conspicuous in the map pattern.

The foreland basins contain notable quantities of mineral fuels. The Allegheny synclinorium preserves the great coal deposits on which the industrial complex of Pittsburgh and surrounding areas is based; it was the site of the first oil test in the United States, the Drake well of 1859. It was here also that John D. Rockefeller in the latter part of the 19th Century put together his Standard Oil Company which later spawned a whole family of corporate giants. The foreland basins of Oklahoma and Texas have likewise been established as oil provinces for many years, and the Williston and Alberta basins in the far northwest have become so since the Second World War.

MEANS OF ANALYSIS OF DOME AND BASIN STRUCTURE. The domes and basins in the Paleozoic rocks, so strikingly shown on the geologic maps, are clearly of pre-Mesozoic age; where rocks of the latter overlie them, they truncate their eroded surfaces.

It has been assumed that the Appalachian mountain belt farther southeast was created by an Appalachian Revolution toward the close of Paleozoic time; we will see later that this idea will require modification (Chapter IV, Section 5). But granting for a moment that it is so, did the more gently dipping structures in the Paleozoic rocks of the Interior Lowlands similarly originate in one great period of movement?

Various means have been devised for analysis of this and other structural problems of gently dipping strata:

(A) *Structure contour maps* show configuration of the surface of a stratum. Imagine instead of a normal land surface, that all the overburden had been stripped away down to a single stratum, and that all its eroded parts had been restored. Contours of such a surface would show its topography, which would also be its structural configuration. In practice, of course, it is unnecessary to strip away or restore the surface of the stratum. Altitudes on the stratum from which contours can be constructed are obtainable from surface outcrops, from well records, and by calculations based on underlying strata where the stratum has been removed by erosion.

The accompanying Figure 16 shows structure contours on successively higher strata in the Michigan

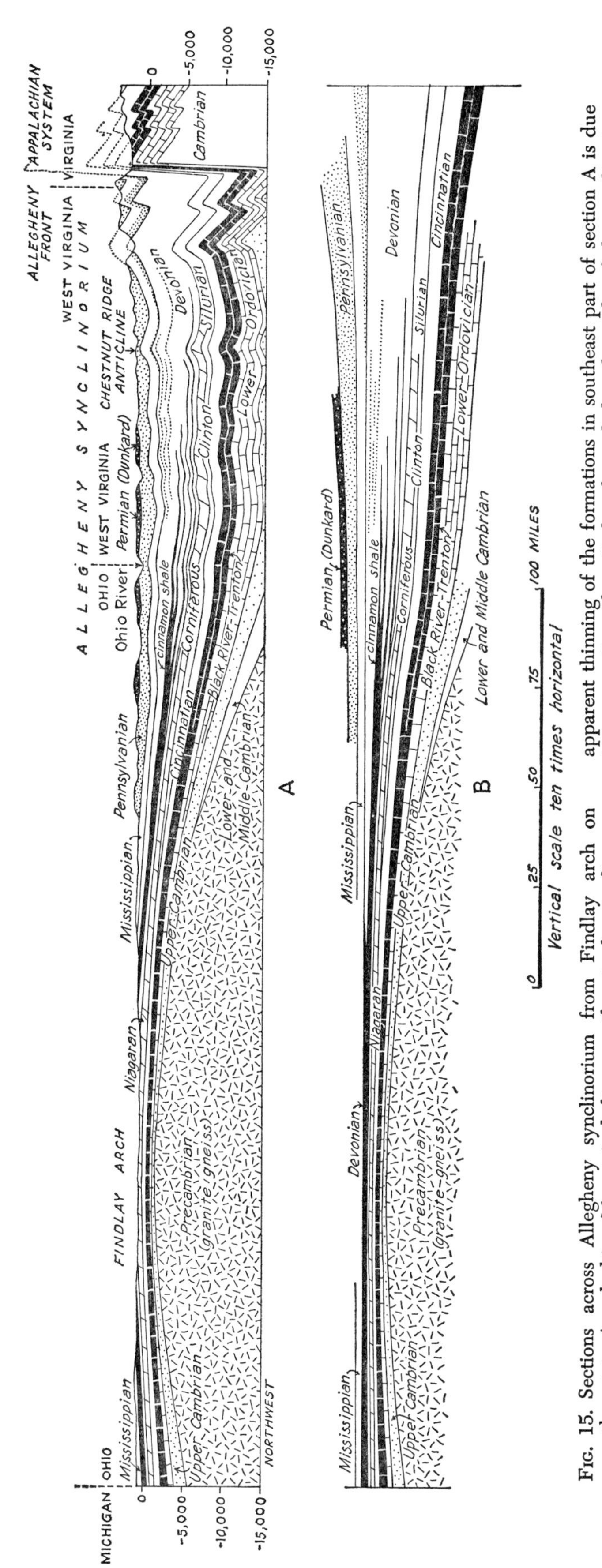

FIG. 15. Sections across Allegheny synclinorium from Findlay arch on northwest to Appalachian Mountain belt on southeast, along a line extending from northwestern Ohio to eastern West Virginia. (A) Structure section. (B) Stratigraphic diagram along same line as structure section, with effects of deformation eliminated. Section B demonstrates that the apparent thinning of the formations in southeast part of section A is due to exaggeration of vertical scale, and that the rocks of the synclinorium were not laid down in an independent depositional basin. Compiled from Wasson and Wasson, 1927; Lafferty, 1941; and other sources.

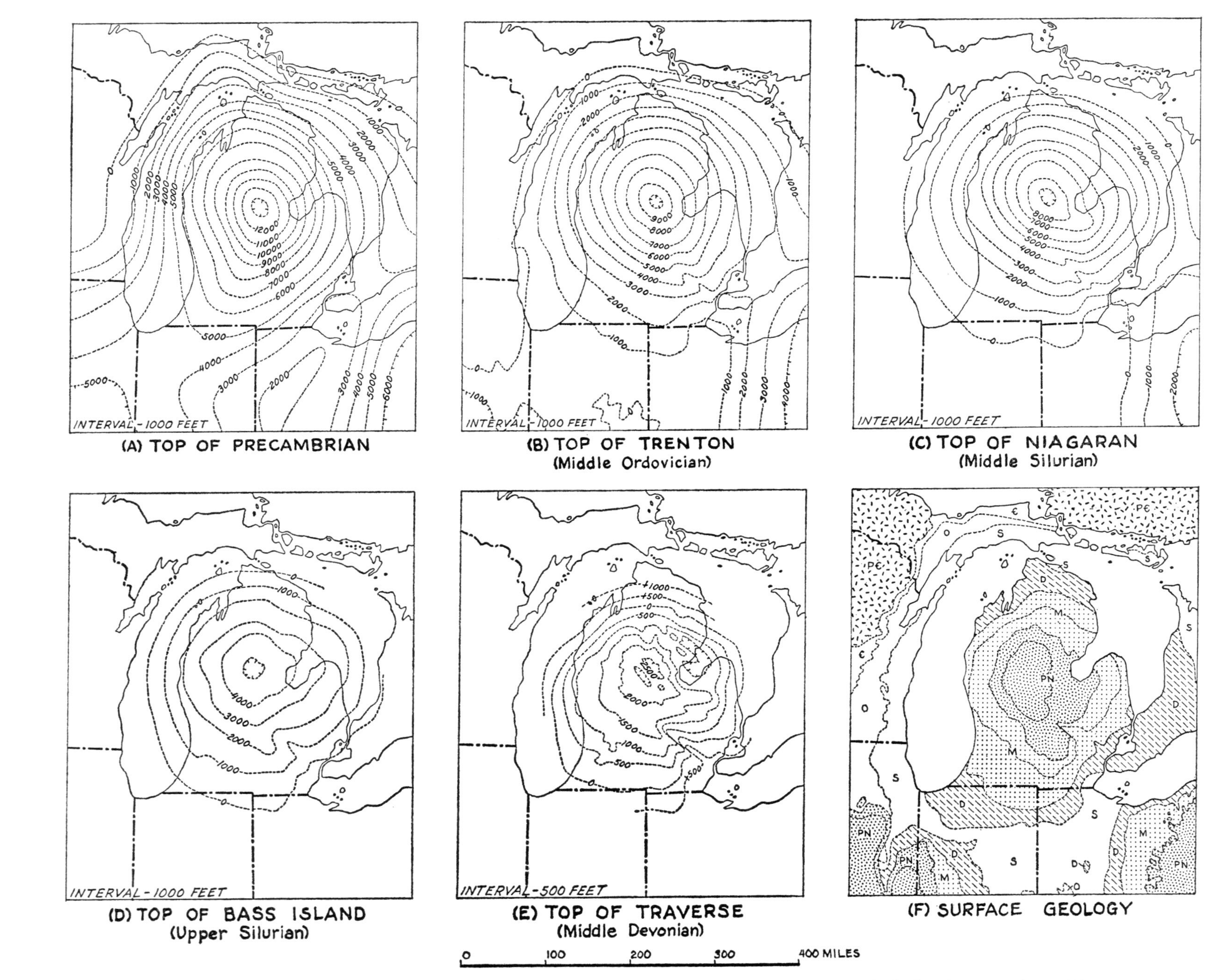

FIG. 16. Maps of Michigan basin and adjacent areas, showing structure contours on successively higher horizons. (A to E), and distribution of geologic systems at surface (F). After Cohee and Landes, 1945, 1947, 1948.

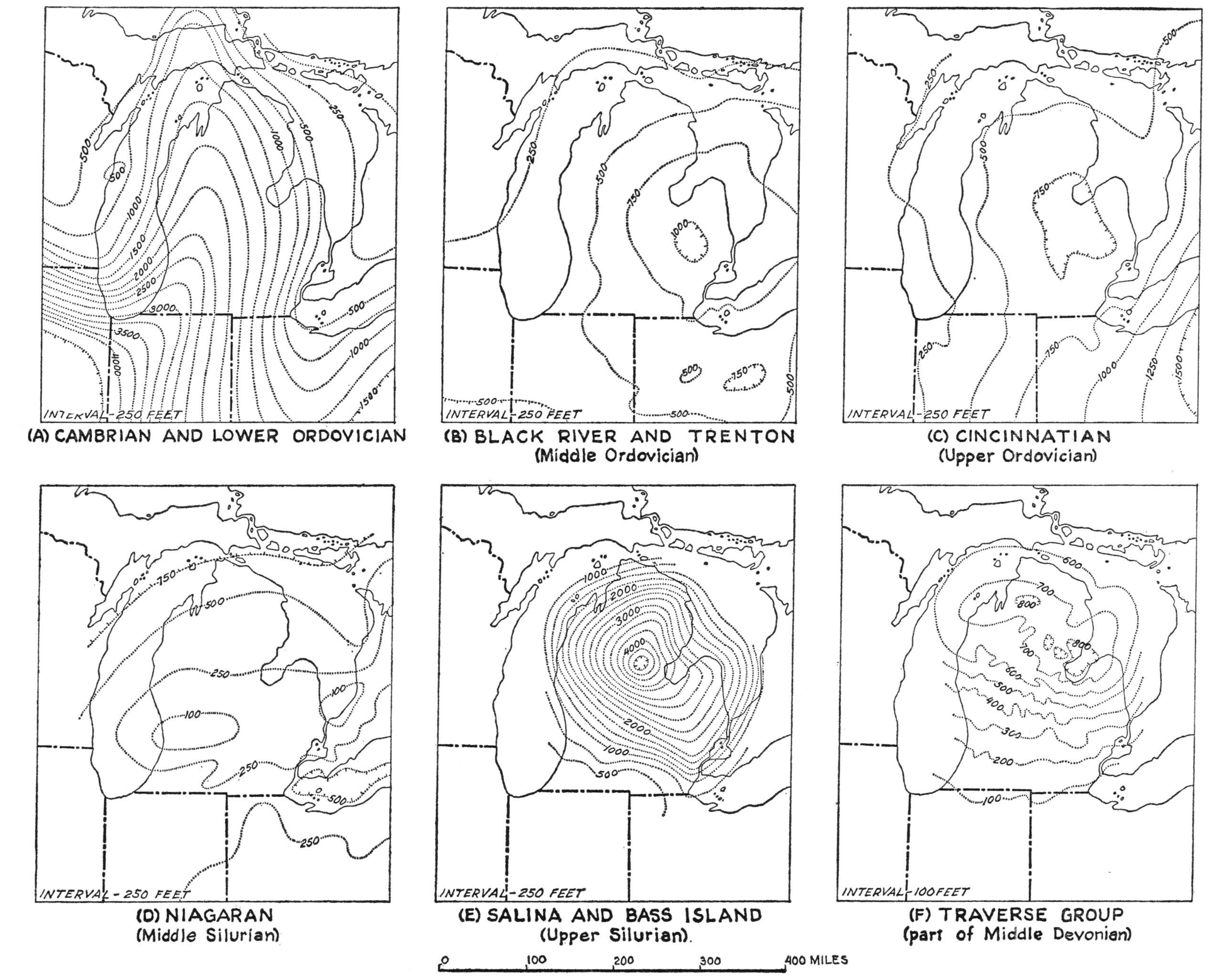

FIG. 17. Maps of Michigan basin and adjacent areas, showing by isopachous lines the variations in thickness of strata laid down there during certain Paleozoic epochs. Note the well-marked basin structure developed during some epochs (E) and its absence from others (D). After Cohee and Landes, 1945, 1947, 1948.

basin. All bring out the nearly circular form of the basin; some hint of its evolution is afforded by progressive flattening of the structure of each younger and higher stratum.

(B) A more subtle form of analysis is the *isopach map*, which shows variations in thickness of a given sedimentary unit. Here we are no longer contouring any real surface, but are representing graphically a body of statistics. Thicknesses of the unit are determined in various places from outcrop sections and drill records, and contours are prepared from these figures.

Our next illustration (Fig. 17) shows isopach lines drawn on successive rock units in the Michigan basin. In general, deposits are thicker in the center of the basin than toward its margins, showing that it subsided progressively during Paleozoic time; but isopach lines on some of the units have little relation to the shape of the modern basin—some show no basinal features at all. On the other hand, isopach lines for the upper Silurian or Salina (Fig. 17 E) perfectly reflect the basin shape, and suggest that a significant phase of its growth took place during that epoch.

(C) Another form of analysis is the *paleogeologic map*. Do not confuse this with the more familiar paleogeographic map; the second shows inferred *geography* of former lands and seas; the first shows *areal geology* of an ancient surface over which younger deposits have been laid.

For demonstration let us shift from the Michigan basin to the state of Iowa. Iowa lies on the flank of the Wisconsin arch and its strata dip gently in a broad homocline southwestward toward the Forest City basin and southward toward the Illinois basin.

In the western part of the state Cretaceous strata spread across all the earlier rocks and their structures, from Precambrian on the northwest to Pennsylvanian on the southeast (Fig. 18 A). This is to be expected, as we have already noted that Mesozoic rocks bevel the Paleozoic of the Interior Lowlands. Let us extend the contacts between different Paleozoic units that appear in surface outcrops westward beneath the Cretaceous, marking them with dotted lines to show the areal geology of the surface on which the latter was deposited (Fig. 18 B). The patch of Precambrian on the northwest now comes into focus as the Sioux uplift, another structurally high area like the Wisconsin arch, away from which the Paleozoic rocks are tilted.

But observe, too, that the Pennsylvanian is discordant on the Mississippian. In central Iowa and farther east in Illinois its basal contact cuts across subdivisions of the latter; outliers of Pennsylvanian in the eastern part of the state lie as well on the Devonian and Silurian, with all the Mississippian missing. The southwestward-dipping homocline of Paleozoic rocks in the state is thus composite; part of it formed after and part before Pennsylvanian time. Let us extend beneath the Pennsylvanian by a dotted line the contact between subdivisions of the Mississippian.

Descending lower in the section, observe how the Devonian truncates the Silurian in the northeastern part of the state, overstepping northward onto Ordovician rocks. Again, the southeast-dipping homocline is composite—part formed after, part before Devonian time. Let us extend beneath the Devonian by a dotted line the contact between the Silurian and Ordovician.

By this demonstration we have not actually made any paleogeologic maps; we would need more data than are afforded by the surface geologic map of the state alone. But we have made a start on three such maps—paleogeologic maps of the pre-Cretaceous, pre-Pennsylvanian, and pre-Devonian surfaces—and have shown how such maps are compiled.

In Iowa our demonstration has indicated that there is a major unconformity above the Paleozoic and two other major unconformities within it, the structures under each being steeper than the structures above. *A. I. Levorsen*, the well-known petroleum geologist, has aptly termed this relation *layer-cake geology*. In Iowa, according to our analysis of the map, the "cake" above the Precambrian has four "layers," or bodies of rock of different structure, each one bounded by unconformities. (Actually, there are five layers, as another unconformity, not apparent on the map, separates Mississippian and Devonian rocks; this acquires greater importance in adjacent states, where it is second in magnitude only to the one beneath the Pennsylvanian.)

Growth of domes and basins. These methods of analysis indicate that the domes and basins in the Paleozoic rocks of the Interior Lowlands, far from having been created by a single period of deformation late in Paleozoic time, have grown progressively through much of the era. Structure contour maps indicate that inclination of the strata steepens as one proceeds downward in the section; isopach maps show that many of the units thicken into the basins and thin toward the domes; paleogeologic maps prove that many of the younger units overstep the older toward the domes on surfaces of unconformity.

Growth of domes and basins is a form of *epeirogeny*, a term used to distinguish this rather passive process from the more positive processes of orogeny or mountain building. It has two guises (Fig. 19):

(1) Greater subsidence of basins and less subsidence of domes, indicated by variable thickness of sediment laid down from place to place in a given

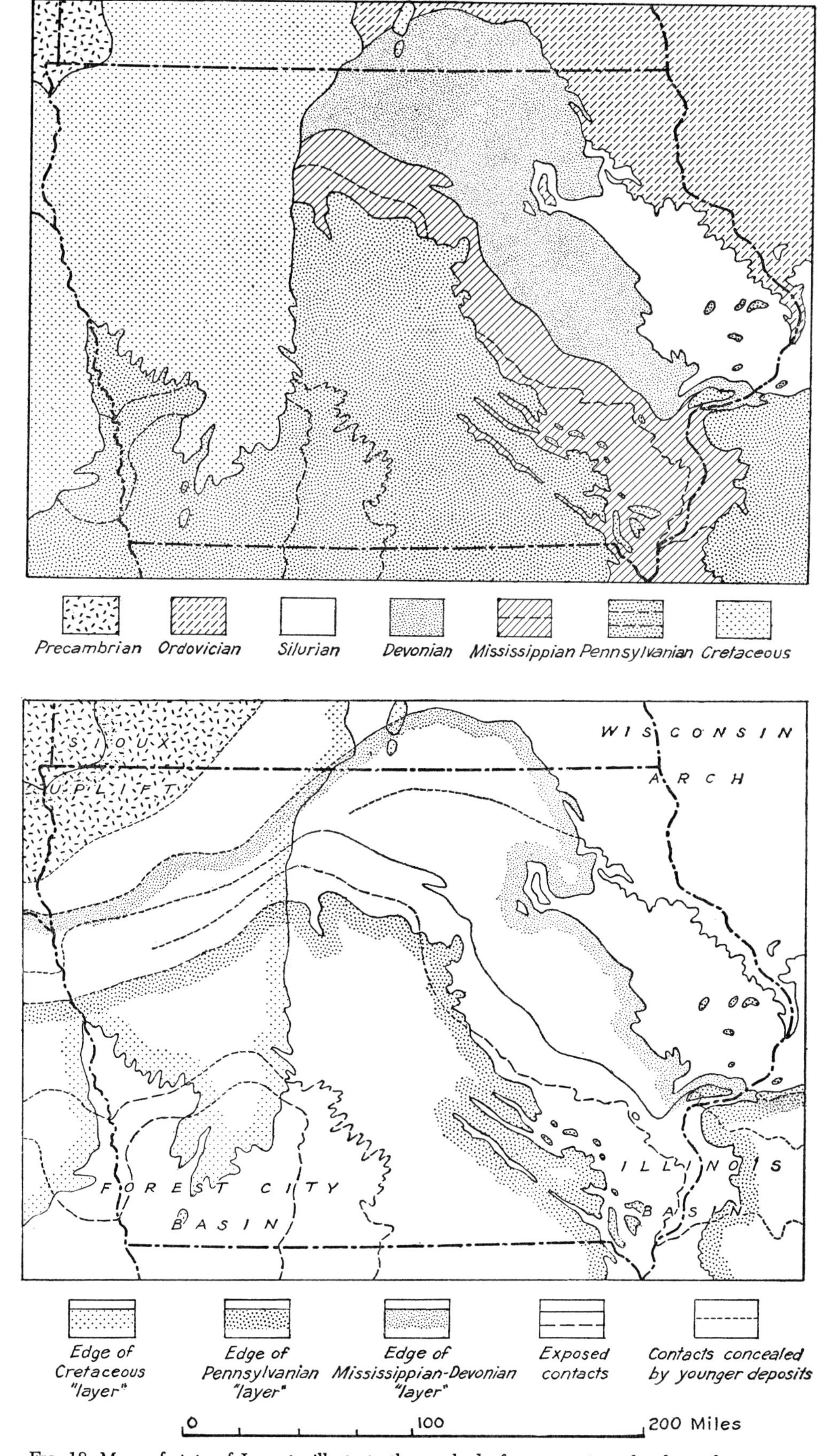

Fig. 18. Maps of state of Iowa to illustrate the method of construction of paleogeologic maps. Above is a conventional geologic map. The map below shows the same exposed geologic contacts, and also contacts which are covered by unconformably overlying deposits. The rocks of the state above the Precambrian are thus divisible into four "layers" of a "layer-cake"—Cretaceous rocks, Pennsylvanian rocks, Mississippian and Devonian rocks, and Silurian and Ordovician rocks. After geologic map of Iowa, 1937; and other sources.

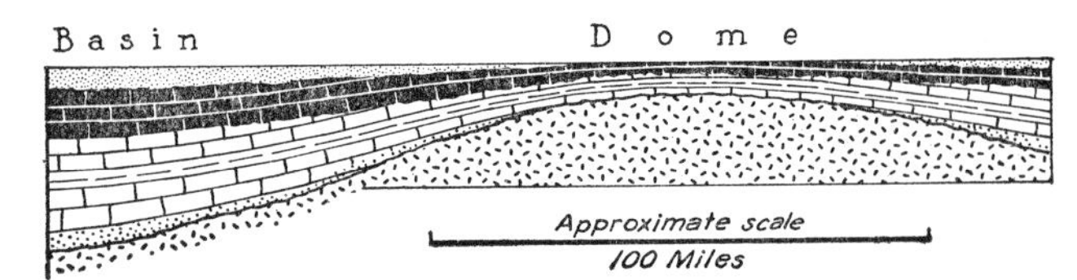

FIG. 19. Idealized section of a dome and basin in the Interior Lowlands, showing thinning of all units away from the basin and toward the dome. This results partly from deposition of a greater thickness of each unit in the basin than on the dome and partly from truncation and overlap along unconformities.

period (see isopach maps of Michigan basin, Fig. 17). This process was secular; that is, it went on more or less without interruption through the Paleozoic era.

(2) Actual uplift of domes during relatively brief periods, which brought about erosion of the earlier units and overlap and overstep of the younger on surfaces of unconformity. (See paleogeologic map of Iowa, Fig. 18; it is also well illustrated on the Ozark and Nashville domes, from which Fig. 19 is generalized.) Some of these times of uplift may correspond to periods of more intense deformation in the mountain belts roundabout; for example, the pre-Mississippian unconformity corresponds roughly to the Acadian orogeny of the Appalachians.

All that we have done here is to furnish some simple examples of the growth of the structures of the Interior Lowlands. The story has been worked out in much detail, especially in the western part where extensive drilling for oil has provided subsurface information. Such drilling has always brought forth surprises; subsurface structure cannot certainly be predicted from outcrops. Gentle surface structures steepen downward due to progressive growth with time. Other structures are encountered, such as the "granite ridge" of Kansas, for which there is no surface indication; they formed early, were eroded and buried, and were little disturbed again.

6. THE WEST TEXAS BASIN

In this book we have sketched the foreland basins briefly, but I cannot close the present chapter without saying more about the basin with which I have been most concerned personally—that in West Texas. Not long ago I had an opportunity to return to West Texas, take part in a field trip, and revisit old geological localities, some of which I had not seen for fifteen or twenty years. On the trip one of the younger men said to me, "You must be pretty smart to know so much about this country and to have worked all this out." My reaction was, "Hells Bells! I was in this country off and on for fifteen years. In that time even a simpleton should have figured out something about it!"

This brought home to me something of the handicap which all younger geologists face in coming into an established geological province where the great principles that control it have all been worked out by geologists who have gone before, in ways they cannot know. During the fifteen years of which I spoke, we geologists who had come early to West Texas had an opportunity to "grow up with the country" just as truly as James Hall had "grown up" with the geology of New York State a hundred years earlier. We saw a major oil province when there were few oil wells and little subsurface information, and when all its complex geology was still a mystery. We were able to probe these mysteries and learn the geological laws of the province the hard way—by slow process of trial and error.

Here, I will depart from the style of the rest of this book and indulge to a greater extent in personal reminiscence, but for the purpose of recreating the spirit of the times in which the modern ideas on the region developed.

WEST TEXAS "PERMIAN" BASIN. The West Texas basin, as the name implies, lies mainly in western Texas, although one corner extends into southeastern New Mexico. For the most part it lies beneath the southern end of the Great Plains, east of the Texas and New Mexico mountains beyond the Pecos River. Southward it is bordered by older mountains—the Ouachita chain of Paleozoic time (or Marathon deformed belt of Fig. 20)—but these are now little evident at the surface, being mostly buried by Cretaceous rocks and emerging only briefly in the Marathon country which we will discuss later (Chapter IV, Section 6). The basin structure of West Texas is indicated at the surface by opposing dips of Paleozoic rocks off the trans-Pecos mountains on the west and off the exposed Paleozoic in north-central Texas to the east, but the basin itself is masked by Triassic, Cretaceous, and Tertiary deposits.

The West Texas basin is known in the trade journals and to all local people as the "West Texas *Permian* basin." Coupling of the word Permian with that of the province connotes the great thickness of Permian rocks there—more than 10,000 feet in places—and the prolific oil production which they have yielded. Permian rocks were the first in the province to produce oil, and although deeper and older oil-bearing horizons have since been found, the former still accounts for more than half the total production.

In fact, our Texas friends have pretty much taken over the word "Permian." In Midland, the oil headquarters town, there is a "Permian Building" and a "Permian Oil Company"; some of the national technical societies have "Permian Basin sections." As the reader may recall, the Permian system was established in 1841 by the British geologist *Sir Roderick Impey Murchison*, who named it for the city and province of Perm, west of the Ural Mountains in European Russia. But the ancient city of Perm is now no more; the Soviets have renamed it "Molotov."* So perhaps the Texans are as qualified as any to appropriate the Permian system.

EARLY YEARS. Like any other college graduate in 1924, I was anxious for a job and uncertain of my future. My professors had written to various alumni of the school in behalf of us graduates. For me, in the week of graduation, came the heartening message from one of them, "Report at once to the Marland Oil Company in Dallas." This I shortly did, being not quite twenty-one at the time. After a few weeks with the company in Dallas I was sent as an assistant to join geological field parties which were being organized by the company in West Texas.

The West Texas country of those days was quite different from what it is now—a backwash of the frontier which had never quite attained prosperity, a land of great cattle ranches joined by ungraded dirt tracks with the raw towns along the railroads. The oil companies were certain that the region had possibilities. It was a sedimentary basin little tested even by the wildcatters, in which oil had already been found in a few places—at Big Lake in Reagan County, and Westbrook in Mitchell County where oil was produced, and in widely scattered test holes where intriguing "oil shows" had been encountered.

But later I came to realize that West Texas at the time was also the "Siberia" of oil geologists—a place where companies sent misfits and amateurs with whom they did not know what else to do. I was certainly one of these; I still remember with embarrassment my foolish errors at the plane table and alidade, my inability to close traverses within the allowable limits of error. Needless to say, West Texas is a "Siberia" no longer, but a great established oil province in which technical employment is an honor rather than a stigma.

I spent a year or so with the oil company, helping to survey structures of the Cretaceous rocks which covered the whole surface. This was all we knew how to do, although we realized that proven oil production was in the Permian, many thousands of feet below the surface, and separated from the surface rocks by at least two unconformities. But the region was one of the first to be explored for oil in which producing horizons were markedly unconformable beneath surface rocks—it was perhaps the first encounter of the oil geologist with "layer-cake geology."

All we knew of subsurface stratigraphy and structure was from drillers' logs, written by men untutored in the fine arts of geology. The logs were sufficiently baffling, as one would report thousands of feet of limestone, and another only a few miles away indicated a great thickness of salt and gypsum or a great thickness of sandstone. A testimony of those days is a Geological Survey Bulletin on the potash resources of the region,† which discusses at length the meaning of these logs; it makes curious reading today. Actually, the drillers' logs were not mendacious; they gave a blurred impression of the different kinds of rocks penetrated and would probably be meaningful to a modern West Texas stratigrapher. Real difficulties were with the rocks themselves, which obeyed no laws with which we were familiar at the time.

While we were struggling with the surface structures of the region, the Gulf Oil Company was leasing great blocks of West Texas territory at low prices and under long-term contracts. Even at terms of a few cents an acre, such contracts meant riches to the local ranch people, each of whom owned hundreds of square miles of barren country, but were "land poor" from a disastrous succession of droughts and from slumps in the cattle market that followed the First World War. There seemed to be no sense or system in land leasing by the Gulf; some of the blocks were in areas that other geologists of the time did not suppose had any possibilities. Years later I learned that the leasing was based on real method, not aimless guessing—a method for which its unsung discoverer deserves honor. It was based on salt thicknesses as reported in drillers' logs. If salt was thin, land was leased; if salt was thick, the area was rejected. As things turned out later, this was as good a rough-and-ready means of blocking out promising territory as one could find at the time. In later years the Gulf Oil Company found itself in possession of leases on some of the most favorable land in the basin.

CENTRAL BASIN PLATFORM. From surface indications and drillers' logs we believed that the deepest part of the West Texas basin was in Winkler County, midway between opposing dips on the east and west sides. Oil should be trapped, we thought, in structures marginal to such a basin; Winkler County in the center should offer the least possibilities of any. But about this time a wildcatter without geological inhibitions began to drill here with no encouragement from the companies (except perhaps the Gulf). He struck oil—spectacularly—irrevocably overturning all previous theories

* Or at least it was so named up to 1957; I do not know what they are calling it now.

† *U.S. Geol. Survey Bull. 780 B*, p. 33-126, 1926.

about the West Texas Permian basin. The reader can picture the excitement—provisions which had to be made for the unexpected flow of oil, the new town of "Wink" hurriedly laid out by land promoters, and the readjustment which geologists had to make in all their thinking.

As drilling in Winkler County progressed, it became evident that oil occurred here on a limestone "high," and geologists began to realize that the West Texas basin was composite. Instead of being a single great geological depression, it consisted of several subbasins separated by structural ridges. One of these, the "high" in Winkler County, came to be known as the *Central Basin platform*; it divided the larger West Texas basin into the *Delaware basin* on the west and the *Midland basin* on the east (Fig. 20).

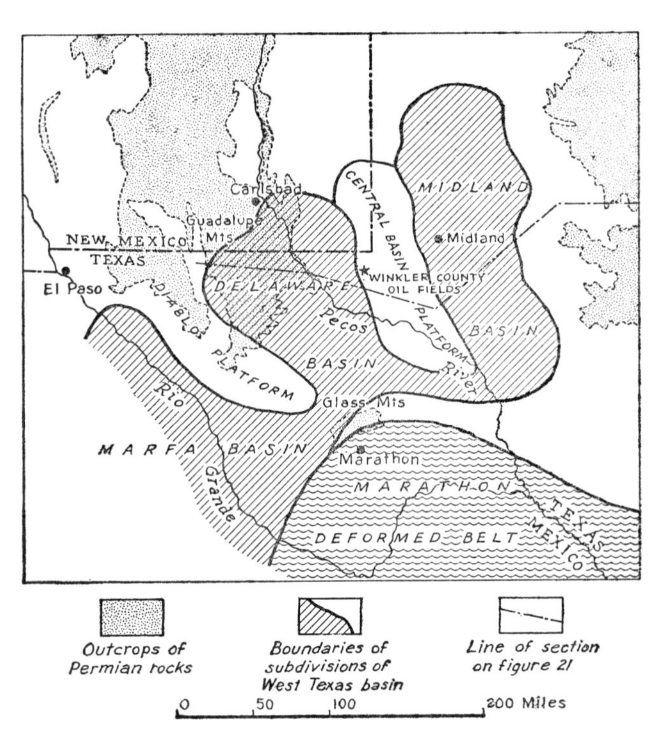

Fig. 20. Map of West Texas basin showing subsidiary basins and platforms into which it is divided. After King, 1942.

The reef theory. But this broad structural pattern did not explain all the anomalies of the region. The geology revealed by drilling was not merely a matter of simple folding or warping of previously deposited strata, but one in which the strata themselves changed in character from one unit to another (Fig. 21). Different rock sequences thus occurred in each unit beneath the surface rocks:

Delaware basin A thick body of salt underlain by a thick body of anhydrite, and this underlain by sandstone.

Central Basin platform A nearly solid sequence of

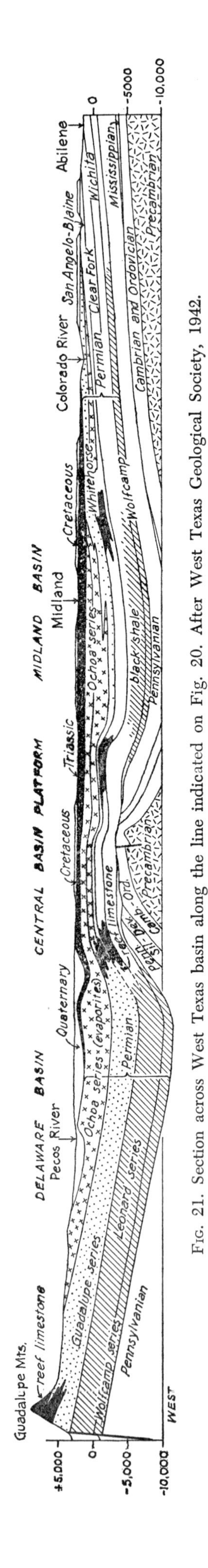

Fig. 21. Section across West Texas basin along the line indicated on Fig. 20. After West Texas Geological Society, 1942.

limestone down to basement rocks, which had been reached in a few places by 1929.

Midland basin Interbedded salt, anhydrite, and red beds underlain by shales and thin limestones.

At about this time I left the oil company and began work on the outcrop areas of Paleozoic rocks southwest and west of the basin—first as a graduate student for the Texas Bureau of Economic Geology in company with R. E. King, later for the U.S. Geological Survey. This work was not as remotely related to the geological work by the oil companies as one might suppose. The modern West Texas basin, it is true, is blocked off by mountain uplifts on the west, but these are of Laramide or later age; during Permian time when the critical features of the basin developed, it extended much farther west.

My first work was in the Glass Mountains on the southwest, where Permian rocks emerge at the edge of the basin (Fig. 20). Later it was in the Marathon lowland to the south, in folded earlier Paleozoic of the Ouachita mountain belt. Still later, it was in the Sierra Diablo and Guadalupe Mountains west of the basin, where Permian rocks again come to the surface.

On the outcrops we began to find the same stratigraphic puzzles as in the subsurface sections. In the Glass Mountains lateral changes in the Permian stratigraphy were extreme; the rocks in the section on the west were almost totally unlike those on the east—on the west interbedded shales, sandstones, and limestones; on the east nearly solid thin-bedded limestone.

Professor Charles Schuchert, who was supervising our work, told me that the geologists who had been in the area earlier—*J. A. Udden* and *Emil Böse*—had talked of an idea that the thick limestone bodies might be reefs; apparently they did not know quite how, and the idea was so vague that it had never appeared in any of their publications. Schuchert urged me to look into the possibility. I was unimpressed, and to bolster my objections I looked up the subject of fossil reefs in Grabau's "Principles of Stratigraphy." There, a drawing which I had merely glanced at before—a view of Triassic reef structures in the Tyrolean Alps—suddenly took on new meaning. The lay of the strata on the Alpine mountainside was a near replica of oddities we had seen on the slopes of the Glass Mountains, and for which we had not found an explanation. Perhaps the odd structures were marginal to a great reef! Perhaps (but the idea seemed too outrageous) reef barriers were responsible for the unlike stratigraphic sequences at the two ends of the Glass Mountains!

Amazingly, as has happened with many other great scientific discoveries, nearly every geologist who was working in West Texas came independently to the same conclusion at about the same time; by 1929 the "reef theory" as an explanation of the Permian stratigraphy of the region was in full flower. Through the years that followed, further evidence has accumulated, so that reef theory has become reef fact, and is now established as one of the controlling principles of West Texas Permian geology.

PERMIAN OF GUADALUPE MOUNTAINS. Although we could sense the dominant control of the reefs on the stratigraphy of the Glass Mountains, structural complications and the remoteness of the area from country that was being intensively drilled prevented us from grasping the regional implications. Other geologists had better success, notably *E. Russell Lloyd*, who was examining the same rocks in the Guadalupe Mountains farther northwest. It was my privilege to work in these mountains much later.

The Guadalupe Mountains begin near the Pecos River in New Mexico and extend southwest into Texas with steadily increasing altitude, terminating in a great point or cliff. On this is exposed a wonderful section of about 4,000 feet of strata (right-hand end of Fig. 22 B), all of Permian age and highly fossiliferous. This was first discovered by *George G. Shumard*, one of the geological explorers before the Civil War, at the time of the Pacific Railroad Surveys. It consists of the following, in descending order:

Capitan limestone White, massive or poorly bedded.

Delaware Mountain formation (or group) Sandstone, mainly fine-grained and thin-bedded with some coarser and more massive layers, interbedded with shaly sandstone and thin-bedded dark limestone.

Bone Spring limestone Black, thin-bedded limestone, shaly in part.

But this fine sequence, which one would think should provide a stratigraphic key to the mountains, does not extend far in any direction. Bone Spring limestone and Delaware Mountain formation continue to the southeast, but the latter is overlain by Castile gypsum—where is the Capitan limestone? Sands of the Delaware Mountain formation extend only a few miles to the northwest, beyond which most of the section is limestone, but limestone which has lost the distinctive features of the Capitan and Bone Spring (left-hand end of Fig. 22 B).

Despite its short extent to the southeast and northwest, the Capitan limestone forms a massive front or escarpment that can be followed fifty miles northeastward to the Pecos River, where it plunges beneath the surface. But it can be traced by drilling beyond, curv-

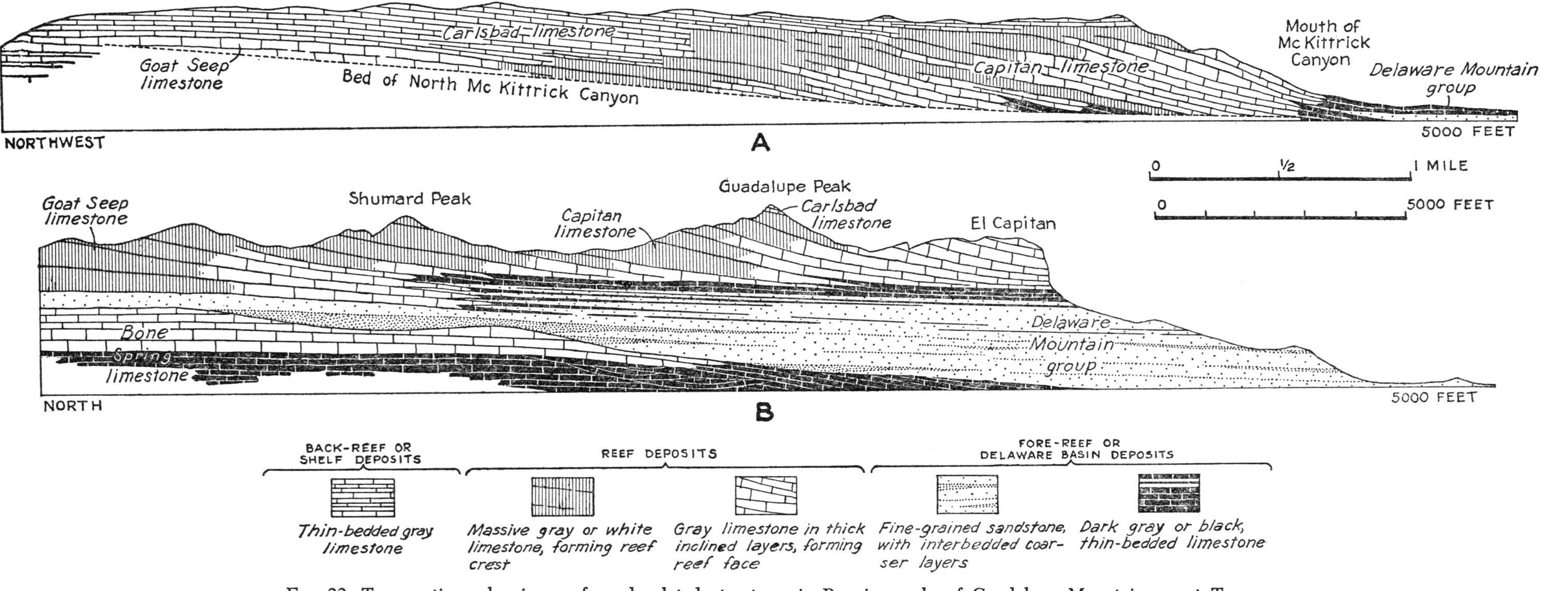

FIG. 22. Two sections showing reefs and related structures in Permian rocks of Guadalupe Mountains, west Texas. (A) Along a canyon wall in the mountains showing upper part of succession. (B) Along escarpment on west side of mountains showing lower part of succession. After King, 1948.

ing in a semi-circle around the north end of the Delaware basin, to pass through the Winkler County "high" and thence into the Glass Mountains. The sands of the Delaware Mountain formation are thus a body confined in their extent to the Delaware basin, and are surrounded by partly contemporaneous limestone deposits (Fig. 20).

Geologists imbued with the reef theory began to discover strange things about the Capitan limestone. Instead of being a well-behaved, flat-lying deposit, all its layers are inclined to the southeast toward the Delaware basin (Fig. 22 A). At their lower southeastern ends each flattens and thins abruptly changing from white massive limestone into thin-bedded limestone or sandstone of the facies of the Delaware Mountain formation. At their upper ends the layers pass into exceedingly massive limestone which, farther northwest, intergrades with flat-lying, thin-bedded limestone. Structures of the Capitan are thus very much like an ideal diagram of a delta, with topset, foreset, and bottomset beds (Fig. 13 B).

We now know that the massive limestone at the top of the slope was a growing barrier reef built by sponges, algae, bryozoans, and other wave-resistant, lime-secreting organisms. The inclined beds below were reef talus, and the sands at their bases were laid down on a sea floor in front of the reef, probably in a thousand feet or more of water.

The thin-bedded limestones to the northwest were laid down in a shallow lagoon behind the reef barrier. Farther in this direction they intergrade with dolomites, evaporites, and red beds formed in evaporating pans toward the shore. The Castile gypsum, which fills the Delaware basin in front of the reef, lies at the same level as these back-reef deposits, but is a later evaporite formed after the Capitan was deposited, when the Delaware basin itself was cut off from free access to the sea.

The substructure of the Capitan reef is laid bare in cross-section in the great fault scarp on the western side of the mountains (Fig. 22 B). A short distance back from the front of the Capitan reef the middle part of the Delaware Mountain formation changes into massive limestone of reef facies, just as the upper part has changed into Capitan. This precursor of the Capitan reef (the Goat Seep) also has its reef talus and growing reef, and rises to the same height as the crest of the Capitan deposits, but lies several miles behind it.

But the lower third of the Delaware Mountain disappears northwestward in another manner. The top of the black Bone Spring limestone rises in this direction in a flexure, crowned at its upper end by a bank of gray limestone. The top of the flexed Bone Spring is a local unconformity within the Permian, and the sands of the lower Delaware Mountain overlap upon and wedge out against its sloping surface.

This assemblage of complex stratigraphic and structural features—barrier reefs, banks, and flexures—has caused the contemporaneous deposits in front of, within, and behind the reef zone to be so very different. And our sections (Fig. 22) show the features the way they really are. In other sketches and figures in this book it has been necessary to represent features on an exaggerated vertical scale, or to assemble the items from exposures scattered over a considerable distance. Here, the features are shown on true scale, and are all laid bare in cross-section on single canyon walls and mountain sides.

Relation of Permian stratigraphy to oil accumulation. All this is interesting enough, the reader may say, but how does it relate to the oil of the West Texas basin? Remember, however, that our cross-sections (Fig. 22) are exposed samples of a structure traceable for hundreds of miles around the rim of the Delaware basin, the eastern half of which, in Winkler and adjacent counties, is buried, and followed by a chain of oil fields along the west edge of the Central Basin Platform. The same substructure, or something very much like it, must underlie the oil fields.

The Capitan reef has not itself entrapped oil in this eastern segment, but deposits adjacent to it and controlled by it have done so; many of the fields are in porous lagoonal beds behind the reef. One should remember, too, that the reef, a narrow barrier, is not the only Permian deposit in the West Texas basin, but merely the most spectacular part of a complex of deposits spread over a vast area. Many different parts of this complex, when porosity and structure were right, have become oil reservoirs.

As exploration of the West Texas Permian basin continued; the more venturesome test wells were extended through the Permian into earlier Paleozoic strata, and these strata were found to contain oil reservoirs down into the Lower Ordovician. In the earlier Paleozoic rocks we have entered a lower slab of the West Texas "layer cake," a slab with its own primitive set of structures that was the foundation on which the reefs and other Permian deposits were built.

At present, the only foreseeable downward limit of oil possibilities in the region is the last and deepest slab of the "layer cake," the Precambrian floor of the basin (Fig. 21). The Precambrian has been reached by the deepest test wells, and consists of granitic plutonic rocks, roofed in places by altered lavas and continental sediments; these would not have generated oil in any case, even had they been much younger.

Now the West Texas basin has been probed to its depths, and perhaps we can be content. We have learned that it is a great depression whose construction occupied much of Paleozoic time. It was covered by successive seas, each of whose organisms generated oil that was caught in a myriad of stratigraphic and structural traps, where it was ultimately discovered by the eager search of man.

REFERENCES

2. *Geological investigations*

Merrill, G. P., 1924, *The First One Hundred Years of American Geology*: Yale Univ. Press, chaps. 3 and 4, p. 127-292, especially p. 223-235.

3. *Precambrian basement of the lowlands*

Burwash, R. A., 1957, Reconnaissance of subsurface Precambrian of Alberta: *Am. Assoc. Petrol. Geol. Bull.*, v. 41, p. 70-103.

Dake, C. L., and Bridge, Josiah, 1932, Buried and resurrected hills of the central Ozarks: *Am. Assoc. Petrol. Geol. Bull.*, v. 16, p. 629-652.

Flawn, P. T., 1956, Basement rocks of Texas and southeast New Mexico: *Texas Univ. Publ. 5605*, 261 p.

Thiel, Edward, 1956, Correlation of gravity anomalies with the Keweenawan geology of Wisconsin and Minnesota: *Geol. Soc. America Bull.*, v. 67, p. 1079-1100.

4. *Sedimentary cover of the lowlands*

Dake, C. L., 1921, The problem of the St. Peter sandstone: *Missouri School of Mines and Met., Tech. ser.*, v. 6, p. 1-228.

Cloud, P. E., Jr., and Barnes, V. E., 1948, The Ellenburger group of central Texas: *Texas Univ. Publ. 4621*, 473 p., maps.

Lowenstam, H. A., 1950, Niagaran reefs of the Great Lakes area: *Jour. Geology*, v. 58, p. 430-487.

Sloss, L. L., 1953, The significance of evaporites: *Jour. Sed. Petrol.*, v. 23, p. 143-161.

Weller, J. M., 1930, Cyclical sedimentation of the Pennsylvanian period and its significance: *Jour. Geology*, v. 38, p. 97-135.

5. *Structures of the sedimentary cover*

Kay, Marshall, 1942, Development of northern Allegheny synclinorium and adjoining regions: *Geol. Soc. America Bull.*, v. 53, p. 1601-1658.

Levorsen, A. I., 1927, Convergence studies in the mid-continent region: *Am. Assoc. Petrol. Geol. Bull.*, v. 11, p. 657-682.

———, 1933, Studies in paleogeology: *Am. Assoc. Petrol. Geol. Bull.*, v. 17, p. 1107-1132.

Moore, R. C., 1936, Stratigraphic evidence bearing on problem of continental tectonics: *Geol. Soc. America Bull.*, v. 47, p. 1785-1808.

Powers, Sidney, 1931, Structural geology of northeastern Oklahoma: *Jour. Geology*, v. 39, p. 117-132.

Wilson, C. H., Jr., and Born, Kendall, 1943, Structure of central Tennessee: *Am. Assoc. Petrol. Geol. Bull.*, v. 27, p. 1039-1059.

6. *The West Texas basin*

Adams, J. E., and Frenzel, H. N., 1950, Capitan barrier reef, Texas and New Mexico: *Jour. Geology*, v. 58, p. 289-312.

King, P. B., 1948, Geology of the southern Guadalupe Mountains, Texas: *U.S. Geol. Survey Prof. Paper 215*, 183 p., maps.

Lloyd, E. R., 1929, Capitan limestone and associated formations of New Mexico and Texas: *Am. Assoc. Petrol. Geol. Bull.*, v. 13, p. 645-658.

Newell, N. D., and others, 1953, *The Permian Reef Complex of the Guadalupe Mountains Region, Texas and New Mexico*: W. H. Freeman & Co., San Francisco, 236 p.

West Texas Geological Society, 1942, Résumé of geology of the south Permian basin, Texas and New Mexico: *Geol. Soc. America Bull.*, v. 53, p. 539-560, with cross-section. (Other and more detailed sections of the West Texas basin have been issued later by local societies, but are not in readily accessible periodicals, and do not greatly change the general picture here presented).

CHAPTER IV

APPALACHIAN AND RELATED SYSTEMS; PALEOZOIC STRUCTURES SOUTHEAST AND SOUTH OF CENTRAL STABLE REGION

1. TOPOGRAPHY AND STRUCTURE

We pass now to the younger mountain belts which border the Central Stable Region, to which we will devote much of the remainder of the book. Attention will be given first to those structures formed in Paleozoic time that lie southeast and south of the stable region (Plate 1).

Topographic mountains. We have referred to these Paleozoic structures as a "mountain belt," but this is true more in a geological than a geographical sense. While the system was once truly as mountainous as the Cordillera of the west is today, that was long ago. Since then, deformation of the region has been inconsequential; parts have gone through several cycles of upwarp and erosion; other parts, of subsidence and burial bequeath younger deposits. The mountains we see today are only the deeply eroded stumps of a fraction of the original mountain belt.

The word "Appalachians" can be used broadly for that segment of the former mountains east of the Mississippi River, although topographically the term "Appalachian Mountains" applies more strictly to high ridges which extend along its axis from Pennsylvania to Alabama. The New England Upland is also geologically part of the Appalachians, although it is largely separated from the part to the southwest by a depressed segment near New York City, partly covered by younger deposits and partly by the sea. Canadians again use the term "Appalachians" for ridges northeast of New England in southeastern Quebec and in the Maritime Provinces, i.e. New Brunswick, Nova Scotia, and Prince Edward Island.

West of the Mississippi River in Arkansas and Oklahoma other stumps of the once lofty system project as the Ouachita, Arbuckle, and Wichita Mountains; a final fragment emerges far to the southwest in the Marathon region of western Texas. The fragments west of the Mississippi are now so widely scattered that their original plan is no longer evident either from surface topography or exposed structures; we will reserve analysis of them for the last part of this chapter (Section 6).

Mountains worn down to plains. Parts of the system have been so deeply eroded that they are no longer topographically mountainous.

Between the present Appalachian ridges and the edge of the younger deposits on the southeast in the Atlantic Coastal Plain is a belt of low country about one hundred miles broad which has been called the Piedmont Plateau. This is not even a very good topographic "plateau," as much of it in Virginia, the Carolinas, and Georgia is as featureless and heavily farmed as the Interior Lowlands of the Middle West in Illinois and Iowa. Unlike the latter which are underlain by glacial drift and gently dipping bedrock strata, the lowlands of the Piedmont are formed on the edges of steeply upturned and altered rocks that were originally part of the Appalachian mountain system.

Buried mountains. Elsewhere the system is not only worn down but lost to view, as it is buried by Mesozoic and Tertiary deposits of the Atlantic and Gulf Coastal Plains (Fig. 31). These deposits were laid over the eroded edges of the ancient mountain structure, and dip seaward with little deformation. While they are of much interest in themselves, for our present purpose they are merely an inconvenient cover that must be probed by drilling and other means to determine the substructure.

Southeastward across the strike of the southern Appalachians the worn down rocks of the Piedmont Plateau pass beneath deposits of the Atlantic Coastal Plain; the southeastern edge of the Appalachian system, whatever it was, is now no longer visible. Also, southwestward along the strike the whole mountain structure plunges beneath deposits of the Gulf Coastal Plain. It is thus not possible to determine exactly the relations of the Appalachians to the similar Ouachita system across the Mississippi Embayment on the west. Much of the Ouachita and other systems in Oklahoma and Texas are similarly buried by Cretaceous or even by later Paleozoic deposits.

SUBMERGED MOUNTAINS. Other parts of the former Paleozoic mountain chain are still less accessible, as they lie submerged beneath the sea under the continental shelves.

Much of the connection between the central Appalachians and the part in New England is thus submerged in the segment near New York City. Northeastward beyond New England the Appalachian folds of Gaspé and the Maritime Provinces also run out beneath the Gulf of St. Lawrence. They emerge briefly again in the island of Newfoundland, but on its northeastern side they pass beneath the waters of the Atlantic Ocean and are gone from our sight for good (Plate 1).

No one knows for certain what becomes of the Appalachian structures northeast of Newfoundland. On the opposite side of the Atlantic in Ireland, Great Britain, and Brittany the Hercynian or Variscan system—a worn down mountain chain similar to the Appalachians—passes westward out to sea. It is made up of much the same kinds of rocks and structures as the Appalachians, and was formed in the latter part of Paleozoic time. Somewhat farther north in Scotland is the Caledonian mountain system, which was formed earlier in Paleozoic time.

It is tempting to connect the Appalachians with these far-away systems, yet they are separated from it by thousands of miles of deep ocean, probably floored by simatic crust, on which no topographic trace of Paleozoic deformation has been detected. Proponents of continental drift would have us believe that the broken-off ends of the Appalachians were once directly connected with the Hercynian and Caledonian systems, and have now been separated by displacement of the continental blocks.

As an alternative, the Appalachians might once have joined the fold belt on the east side of the island of Greenland (Plate 1), so that the two formed a continuous border along the North American continent. Much of the deformation in eastern Greenland was completed by middle Paleozoic time, but such early deformational phases are also well displayed in the Appalachians of Newfoundland. Even the fragments of former mountains in Newfoundland and Greenland are separated by a wide expanse of open water, so that the suggested connection is conjectural.

The problem of northeastward linkage of the Appalachians with other structures in Europe or Greenland remains unsolved, and we will carry our speculations no further in this direction.

STRUCTURAL PATTERN. So much for preservation of the old mountain system in modern topography and its possible extensions overseas. Returning again to North America, let us consider its pattern.

On the surface the Appalachians extend with interruptions from Newfoundland to Alabama, a distance of nearly 2,000 miles, disappearing northeastward under the Atlantic and southwestward under the Gulf Coastal Plain. In this long expanse the course of the system is sinuous rather than straight, so that it forms a series of salients and recesses, each salient being 400 to 600 miles across (Plate 1).

One salient occupies most of New England and the Maritime Provinces; to the northeast is a recess in the Gulf of St. Lawrence and to the southwest another near New York City. The structures in Newfoundland northeast of the Gulf of St. Lawrence may be part of another salient. Southwest of New England less strongly convex salients are visible on the map, one centering in Pennsylvania, another in Tennessee and the Carolinas.

West of the Mississippi the pattern of salients and recesses continues, although more obscured by cover of younger deposits. The Ouachita Mountains of Arkansas and Oklahoma seem to lie on the apex of one salient, the Marathon area of western Texas on another, with a deep recess east of the Ouachita salient beneath the Mississippi Embayment and another between the Ouachita and Marathon areas southeast of the Llano uplift in central Texas.

There has been much speculation as to the meaning of these salients and recesses. They might be accidental sinuosities of the former mountain chain, or they might have deeper meaning. Some geologists have suggested that the salients were loci of most active deformation of the system, and of greatest crowding of its rocks northwestward toward the continental interior. Perhaps this is so, yet the evidence adduced for the existence of such loci can be otherwise interpreted.

WIDTH OF EXPOSURE. Recall that east of the Mississippi River the Appalachian system not only passes from view at its ends, but that its southeastern edge is either buried or submerged, so that a variable width of the former structure is now visible.

Broadest exposure is toward the northeast in Gaspé and the Maritime Provinces where a section 400 miles wide is exposed. Southwestward the visible section narrows toward New York City, but widens again in the salient of Tennessee and North Carolina where a section 250 miles wide is exposed (Fig. 30 B). To the southwest exposures narrow once more toward the Gulf Coastal Plain, but in Georgia, Alabama, and Florida drilling has been carried through the coastal plain deposits in so many places that we know much about a cross-section, exposed and buried, that is nearly 350 miles wide.

2. GEOLOGICAL INVESTIGATIONS

Before going further, let us pause to say a few words on the birth of our knowledge of the Appalachian region:

A century and a half ago, in the days of William Maclure, the meaning of steeply inclined strata was only imperfectly understood. Ideas of the 18th Century German "geognocist" *Abraham Gottlob Werner* were still in vogue—that the inclination of strata originated during formation of the sediments, when they were precipitated from waters of a universal ocean onto the slopes of the "primitive" or initial mountains. Inclined strata of this sort were supposed to have formed during a "transition" era, and most of the sedimentary rocks of the Appalachians were assigned to this era by Maclure.

Realization of the true state of affairs came slowly. While the nature of rock deformation had been apprehended by *James Hutton* in Scotland in the latter part of the 18th Century, it was also worked out independently by various American geologists in the decades immediately succeeding Maclure's pioneer publication, as a result of patient field work in the Appalachian chain.

We have said that James Hall and his colleagues of the New York Geological Survey laid the groundwork for American stratigraphy. Fundamentals of American structural geology were being established at about the same time by state surveys to the south, especially in the beautifully folded Appalachian region of Pennsylvania. Here in 1838 *Henry Darwin Rogers*, state geologist of Pennsylvania, and his assistants, found the stratigraphic key of the region—a sequence of formations in their proper order, laid out on one mountain side—and used this to unlock the structure of rocks that elsewhere had been disordered by folding and faulting. The same methods were extended by his brother *William Barton Rogers* into Virginia (which at that time included West Virginia). Results of their work were presented in an epoch-making paper of 1842, in which the Appalachian structures were likened to a series of wave-like undulations moving from the southeast toward the little-deformed continental interior. Details of the structure as there described are still valid—the great linear extent of the folds, the manner in which they are overturned and broken by faults, and the meaning of slaty cleavage.

On this foundation many other geologists have built —this account is too brief to give them the honor they deserve. Of the many we can mention briefly: *J. Peter Lesley*, ordained Presbyterian minister turned geologist and engineer, who carried forward the work of Rogers as state geologist for the Second Pennsylvania Geological Survey, and whose maps are models of technical skill; *James Merrill Safford*, who laid out single-handed the geology of Tennessee during the trying times of the Civil War period; and *Arthur Keith*, who in the latter part of the century produced a great series of geological folios of the southern Appalachian region for the U.S. Geological Survey. It was Keith and his colleagues of the Geological Survey (*Bailey Willis*, *C. W. Hayes*, and *M. R. Campbell*) who discovered the one remaining major feature of Appalachian structure that had escaped the Rogers brothers—the great low-angle thrust faults.

3. APPALACHIAN CROSS-SECTION

Let us now examine the characteristic structures of the Paleozoic mountain system. These can best be illustrated by a cross-section of the Appalachians from northwest to southeast—from the little deformed rocks of the continental interior to the complex, highly altered rocks nearer the coast.

Treatment in terms of such a cross-section is more appropriate for the Appalachians than for many other mountain systems. The Appalachians have been little confused or modified by later structures, and maintain a remarkable similarity throughout their length; they are characteristically divisible into several narrow belts of like rocks and structures that are much the same from Canada to Alabama.

FORELAND AREA. To begin our cross-section let us recall the features of the foreland area on the northwest (Fig. 14)—that this is part of the Interior Lowlands and became the edge of the continental platform early in Paleozoic time by inland overlap of the Cambrian deposits; that during growth of the mountain system the foreland was covered by wedges of sediments eroded from the system; that during closing stages of the deformation the foreland was mildly folded and warped down into a basin.

Largest of these foreland basins is the *Allegheny synclinorium* (Fig. 15) which extends southwestward from Pennsylvania into Kentucky. Topographically it forms the Allegheny Plateau and its extension to the southwest, the Cumberland Plateau—a broad, high-standing region, everywhere deeply cut by rivers and streams, the intervening ridges rising to accordant heights and without pattern other than that imposed by erosion. These plateaus, much more than the ridges of the true Appalachian Mountains on the southeast, were the great barrier to westward expansion of the colonists along the eastern seaboard. Here, as before, we can point out a discrepancy between "topographic mountains" and "geological mountains."

Surface rocks of the plateau and synclinorium are largely Pennsylvanian continental and coal-bearing beds. They lie conformably above Mississippian and Devonian strata and are succeeded conformably in the central part of the synclinorium in West Virginia by a small thickness of similar strata of early Permian age. The Pennsylvanian and associated rocks have been warped into a series of anticlines and synclines by Appalachian movements, but most of the deformation is so light that over wide areas the strata appear to lie nearly flat. Presence of the early Permian rocks in this conformable sequence is of interest in demonstrating that at least the northwestern part of the Appalachian area was deformed later than early Permian time. This fact has led to a further widely held assumption that the whole Appalachian orogeny occurred toward the end of the Paleozoic era. We will see later in the chapter (Section 5) why this idea should be modified.

On the southeast the Allegheny Plateau breaks off along the *Allegheny Front* (Fig. 25 A), an imposing escarpment which overlooks the more varied, linear ridges and valleys of the true Appalachians. The front marks an abrupt change in style of deformation; the strata now turn up abruptly, and beyond they are heavily folded and faulted; we pass here from the foreland into the main deformed belt.

SEDIMENTARY APPALACHIANS (VALLEY AND RIDGE PROVINCE). Broadly speaking, the deformed belt of the Appalachians is divisible into two parts which I will call, for want of better terms, the *sedimentary Appalachians* and the *crystalline Appalachians*; the former is encountered first on proceeding toward the southeast.

The sedimentary Appalachians are built of sedimentary rocks of Paleozoic age, a thick mass that was laid down in the Appalachian geosyncline southeast of the continental platform; this geosyncline will be discussed later in the chapter (Section 5). The Paleozoic begins with Lower and Middle Cambrian deposits which, as we have seen, wedge out toward the continental interior. Most of the first half of the succession, up to the middle of the Ordovician, consists of carbonate rocks (that is, limestone and dolomite); higher parts above the middle of the Ordovician are mainly sandstones and shales—the significance of this arrangement will be seen farther on.

In the humid climate of the Eastern States the limestones and dolomites are more susceptible to erosion than the sandstones and shales; wherever deformation has raised them to view they are worn down to low ground, whereas the adjacent sandstones and shales project as ridges. Characteristic topography of the sedimentary Appalachians is thus a succession of parallel valleys and ridges which form the *Valley and Ridge province.* As the underlying limestones and dolomites are raised highest on the southeast, they occupy their widest surface extent in this direction and form a broad expanse of lowland, the *Appalachian Valley.*

Strata of the sedimentary Appalachians have been flexed into a succession of anticlines and synclines, each a few miles across, one fold crowded closely against another. Most folds are *asymmetrical* and have been pushed over toward the northwest relative to the rocks beneath them, so that on this side of each anticline the strata are steep or overturned, whereas on the southeast side they dip gently. With increase of the northwestward push the strata on the steeply dipping sides have been broken by southeast-dipping faults on which the rocks above have been thrust toward the northwest.

Folds dominate the structure of the Valley and Ridge province in Pennsylvania, but southwestward across Virginia into Tennessee faults increase in number and magnitude until few unbroken folds remain (Fig. 23 B). Many of the faults are more than simple breaks on the flanks of anticlines, and are great *low-angle thrusts* which have carried sheets of rock for miles northwestward over rocks beneath—as shown where the thrust sheets have been warped and breached by erosion, revealing the overridden rocks in *windows.* Some great thrusts, such as the *Saltville* and *Pulaski faults,* are traceable for hundreds of miles along the strike, although they eventually die out at their ends.

Like the folds, the low-angle thrusts result from a relative push of the rocks toward the northwest during the Appalachian deformation. The thrusts were formed by the tearing loose of higher parts of the mass of sedimentary rocks from lower parts along weak, shaly layers interbedded in the succession.

This principle is illustrated by a great thrust block

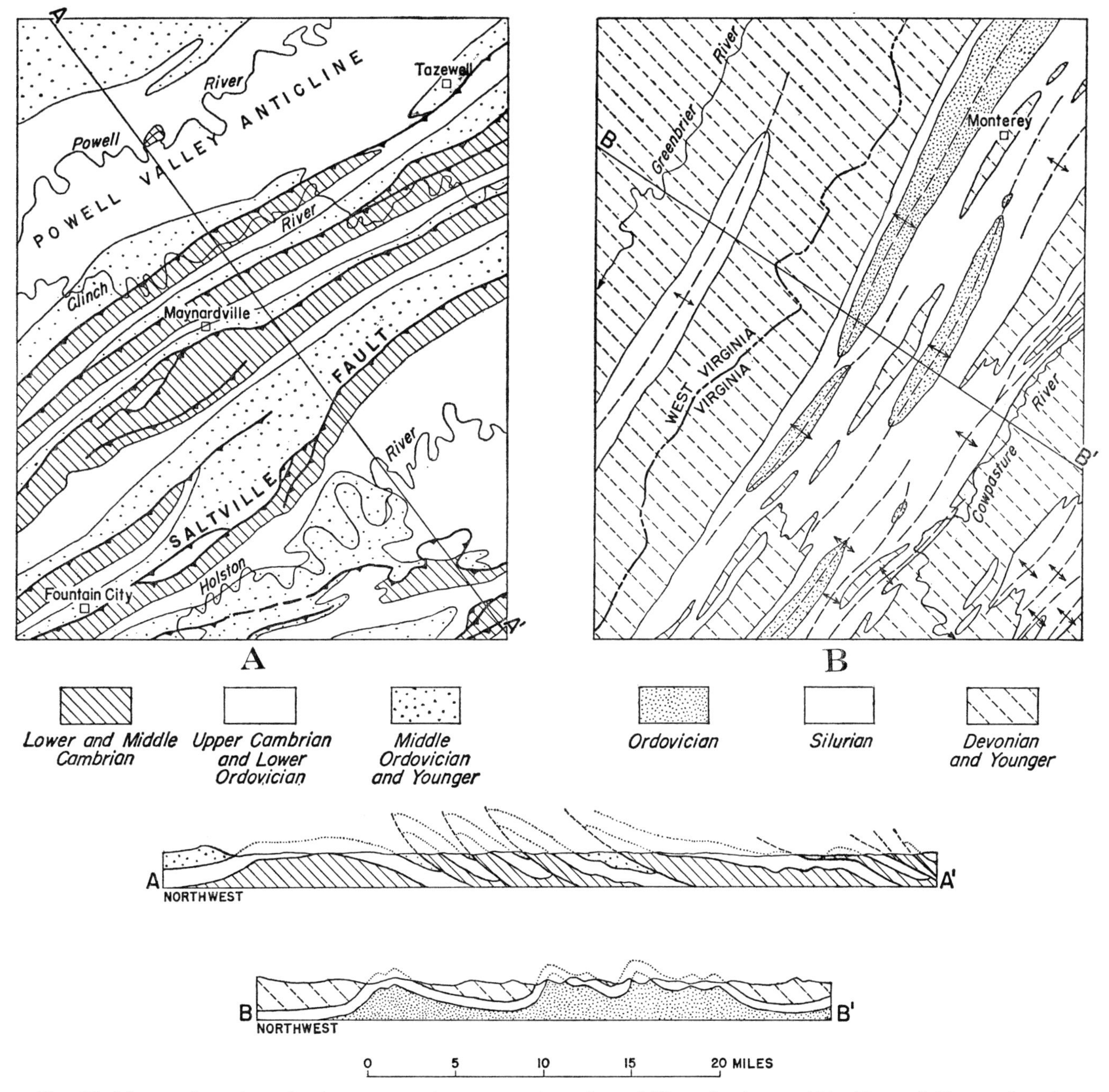

Fig. 23. Maps and sections showing contrast in structure in Valley and Ridge province between the Southern Appalachians where faulting dominates, and Central Appalachians where folding dominates. (A) Maynardville quadrangle, eastern Tennessee. (B) Monterey quadrangle, Virginia and West Virginia. After Keith, 1901; and Darton, 1899.

in the Cumberland Mountain area near the common corner of Virginia, Tennessee, and Kentucky; the block lies along the boundary between the Valley and Ridge province and the Allegheny Plateau, and its initial structures are better preserved than those of the more complex thrusts to the southeast (Fig. 24 A). The block was separated from the rocks beneath along two zones of weak strata: shales in the Lower and Middle Cambrian (Rome and Conasauga formations), and shales at the base of the Mississippian (Chattanooga formation) (Fig. 24 C). When a relative push of the rocks to the northwest occurred, the stronger intervening and overlying strata moved over the shales and also internally along more steeply dipping shear planes; eventually the whole system of fractures united to form a single low-angle thrust, the *Pine Mountain fault.* During northwestward movement along this fault of four or five miles (in the segment shown in Fig. 24) rocks of the overlying thrust block of the Cumberland Mountains became warped as they passed over its irregular, step-like surface (Fig. 24 B).

In the more complexly faulted rocks of the Valley and Ridge province on the southeast, thrust slice after thrust slice brings up weak shaly rocks of the Rome and Conasauga formations, but fails to reveal any older Cambrian or Precambrian rocks. Apparently the thrusts largely originated at the level of the Rome and Conasauga, so that the underlying Cambrian sediments and their basement may have been little disturbed, or at least received a different structure.

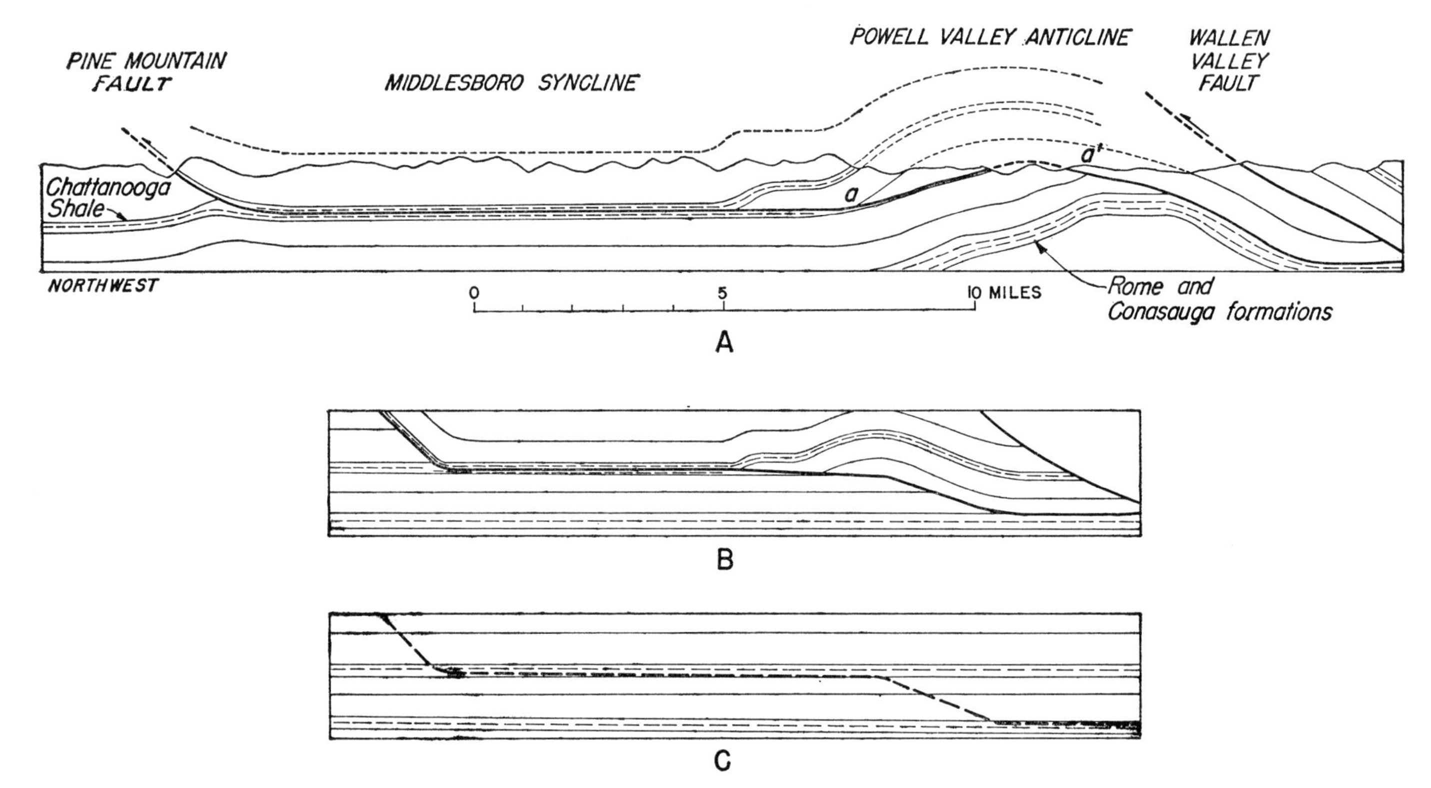

FIG. 24. Sections across Cumberland Mountain thrust block, Virginia and Kentucky. (A) Present structure, with displacement on Pine Mountain fault indicated by points *a* - *a'*. (B) The same, with folding in overridden block eliminated. (C) Area before faulting, showing initial course of break that later became Pine Mountain thrust. Compiled from Butts, 1927; Rich, 1954; and Miller and Fuller, 1947.

Thrusting and probably also folding in the sedimentary Appalachians was thus relatively shallow, confined to the body of sedimentary rocks above the Precambrian basement. It could not have developed independently, but was *marginal* to other structures on the southeast in the Blue Ridge and crystalline Appalachians, whose growth provided a driving force; these we will now examine.

We can summarize that part of the Appalachians so far covered by means of two generalized cross-sections (Fig. 25).

BLUE RIDGE PROVINCE. Southeast of the Appalachian Valley rise other mountains known as the *Blue Ridge* from Pennsylvania southwest to Georgia. The Blue Ridge narrows northeastward and ends in Pennsylvania as South Mountain, but it widens southwestward into Tennessee and North Carolina where it is a massive highland 75 miles across (Fig. 30 B). This includes Mount Mitchell with an altitude of 6,684 feet, the highest summit in the eastern United States, and many other peaks almost as high. These altitudes may not seem impressive by western standards, nevertheless the Blue Ridge is a good rugged mountain area in which relief from ridge top to valley bottom is more than a mile in many places. The Great Smoky Mountains popular in song and legend are part of the Blue Ridge, although the name is restricted by geographers to a small segment in Tennessee and North Carolina. In New England the *Green Mountains* lie in the same position as the Blue Ridge in the cross-section, and have a similar structure (Fig. 30 A).

A structural tendency in the Valley and Ridge province continues into the Blue Ridge—a progressively higher uplift toward the southeast. Here, therefore, rocks beneath the carbonates of the Appalachian Valley emerge—Early Cambrian quartzites, arkoses, and conglomerates that form the bottom of the

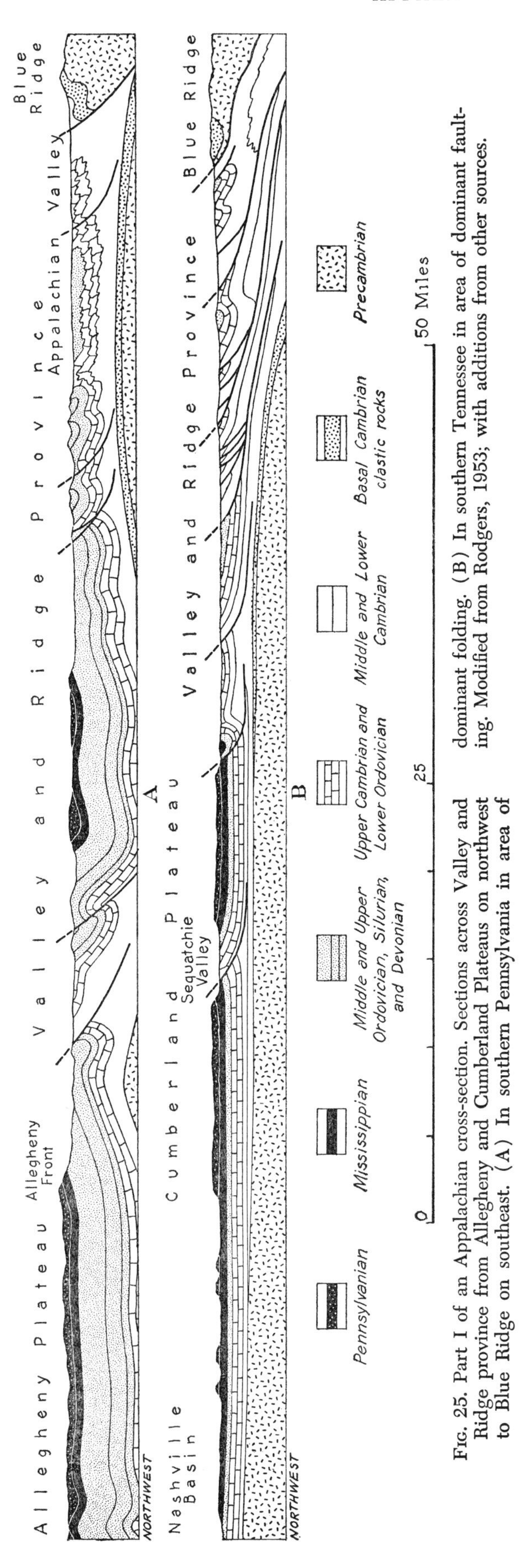

FIG. 25. Part I of an Appalachian cross-section. Sections across Valley and Ridge province from Allegheny and Cumberland Plateaus on northwest to Blue Ridge on southeast. (A) In southern Pennsylvania in area of dominant folding. (B) In southern Tennessee in area of dominant faulting. Modified from Rodgers, 1953; with additions from other sources.

Paleozoic geosynclinal column, and various stratified later Precambrian rocks beneath. The latter consist mainly of basaltic lavas in Virginia, but farther southwest are a great mass of graywackes and other clastics less cleanly washed than the Paleozoic deposits. These stratified rocks—early Cambrian and later Precambrian—lie on a basement of earlier Precambrian granites and gneisses that has been forced up as a welt or anticlinorium in the midst of the Appalachian structure.

Significant changes in style of deformation appear in the Blue Ridge province which foreshadow the kind of deformation in the main crystalline belt of the Appalachians beyond. Folds of the sedimentary Appalachians formed in the strata mainly by *flexing*, distortion being taken up by gliding between the layers, with little development of slaty cleavage. On passing into the Blue Ridge, folding is mainly by *shear*, so that the rocks have been deformed by laminar flow, with great thickening along the axes and thinning along the flanks of the folds, and with extensive development of slaty cleavage; distinctions between stratified overburden and basement become blurred and the whole mass of rocks has been deformed as a unit.

In places deformation of this kind resulted merely in folding, but in others it produced low-angle thrusts which differ from those in the sedimentary Appalachians because they extend as great shears deep into the basement. Such thrusts dominate the border of the Blue Ridge from southwestern Virginia into Tennessee and beyond; their northwestward movement probably provided a driving force which brought about the marginal deformation in this segment of the sedimentary Appalachians.

CRYSTALLINE APPALACHIANS. (NEW ENGLAND UPLAND AND PIEDMONT PLATEAU). Southeast of the Blue Ridge follow the main crystalline Appalachians. Some of the crystalline area projects in mountains quite as high as the Blue Ridge—the New England Upland extends eastward from the Green Mountains; the summit of the White Mountains beyond in New Hampshire is Mount Washington with an altitude of 6,293 feet, the highest peak in the Northeastern States. But farther southwest, because of relatively rapid erosion of the crystalline rocks in the humid climate, wide areas have been worn down to low ground in the Piedmont Plateau, as we have seen.

The crystalline Appalachians are made up of metamorphic schists and gneisses and of various plutonic (or deep-seated igneous) rocks, mainly acidic or granitic, but partly basic or even ultrabasic.

Because the crystalline complex resembles similar altered Precambrian rocks in the Canadian Shield, it has long been inferred that it also is of ancient age.

Over wide areas, it is true, exact ages of the plutonic and metamorphic rocks have not been determined, but radiometric determinations on the plutonic rocks have yielded no such extreme ages as those in the shield. In many places in New England and Canada as well as a few farther south, fossils of Paleozoic age have been found in the metamorphic rocks. Even where Precambrian rocks came to the surface, they no longer acted as a basement but were deformed in the same manner as their overburden.

The schists and gneisses, both Paleozoic and earlier, were formed from what were originally surficial and stratified rocks—argillites, sandstones, tuffs, and lavas. Not only have those of Paleozoic age undergone a different deformational history from those in the sedimentary Appalachians, but their original nature was not the same and they developed in a contrasting environment.

Scattered through the metamorphic rocks are small pods and lenses of *ultrabasic* (or "ultramafic") plutonic rocks, variously termed peridotites, pyroxenites, dunites, or serpentines according to their mineral composition. Although individual bodies are small, they are characteristically grouped in belts or swarms; in both New England and the Southern Appalachians the most prominent belts are in the northwest part of the crystalline area in or near the Green Mountain and Blue Ridge uplifts; occasional bodies occur farther southeast.

Of much greater extent are the *acidic plutonic rocks*, the granites and their allies, which permeate the metamorphic rocks so widely that few areas are wholly devoid of them; in part they replaced or engulfed the metamorphic rocks to such an extent that not much of the latter remains.

The acidic plutonic rocks exhibit a wide variety of relations to the metamorphic rocks (Fig. 26). Some are thoroughly gneissic and form thin to thick lenses or beds concordant with the foliation of the metamorphics; they fade out into the metamorphics at their ends as streaks and lit-par-lit injections. At the opposite extreme are massive granitic rocks which break through and cut off the ends of the metamorphics and the granite gneisses, forming bodies of various shapes and sizes up to scores of miles across, and including in New England a spectacular array of ring dikes.

But these extremes in the acidic plutonic rocks are not sharply separated; between are many bodies that are partly concordant and weakly gneissic, yet which break through the enclosing rocks in many places. The whole assemblage, concordant to discordant, forms a *granite series* which seems to have developed through time, beginning with the former and ending with the latter.

Significance of the plutonic rocks—ultrabasic to acidic—we will examine later (Section 5).

The metamorphic and plutonic rocks of the crystalline Appalachians have a highly complex structure, so that their pattern on the geologic map is one of swirls and knots that is sometimes difficult to resolve into any orderly system of folding (Fig. 27). Foliation mostly stands at high angles, but in places it rolls over the crests of domes or arches and in other places it dips at low angles over wide areas. Where the metamorphic rocks have been mapped in detail, it has been possible to subdivide them into units and resolve their structure, but this structure proves to be very different from that of the sedimentary Appalachians, in that it involves much flowage of the rocks and thickening and thinning of the units with little breaking or faulting. Nearly all the faults of the crystalline area formed after the rocks had been deformed and congealed. Deformation of the crystalline Appalachians must have taken place under greater overburden and at greater pressures and depths than the folding and faulting of the sedimentary Appalachians, in a realm where large supplies of heat and magmatic juices were available.

Conditions in this realm are suggested by various index minerals which grow from the rock materials with increasing temperature and pressure. First traces of metamorphism are indicated by chlorite; progressively higher metamorphic ranks are shown by biotite, garnet, staurolite, and finally by sillimanite. Commonly the metamorphic rank increases near larger acidic plutonic bodies—obvious local sources of heat—but this cannot be the whole story; the metamorphic mineral assemblages as well as the degree of physical metamorphism progresses regionally from the edges toward the interior of the crystalline belt; regional metamorphism seems to result from a combination of all conditions at work in the axial zone of the orogenic system: depth, pressure, heat, and plutonism.

In the Southern Appalachians, as one might expect, metamorphic rank increases progressively southeastward from the Valley and Ridge province across the Blue Ridge and into the Piedmont. But the climax in the sillimanite zone does not come in the extreme southeastern exposures; instead, it appears unexpectedly in a belt about midway between the Blue Ridge uplift and the edge of the Coastal Plain (between Brevard and Kings Mountain belts on Fig. 30 B); southeast of this metamorphic rank again decreases. We will see the significance of this in a moment.

Post-orogenic deposits. Besides the foundation of

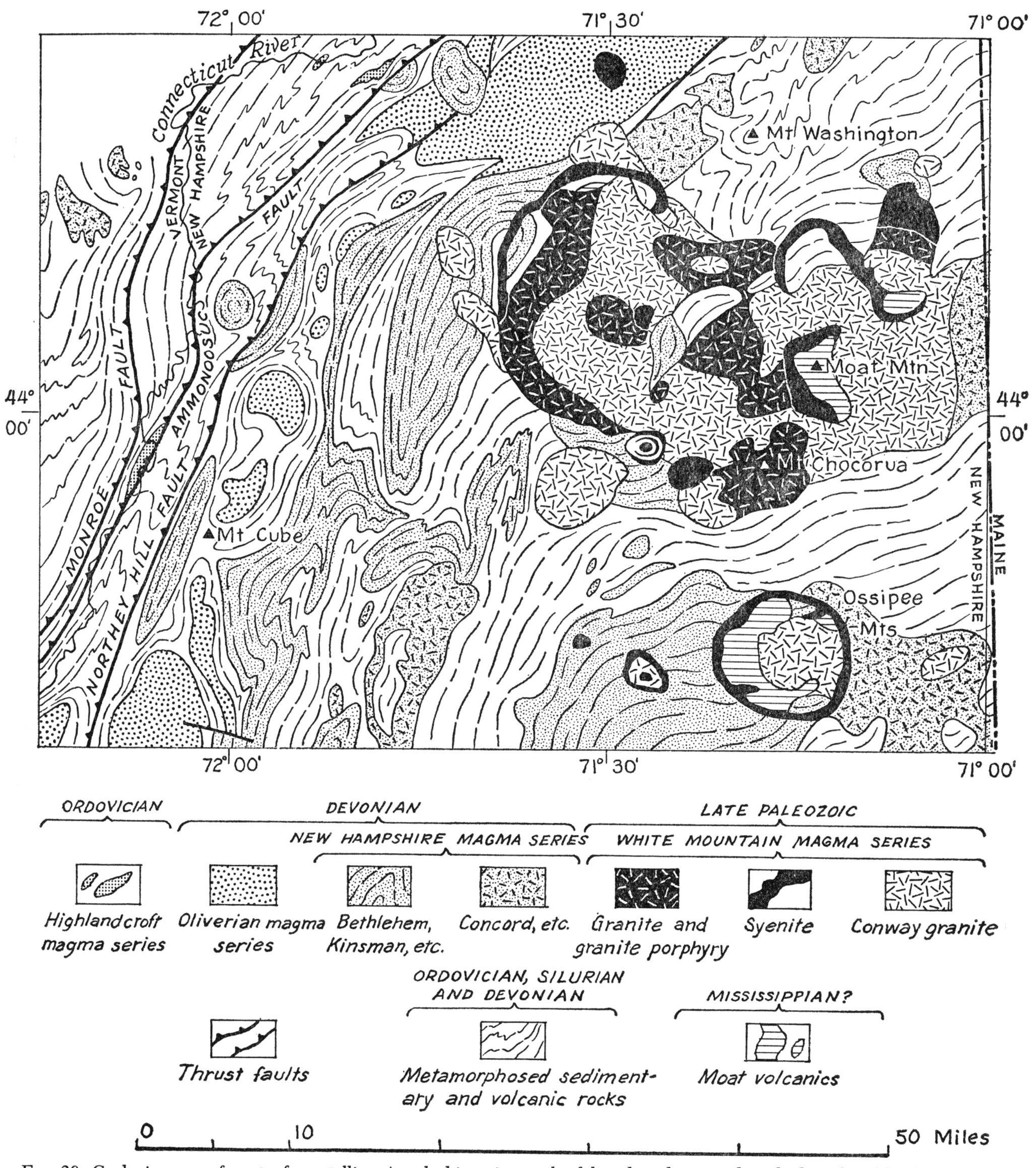

FIG. 26. Geologic map of part of crystalline Appalachians in New Hampshire, showing bodies of acidic plutonic rocks lying in metamorphic host rocks that were originally sediments and volcanics laid down in the eugeosynclinal area. The plutonic rocks have various structures ranging from early foliated and concordant bodies (Highlandcroft, Oliverian, and part of New Hampshire magma series) to later massive and through-breaking bodies (mainly White Mountain magma series); these form a "granite series." After Geologic Map of New Hampshire, 1955; and other sources.

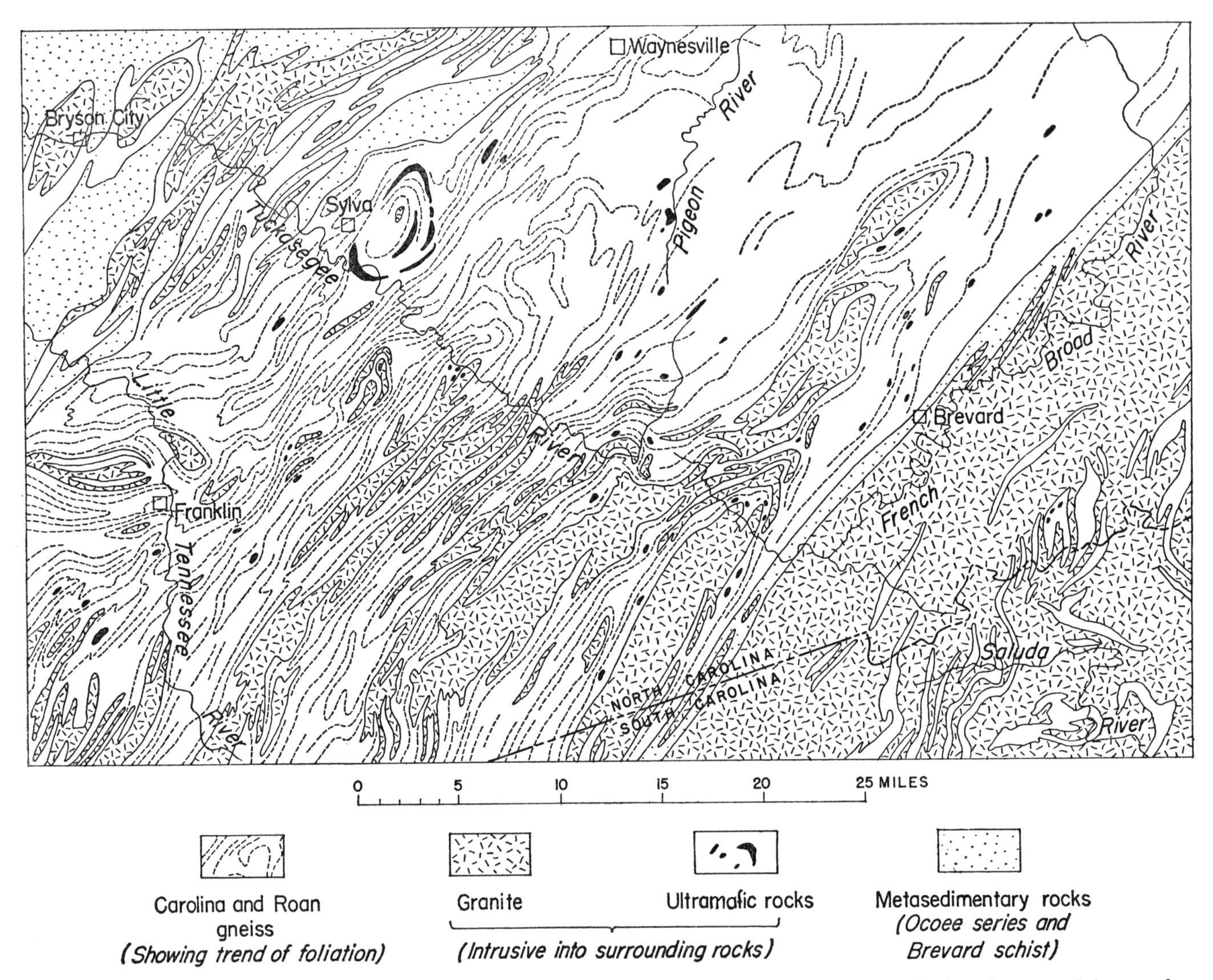

FIG. 27. Map of typical part of crystalline Appalachians in Blue Ridge area of North and South Carolina, showing swirled pattern of deformed metamorphic rocks with small ultrabasic intrusives and larger bodies, knots, and lenses of acidic plutonic rocks. From geological maps of Cowee and Pisgah quadrangles by Arthur Keith.

crystalline rocks, the inner part of the Appalachians contains patches of sedimentary and igneous rocks that are partly or wholly younger than the principal deformation. The sedimentary rocks lie unconformably on the eroded edges of the deformed rocks, and contain fragments of their metamorphosed and plutonized products; they are, therefore, *post-orogenic.*

From Nova Scotia to South Carolina rocks of the *Newark group* of late Triassic age form long strips in the crystalline area; similar rocks have been encountered in wells under the coastal plain deposits as far south as Florida. Newark sediments are mainly continental red sandstones, but include shale, conglomerate, and thin beds of coal. They probably accumulated in downfaulted troughs of little greater dimensions than the present outcrops, after the great deformation of the crystalline foundation had ceased but while the region still possessed considerable relief. The sediments are not metamorphosed or even folded, but have been tilted, warped, and broken by normal faults produced by crustal tension (Fig. 28). Sedimentation was accompanied by igneous and volcanic activity; in places the sediments are interbedded with basaltic lavas and intruded by masses of diabase; diabase dikes of the same age also penetrate the crystalline foundation widely, even far from any areas of Triassic sediments (Fig. 29).

The Triassic rocks provide a significant terminal date beyond which no Appalachian deformation could have taken place; hence, Appalachian deformation is definitely older than the late Triassic. For the present, let us conclude that deformation of the crystalline rocks on which they lie might have been only a little earlier than late Triassic, or might have been much

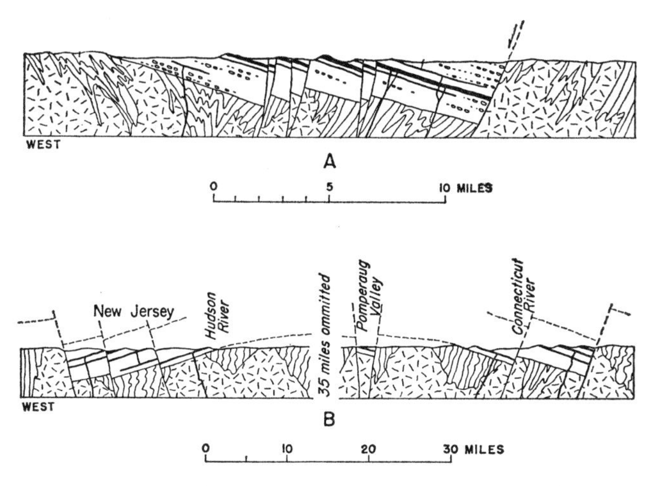

Fig. 28. Sections showing structure of Triassic rocks (Newark group) in southern part of Northern Appalachians: (A) Section across Connecticut Valley near Hartford, Connecticut. (B) Regional section from Connecticut to New Jersey. After Barrell, 1915; and Longwell, 1932.

earlier; we will return to the problem later (Section 5).

Post-orogenic deposits which bridge, to some extent, the lost interval between Triassic deposition and the deformational climax occur in the northeastern part of the Appalachian system in the Maritime Provinces of Canada.

Here are extensive areas of Mississippian and Pennsylvanian sedimentary rocks which are less deformed than the Devonian and earlier rocks beneath, and show no trace of their metamorphism. Although some marine beds occur in the Mississippian, most of the deposits are continental, and the Pennsylvanian contains beds of coal that are mined on Cape Breton Island and the mainland of Nova Scotia. These later Paleozoic rocks are nearly flat-lying in places, but in others are mildly to steeply folded. Their structure indicates that the major deformation of the northeastern segment of the crystalline Appalachians had been completed before Mississippian time, although

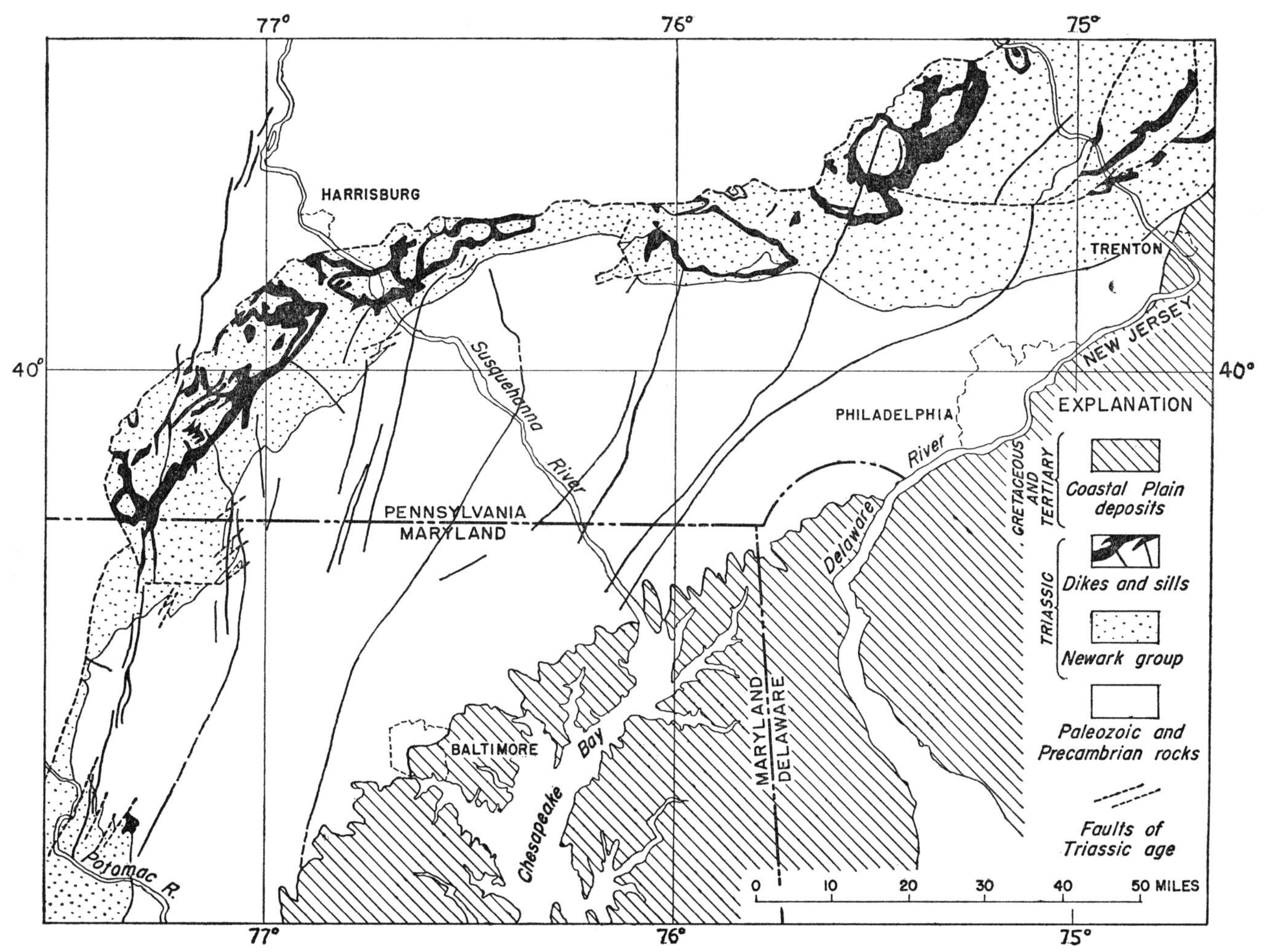

Fig. 29. Map of southeastern Pennsylvania and eastern Maryland showing sedimentary rocks of Newark group, and dikes and sills of diabase which intrude both the Newark group and the earlier crystalline and sedimentary rocks of surrounding areas. After Stose and Stose, 1944.

milder deformation continued nearly to the end of the Paleozoic era. The Mississippian and Pennsylvanian post-orogenic deposits, like those of the Triassic, are basinal and mainly continental, but they differ in that they bear the marks of the last crustal compression.

We have now carried our exposition of the Appalachian system far enough to present a second part of our cross-section (Fig. 30).

SOUTHEASTERN BORDER OF APPALACHIAN SYSTEM. What happens on the southeast side of the Appalachians?

As we have seen, the tilted, metamorphosed, and plutonized rocks of the crystalline Appalachians pass from view in this direction beneath deposits of the Atlantic Coastal Plain and waters of the continental shelf, so that this flank of the system is concealed. But the deformed belt cannot continue indefinitely southeastward, since beyond the continental shelf is the deep Atlantic Ocean, floored by simatic crust—a quite different structural element.

Although much of the southeastern border of the Appalachian system is not visible, we nevertheless can make many educated guesses about it from various lines of evidence—from rocks and structures exposed immediately northwest of the edge of the coastal plain, from wells which have been drilled through the coastal plain deposits, and from geophysical data on the land and at sea.

Surface outcrops suggest that the axis of greatest deformation in the Appalachians does not lie at some undetermined point beneath the coastal plain deposits but within the exposed area of the Piedmont. As already noted, the main sillimanite zone of the Piedmont, or metamorphic climax, lies well northwest of the edge of the coastal plain. In many places southeast of it, cleavage, axial planes of folds, and occasional thrusts dip northwest. From Virginia through the Carolinas into Georgia between the sillimanite zone and the edge of the coastal plain is the *Carolina slate belt* (Fig. 30 B). This is formed of slates, graywackes, pyroclastics, and lavas which are only moderately folded or metamorphosed except near occasional granitic bodies. Age of these rocks is unknown, as they have not yet yielded fossils, but they are generally believed to have been formed during some part of Paleozoic time.

Proven Paleozoic rocks have been discovered by drilling beneath the coastal plain deposits of southern Georgia and northern Florida, again southeast of the crystalline Appalachians. Well cores show that these are shales and sandstones which are nearly flat-lying and quite unmetamorphosed; fossils of Ordovician to Devonian ages have been collected from the cores.

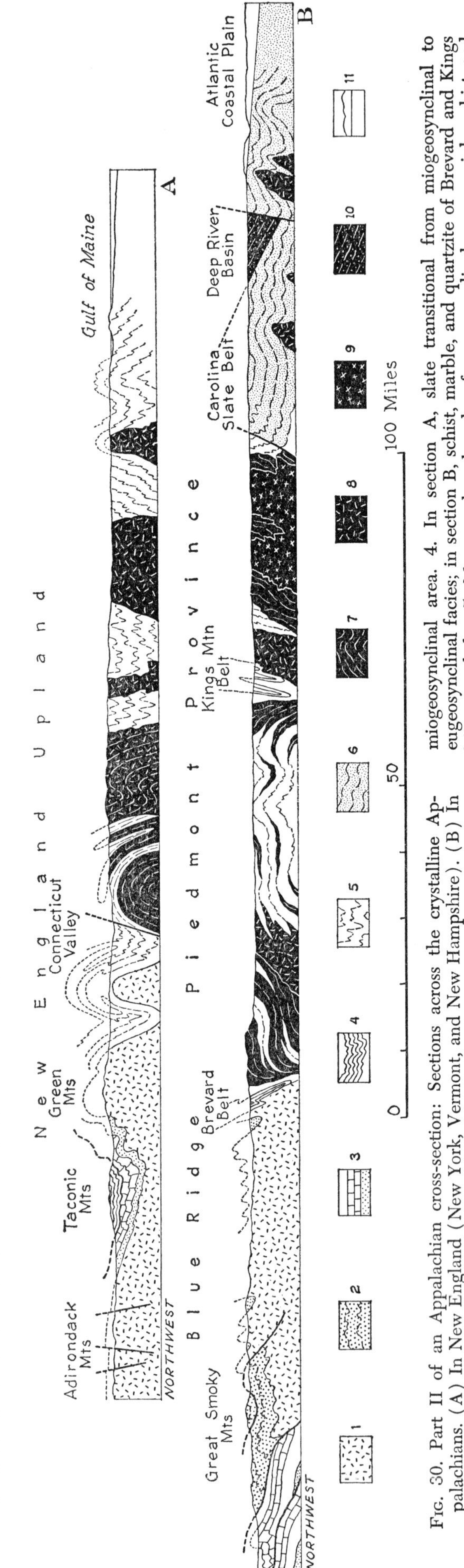

FIG. 30. Part II of an Appalachian cross-section: Sections across the crystalline Appalachians. (A) In New England (New York, Vermont, and New Hampshire). (B) In Tennessee and North Carolina. Note that these sections are on a smaller scale than Part I of the cross-section (FIG. 25). Generalized from sections in Geological Society of America guidebooks, 1952 and 1955.

Explanation of symbols. Precambrian rocks: 1. Basement rocks, mainly granite and gneiss. 2. In section B, Ocoee series, altered sedimentary rocks of late Precambrian age. *Paleozoic rocks of surficial origin:* 3. Limestone, quartzite, and shale, formed in miogeosynclinal area. 4. In section A, slate transitional from miogeosynclinal to eugeosynclinal facies; in section B, schist, marble, and quartzite of Brevard and Kings Mountain belts. 5. Metamorphosed rocks of eugeosynclinal area, mainly schist and gneiss; in section B, may include some Precambrian. 6. In section B, slightly metamorphosed clastic and volcanic rocks of Carolina slate belt. *Paleozoic plutonic rocks:* 7. Foliated acidic rocks. 8. Massive or weakly foliated acidic rocks. 9. In section B, diorite and gabbro. *Younger sedimentary rocks:* 10. Triassic. 11. Cretaceous and Tertiary.

In this region we clearly have passed beyond the axis of greatest deformation and metamorphism in the crystalline Appalachians and come out on the other side; here, at least, the Appalachians are a two-sided mountain system.

These unaltered Paleozoic sediments southeast of the crystalline axis of the Appalachians may be unique; this segment of the system seems to be bordered by a small plate of continental rocks that underlies Florida and the Bahama Islands, whereas elsewhere its southeastern side is flanked by simatic ocean bottom. At this time, at least, we cannot say whether such little altered rocks continue northeastward; if present they would be deeply buried under the coastal plain and continental shelf.

COASTAL PLAIN AND CONTINENTAL SHELF. Further inquiry regarding the southeastern border of the Appalachian system must be by indirection—by deduction from geophysical observations that have been made from the coast across the continental shelf into the deep Atlantic Ocean basin.

Much of this information we owe to *Maurice Ewing* and his associates of Lamont Geological Observatory at Columbia University, who in the last twenty years have made a series of seismic refraction profiles in the segment between Newfoundland and southern Virginia. Means by which this work was done, especially the geophysical observations made at sea, is a fascinating story in itself which we cannot go into here; we will confine ourselves to a summary of the geological results.

Seismic refraction profiles indicate that the Atlantic Coastal Plain and its continuation beneath the continental shelf are underlain by a southeastward thickening wedge of sediments, unconsolidated above (mainly Tertiary) and semiconsolidated below (mainly Cretaceous). Toward their outer edge the deposits are somewhat more than 15,000 feet thick; they form a sort of embankment which breaks off abruptly at the outer edge of the shelf; sedimentary cover of the deep ocean bottom beyond is relatively thin.

The surface of the crystalline Appalachians—the substructure of the embankment of coastal plain and continental shelf deposits—descends gently seaward. At its inner edge, at least, the surface was one of pre-Cretaceous subaerial erosion; why it was downflexed toward the ocean raises interesting questions regarding behavior of the continental margins which we cannot pursue now. Thickest continental shelf deposits are somewhat northwest of the edge of the shelf, and the surface of the foundation rises again toward the ocean basin.

The exposed schists, gneisses, and acidic plutonic rocks, and probably those well out beneath the sedimentary cover form a typical sialic crust. Somewhat beyond the edge of the continental shelf these thin out and give place to the simatic crust of the ocean bottom. A curious feature of the geophysical results is that the sialic rocks in many profiles show a decrease in velocity properties near the edge of the continental shelf, which may mean that such rocks are less metamorphosed than those farther inland—an inference similar to our conclusion from outcrops and well data in the Carolinas, Georgia, and Florida.

Geophysical observations permit us to add a final part to our cross-section of the Appalachian system, as shown in Figure 31.

SUMMARY. We have now reviewed the topographic provinces and structural units of the Appalachians. Later we will consider the sedimentary and other environments in which these different units were formed. It is our misfortune that it is necessary to use for each of these—the topographic, structural, and sedimentary units—a different set of names, many of which are hallowed by usage. To prevent confusion and make clear how these are interrelated, the following table is inclined for reference. The terms given under "rock facies" are defined in Section 5 of this chapter.

COMPARISON BETWEEN DIFFERENT NAMES FOR LARGER UNITS IN THE APPALACHIAN AREA

Physiographic province:	*Structural unit:*	*Rock facies:*
Allegheny and Cumberland plateaus	Allegheny synclinorium	Foreland
Valley and Ridge province	Sedimentary Appalachians (folded and faulted sedimentary rocks)	Miogeosyncline (see Section 5)
Blue Ridge province (including Green Mountains in north)	Anticlinorium in some places, faulted uplift in others	Exposures of basal Paleozoic and of Precambrian basement
New England Upland and Piedmont province	Crystalline Appalachians (metamorphic and plutonic rocks)	Eugeosyncline (see Section 5)
Atlantic Coastal Plain and continental shelf	Cretaceous and Tertiary sediments, lying on crystalline rocks of eugeosynclinal origin	
Atlantic Ocean basin	Sea floor at depths of 15,000 feet, with sima not far beneath sedimentary cover.	

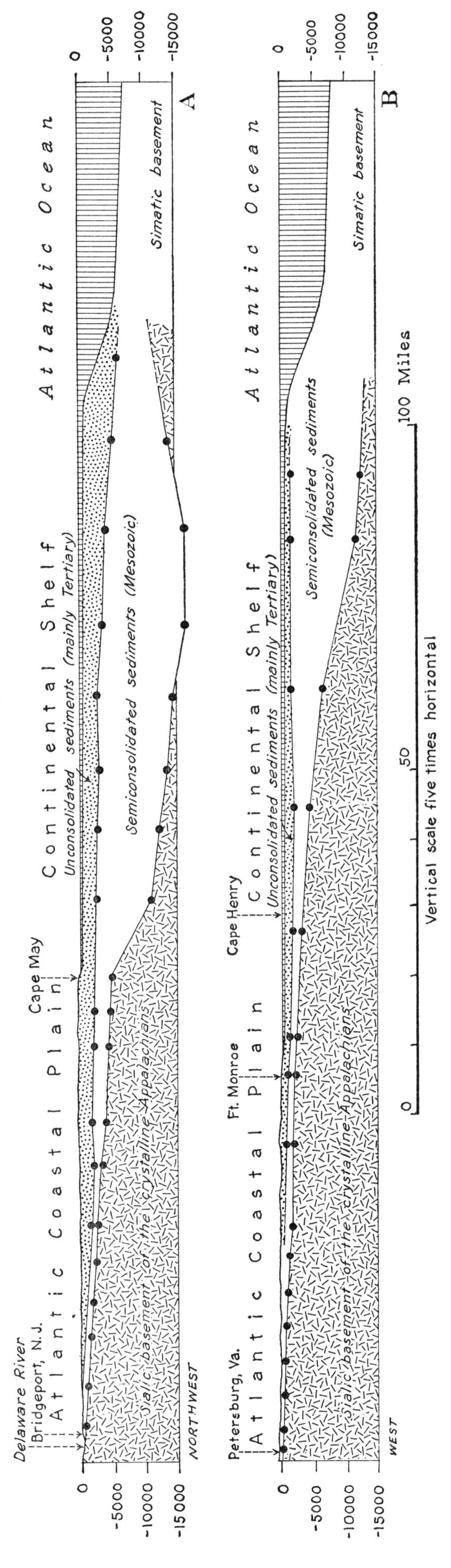

FIG. 31. Part III of an Appalachian cross-section: Sections across Atlantic Coastal Plain and continental shelf to Atlantic Ocean basin, based on seismic refraction profiles. (A) Near Cape May, New Jersey. (B) In southern Virginia, Black circles indicate points located by geophysical means. After Ewing and others, 1950 and 1937.

4. GEOSYNCLINES

In preceding parts of this chapter we have placed the Appalachians in *space* by summarizing their surface form, their plan, and their cross-section. Although we have been able to draw some inferences, we have not yet presented a comprehensive story, and many of the facts given may have seemed unrelated. The next step in our analysis will be to arrange these facts in a sequence and review the Appalachians in *time*—how they originated, how they grew, and how they were built into the structures we see today. None of this story took place overnight or even in a geological period or two; deformation was preceded by a long time of preparation.

During much of the time of preparation the Appalachian region was a geosyncline. Before discussing the growth of the system itself it will be profitable to discuss geosynclines in general—a subject to which we have alluded briefly in several of the earlier chapters.

HALL'S IDEAS. First, it is worth setting forth how the idea of geosynclines originated in the minds of geologists.

The notion of what we now call a geosyncline probably began with *James Hall*, whom we have encountered already as one of the pioneers of American paleontology and stratigraphy (Chapter III, Section 2). In making his observations on strata and fossils he traveled widely in New York State, both in the folded rocks of the Appalachian area and farther west, besides which he extended his studies into the interior region where he at one time made a geological survey of Iowa. From these experiences he came to significant conclusions regarding the lay of the strata and their bearing on the formation of geologic structures which he summarized in an address in 1859.

In brief, Hall observed that Paleozoic sediments of the interior region are thin, whereas those of the Appalachian area are thick; yet he observed that each had formed during about the same span of time. He reasoned that there must be some relation between the greater thickness of sediments in the Appalachians and their mountainous character, and concluded that all large mountain chains must represent areas of greatest acccumulation of sediments.

Sediments in the part of the Appalachians which Hall had studied, like those of the interior region, were laid down in shallow water. The greater thickness of shallow water sediments in the Appalachians was due to greater subsidence of the crust there than in the interior; this, in turn, must have resulted from gradual yielding of the crust beneath the weight of the sediments themselves. Later, these great sedimentary accumulations were uplifted, then eroded into the mountain belts we see today.

Mountain formation was thus largely a matter of thick accumulation of sediments, subsidence of the crust, and subsequent uplift; deformation was incidental to the accumulation and a consequence of the subsidence. Along the axis of the subsiding area sediments became folded and in places were so ruptured as to allow the ascent of igneous magmas. Here, then, was a theory of mountain formation in which the great transformations on and within the crust were ascribed to slow surficial processes—those of erosion and sedimentation.

DANA'S IDEAS. The significance of Hall's observations and speculations gradually came to be appreciated among geologists, but so far the ideas were a theory without a name.

It remained for another geologist, *James Dwight Dana*, to give them a terminology. Dana was a contemporary of Hall's, and for many years was professor of geology at Yale University. He originated the name *geosyncline* (which he called a "geosynclinal"), as well as the less commonly used term *geanticline* (in the form "geanticlinal"), applied to the opposite of a geosyncline—a part of the crust that is gradually rising and being eroded, rather than subsiding and being covered by sediments.

While Dana agreed with the facts as presented by Hall, he was more critical of his explanation, and appraised it as "a theory of the origin of mountains with the origin of mountains left out." Dana had theories of his own and was one of the first geologists to point out the fundamental difference in geological character between continents and ocean basins, and to suggest the possibility of continental accretion—matters which we have discussed earlier (Chapter I, Section 3). The earth's crust he thought to be under compression because of a general contraction of the earth's interior; greatest yielding to such compression was between the continents and the oceans along belts warped down to form the "geosynclinals." Here, thick masses of sediments accumulated because a place was made for them by forces within the earth. Eventually the "geosynclinals" were destroyed by "a catastrophe of plications and solidifications."

To sum up, observe that both Hall's and Dana's theories were based on parts of the Appalachian area; hence, that the Appalachians are the site of the type geosyncline; also, that, although certain implications of mountain formation were inherent in both theories, emphasis was primarily on the sediments themselves; finally, that the opinions of the two men diverged greatly as to the nature of the forces involved—whether thick accumulations of sediments resulted from loading of the crust, or whether the crust was bent down by other processes to form a receptacle in which the sediments could accumulate. We will see that each interpretation has its adherents to this day.

LATER IDEAS. Since the time of Hall and Dana, nearly all geologists have given some thought to geosynclines, but there has been a wide range of opinion as to what geosynclines are, what caused them, and what their effect has been on the subsequent history of an area.

Some geologists have rejected the idea entirely, preferring to consider what it seeks to explain in other ways and by other terms. But most geologists have found the name and concept useful.

To some, a geosyncline is almost any area of relatively thick sedimentation, including not only sedimentary troughs that later became deformed mountain belts, but also sedimentary basins in the interior which were never more than slightly deformed, and basins where sedimentation is now actively in progress. This view has the virtue of treating sedimentary accumulations for their own sake, without implications as to what caused them or what they might evolve into; possibly only in this way can we discover any modern analogues of geosynclines, a topic which we will consider in Chapter V.

To most geologists, nevertheless, the problem of geosynclines is inextricably interwoven with that of orogenic belts and the origin of mountains. A recent textbook of structural geology defines geosynclines as "those accumulations of sediments of great thickness which have been severely folded,"* and excludes those accumulations which are not. While this has some virtue for operational purposes, it seems merely to turn our thinking inside out. What were the unique qualities of the sedimentary accumulations and their subcrust that conditioned them for such severe folding later on?

These elusive qualities have been widely sought. Some geologists, such as those who have labored long in the European Alps, believe that geosynclinal sediments are fundamentally different from sediments of other areas—that they are bodies now found only in heavily deformed regions made up of shaly and slaty rocks rhythmically interbedded with sandstones and associated with peculiar limestones, cherts, and submarine volcanics. We will see presently that this sedimentary facies, while real, is probably only a part of the whole mass of sediments in any geosyncline. Such special definitions of geosynclinal sediments (amusingly enough) would make it necessary to ex-

* De Sitter, L. U., 1956, *Structural Geology*, p. 351.

clude from the geosyncline most of the sediments of the Appalachian area—the typical and original geosyncline as envisioned by Hall and Dana.

Some geologists have elaborated on Hall's belief that geosynclines originated from loading of the crust by the weight of the sediments themselves. Arguments have been brought forward, based especially on the Gulf Coast, that the great thicknesses of Cenozoic sediments preserved there resulted from an inordinate volume of material brought down from the land by its many rivers; it is thought that such material could only have been entrapped near the river mouths if the crust were depressed by the sediments. It has been suggested, besides, that sediments can accumulate only up to a limiting thickness before they become folded; hence, that "drainage patterns determine the future courses of mountains."

In extreme form such ideas are rather bizarre, but at least they have a partial basis in *isostasy*. We now know that the crust has limits of strength beyond which it will subside under load—we have seen this in our discussion of the Pleistocene ice caps of the Canadian Shield (Chapter II, Section 1). Great bodies of sediments must therefore depress the crust to some extent, but most geologists do not believe that light unconsolidated sediments can displace an equal volume of heavier consolidated crust beneath.

Instead, these geologists seek the formation of geosynclines and subsequent mountain building in forces at work deeper within the crust. The geosyncline is, then, perhaps a milder initial manifestation of a long sequence of crustal mobility which has resulted in such varied structures as those which we have just reviewed in our Appalachian cross-section.

Geosynclinal attributes. In the face of such wide diversity of opinion about geosynclines it is a brave spirit who dares define them explicitly. Perhaps it would be more profitable if I listed a series of "attributes" which I believe are possessed by geosynclines. Some are fairly obvious; others differ more or less from beliefs of various other geologists.

(1) Geosynclines formed during sedimentation rather than afterwards, and are more sedimentary than deformational features. Characteristically, they received a greater thickness and volume of sediments during their life than non-geosynclinal areas.

(2) Geosynclines are not ordinary synclinal folds or even groups of folds like the Allegheny synclinorium. Synclines and synclinoria were imposed on the strata after they had been deposited, and are not inherent in the deposit.

(3) Thick accumulations of sediment are caused by subsidence of the crust beneath them, but this is a gradual process. Ideally, at the end of a geosynclinal phase the floor is bowed down in the manner of a syncline, but the last sediments to be deposited are nearly horizontal. Such downbowing is very gentle —the trough is likely to be a hundred miles or more wide, but maximum depth of the depression as expressed by thickness of sediments is merely six to ten miles.

(4) Geosynclines commonly occur along the margins of continental platforms, where they form belts which are much longer than wide; they thus resemble ordinary synclines in their linear form. Basins within the continental platform and basins of non-linear form are not properly geosynclines, although they have been so called by some authors. Nevertheless, one type of basin grades into the other so that distinctions are not absolute.

(5) In detail, sediments of geosynclines are varied; no one suite can be set off exclusively as "geosynclinal." Several different sedimentary facies occur in geosynclines, some sufficiently distinct as to warrant separate names (see below). These facies may occur in different parts of the geosyncline, where they persist through small to great thicknesses of strata, or they may succeed each other in the same part of the geosyncline in one sequence of rocks.

(6) Sediments of some of these facies were laid down in relatively shallow water, as observed long ago by Hall. Such sediments differ little in kind from those of non-geosynclinal regions, but they did accumulate to greater thickness and with less interruption. In areas of shallow water sedimentation the geosynclinal trough filled as it subsided; its base was bowed down gradually, yet its upper surface at no time had a synclinal form.

(7) Sediments of some other facies were probably laid down in much deeper water. The Los Angeles and Ventura basins of southern California were deep troughs at the beginning of Pliocene time, and were filled by sediments that accumulated in progressively shallower water. Although evidence is more elusive in the earlier geosynclines, sediments of parts of them probably also formed in deep water. In those parts of geosynclines where the water was deep, sedimentation did not keep pace with subsidence; thus, subsidence is not a necessary consequence of sedimentation but an independent process.

(8) Subsidence of the crust beneath the geosyncline must have resulted primarily from forces at work within the earth. The floor of the geosyncline was bent down by these forces and sediments accumulated thickly because a place was made for them. Weight of the sediments themselves depressed the crust still farther, but not as far as the thickness of strata deposited. Geosynclines, therefore, are sedi-

mentary basins that formed under marked tectonic control.

(9) A geosyncline is a mobile belt, but degree of mobility varies—from one geosyncline to another, between parts of one geosyncline, and with time. Some geosynclines had only low-order mobility throughout their history; others show a steadily mounting crescendo of mobility, so that the geosyncline was finally transformed into a consolidated orogenic belt. Crustal unrest in the belt may become manifest long before the end of the geosynclinal stage itself.

(10) If geosynclines are mobile belts in which sedimentation took place under marked tectonic control, it follows that they are merely the surface expression of more deep-seated crustal structures. What these substructures are has been much debated; considerable fact and much fancy has accumulated regarding them, as we will see in Chapter V.

Geosynclinal terminology. These "attributes" make clear, I think, that the geosynclinal concept is useful for bracketing larger sedimentary features and structures that have a significant bearing on the evolution of a continent. Nevertheless, the attributes show that geosynclines have many guises not expressed by the word "geosyncline" alone. Many geologists have decided that analysis would be sharpened if names were given to these species of geosynclines; an extensive terminology has therefore grown up. Unfortunately, attempts to name such parts have resulted merely in a confusion of tongues, as each author has proposed his own set of names. Some thirty-seven names were listed in a compilation made about ten years ago,* and others have probably been created since.

It is unnecessary for our purpose to review all these names, but it is worth mentioning a terminology used by *Marshall Kay*, as his memoir is the standard work on North American geosynclines (see reference at end of chapter). Some of Kay's terms were borrowed from earlier works of the German geologist, Hans Stille, but others are newly coined, mainly from Greek roots. Most of the names apply to valid enough features which are also recognized in this book, but the names themselves seem needlessly complex. I believe that most of the concepts thus expressed can be indicated as effectively by descriptive phrases in English —or if not in good English, at least in a technical jargon whose source in the English language can be traced. In the table below, the terminology of Kay is listed and compared with that used in this book.

* Glaessner, M. F., and Teichert, Curt, 1947, Geosynclines, A Fundamental Concept in Geology: *Am. Jour. Sci.*, v. 245, p. 587-589.

NOMENCLATURE OF GEOSYNCLINES AND RELATED FEATURES

KAY TERMINOLOGY (*Geol. Soc. America Mem. 48*, 1951)	*KING TERMINOLOGY* (This book)
Non-geosynclinal features	*Non-geosynclinal features*
Craton (and hedreocraton)	Central Stable Region
Tectonic lands	Tectonic lands
Orthogeosynclines	*Geosynclines*
Eugeosyncline	Eugeosynclinal part
Miogeosyncline	Miogeosynclinal part
Parageosynclines	*Basins and troughs*
Intracratonic	*Mainly in Central Stable Region*
Exogeosyncline	Clastic wedge
Autogeosyncline	Basin in Central Stable Region
Zeugogeosyncline	Trough in Central Stable Region
Other	*Mainly in deformed belts, or laid over earlier deformed belts*
Epieugeosyncline	Area of post-orogenic deposits
Taphrogeosyncline	Area of fault-trough deposits
Paraliageosyncline	Area of coastal plain deposits

5. GROWTH OF THE APPALACHIANS

Appalachian miogeosyncline. With the general character of geosynclines in mind, we can now examine the time relations of the Appalachian structures. We will begin with the miogeosynclinal area, from whose rocks the sedimentary Appalachians and the Valley and Ridge province were created.

The miogeosynclinal area borders the edge of the Interior Lowlands and continental platform; the eugeosynclinal area is farther out, away from the continent. Miogeosynclinal deposits are much like those laid down over the Interior Lowlands—marine carbonates and clastics passing upward into continental beds—and like them, were laid down in shallow water or on low-lying surfaces. But unlike the interior area where initial deposits are no older than Late Cambrian, deposition began in the miogeosynclinal area in Early Cambrian time. Besides, the whole sequence from base to top of the Paleozoic column is much thicker, amounting to some five or six miles. Throughout the long period during which these deposits accumulated, the miogeosyncline was a region of quiet sedimentation and slow subsidence with little other crustal activity and no volcanism.

Let us now analyse the miogeosynclinal deposits in several dimensions: first in vertical sequence, then along the trend of the miogeosyncline, finally across the trend from eugeosyncline to foreland.

To demonstrate the character of the deposits in

vertical sequence and their variations along the trend, three sections in representative areas are generalized in Figure 32—in the Southern Appalachians of Alabama and Tennessee, and in the Central Appalachians of Pennsylvania. The thickness of each of the three sections totals about 25,000 feet, but note the considerable differences between them, especially in the upper half.

Initial Paleozoic deposits in all three sections are Early Cambrian marine clastics—conglomerates, arkoses, and shales which pass upward into cleanly washed quartzites. They vary somewhat in thickness

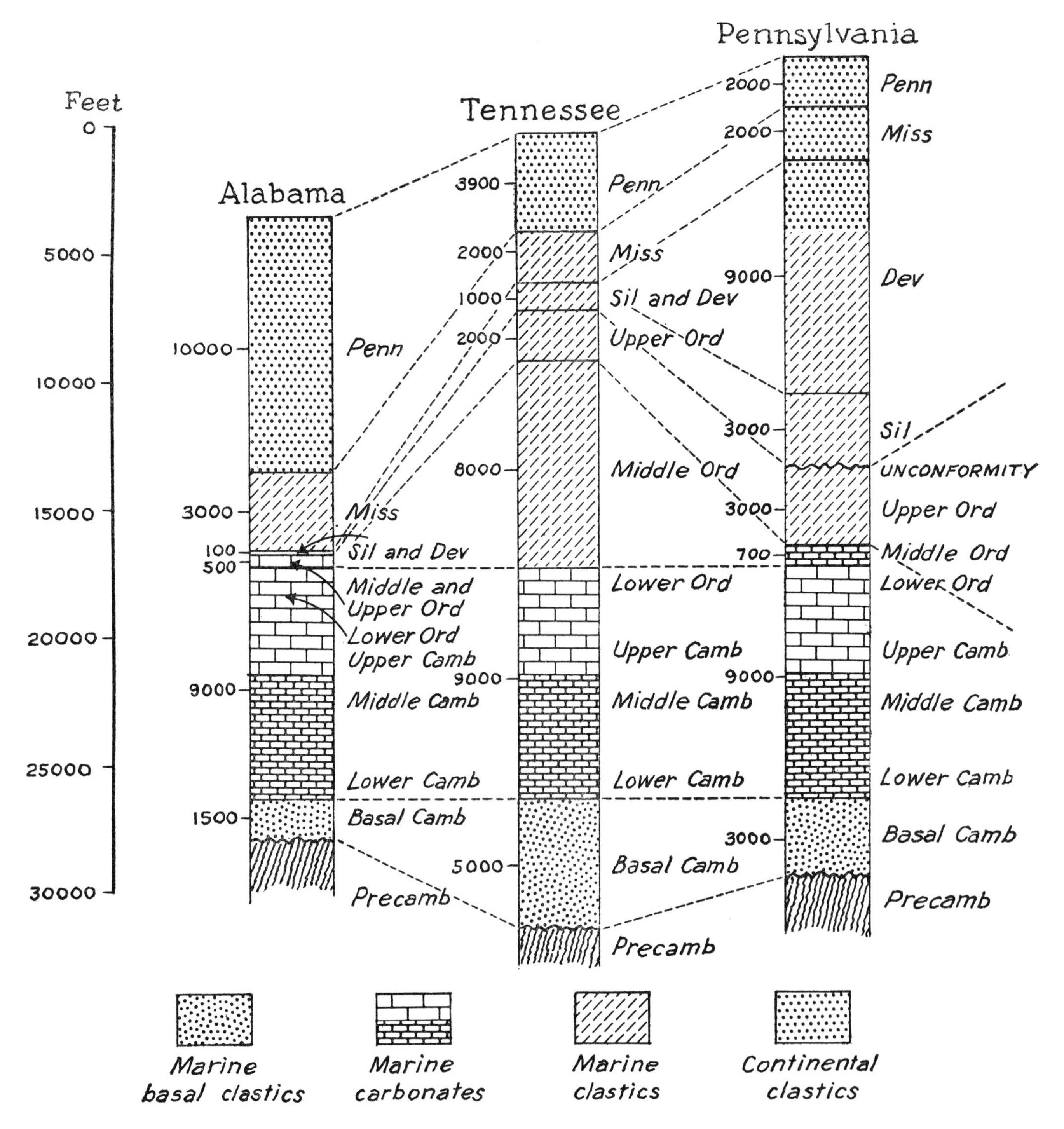

FIG. 32. Generalized columnar sections of the Paleozoic rocks from southwest to northeast in the Valley and Ridge province in Alabama, Tennessee, and Pennsylvania, to show the sequence and character of the deposits along the strike of the Appalachian miogeosyncline.

from place to place because of irregularities of the surface of the Precambrian floor and local differences in subsidence.

Then follows a great carbonate sequence with a nearly constant thickness of 9,000 feet. It embraces the remainder of the Lower Cambrian and extends through Middle and Upper Cambrian and Lower Ordovician; in places it includes Middle Ordovician as well. It is a mass of limestones and dolomites with only occasional interbedded layers of shale and sandstone.

These two components, the basal clastics and the mass of carbonates, form the miogeosynclinal sequence proper. They were laid down during a prolonged time of crustal quiesence on a surface which sloped gently seaward from the continental interior. No nearby lands were being strongly eroded; the sediments bear no indication of any lands to the southeast, and their sandy and shaly layers seem to have been derived from distant parts of the continental interior.

The upper half of the sequence in the miogeosynclinal area formed after a radical change in conditions had taken place; its deposits had best be considered under the succeeding heading.

Clastic wedges. The carbonate sequence gives place upward to marine clastic deposits, beginning in places with the Middle Ordovician, in others with the Upper Ordovician. The marine clastics include thick bodies of shale and sandy shale as well as persistent units of sandstone which are the chief ridge makers of the Valley and Ridge province. They pass upward, in turn, into continental beds that begin in places in the Devonian or Mississippian, in others in the Pennsylvanian.

The clastic deposits, marine and continental, were derived from erosion of lands to the southeast where there seem to have been none before. They spread over the miogeosyncline and far northwestward across the foreland but thinned and tapered in this direction; such deposits are here termed *clastic wedges.*

First, let us consider these clastic deposits along the strike of the miogeosyncline. In Alabama (Fig. 32) the Middle and Upper Ordovician, the Silurian, and the Devonian are all inconsequential; the thick clastic deposits are in the Mississippian and Pennsylvanian. In Tennessee the Pennsylvanian is thinner, the Silurian and Devonian are inconsequential, and the thick clastic deposits are in the Middle and Upper Ordovician. But in Pennsylvania the greatest development of clastics is in the Silurian and Devonian; the Pennsylvanian continental beds which cover the Allegheny Plateau to the west are thin by comparison. If we were to add a fourth column to the figure to show the less completely preserved sequence in New York State, we would find that the Upper Ordovician clastic deposits again thicken.

Consider now the form of the clastic deposits across the strike. In most of them the form in this direction is obscured by deformation and erosion so that only downfolded or downfaulted strips are preserved, extending along the strike rather than across it. Reconstruction of deposits across the strike must be made by correlation between stratigraphic sections that are now separated.

Relations are better shown in the Devonian clastic wedge where it emerges at the northeast end of the Allegheny synclinorium; we can infer that principles proven for it apply to the others. The Devonian crops out along a north-facing escarpment which extends eastward across New York State for some 300 miles, from Lake Erie to the Catskill Mountains overlooking the Hudson River. Throughout this distance the Devonian rocks have been little folded or faulted and beds can be walked out from one end of the wedge to the other, with the results shown in Figure 33.

As shown by the figure, the whole Devonian is only 2,500 feet thick near the west edge of the state, but is more than 10,000 feet thick in the Catskill Mountains where its top has been eroded. Observe that much of this thickening is in the Upper Devonian, some in the Middle Devonian, and very little in the Lower Devonian. Note also that red beds and continental deposits at the thick end of the wedge pass westward through marine sandstones into marine shales. Moreover, these rock facies are not bounded by vertical partitions; continental beds, for example, spread farther west as one ascends the section. When work was first done in the area these westward-spreading facies were thought to be time-stratigraphic units; the thick continental beds of the Catskill Mountains were then thought to be the youngest Devonian deposits of the region, but we now know that they are of early Late Devonian and Middle Devonian age.

If we combine our cross-section of the Devonian clastic wedge (Figure 33) with our earlier data (Fig. 32) we can appreciate its dimensions more fully. The wedge is then found to thin, not only westward toward Lake Erie, but southwestward toward Virginia and Tennessee. Its northeastern part, which once must have covered northern New York State, has now been lost by erosion.

These and other wedges have been likened to fans or deltas, although rather inappropriately, as they are different sorts of sedimentary deposits. Even more fancifully, the whole mass of clastic wedge deposits in the Appalachians can be thought of as a set of gigantic fish scales rooted on the southeast, each

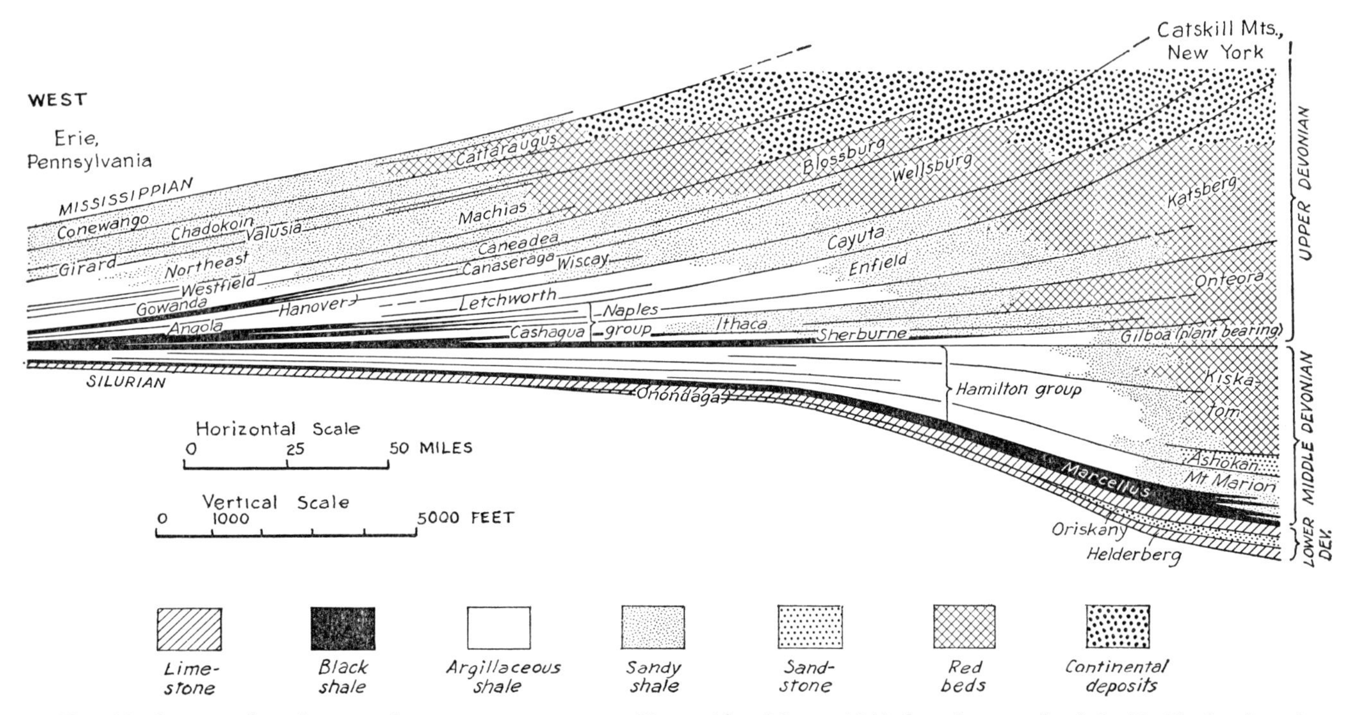

Fig. 33. Stratigraphic diagram from west to east in New York State showing the clastic wedge structure of the Devonian rocks and details of their lithology and thickness. After Moore, 1940, based on work of G. H. Chadwick and G. A. Cooper.

wedge thinning not only northwestward across the strike, but northeastward and southwestward along it. Each wedge (or "fish scale") lies at a different level, and each younger one overlaps the older one adjoining it (Fig. 34).

We have said that the clastic wedge deposits indicate the presence of lands southeast of the miogeosyncline from Middle or Late Ordovician time onward. What were these lands? Two assumptions have commonly been made, which are not entirely compatible.

Taconian and Acadian orogenies. It is thus stated that the Upper Ordovician, Silurian, and Devonian clastic wedge of the northeastern part of the Appalachian miogeosyncline and its foreland were derived from erosion of mountains that had been raised farther back in the geosynclinal area during a *Taconian orogeny* toward the end of Ordovician time and during an *Acadian orogeny* toward the end of Devonian time.

Deformation at such times can be proved at a good many places in the orogenic belt. We have already noted that in the Maritime Provinces Mississippian post-orogenic deposits lie on more deformed Devonian and earlier rocks. In the Valley and Ridge province of New York and Pennsylvania another structural unconformity is evident below the Silurian; it is recorded on Figure 32 in the stratigraphic column for the latter state.

Various lines of evidence—clastic wedges, unconformities, and structural features—thus demonstrate that the geosynclinal rocks and their basement in the Northern Appalachians underwent several periods of deformation in the first half of Paleozoic time. But ordinarily it has been assumed that the Central and Southern Appalachians were deformed much later. Clastic wedge deposits occur here, too, but these have been attributed to a source of a rather different sort—the land of Appalachia.

Appalachia. Appalachia, that mysterious kingdom, was supposed to have been one of the *borderlands* of North America—peculiar land masses, now vanished and in part foundered to ocean depths, which once lay along the present coasts and extended out to sea for two or three hundred miles, or even farther. Like the Canadian Shield, the borderlands were believed to have been composed of basement sialic rocks, but they were thought to have been much more mobile, being raised steadily or spasmodically as geanticlines while the complementary geosynclines (the miogeosynclines of our terminology) subsided on their inner sides. They were, hence, sources of much of the clastic material deposited in such geosynclines.

The notion of Appalachia originated at a time when it was believed that rocks and structures of the crystalline Appalachians, being so different from those of the sedimentary Appalachians, were probably much

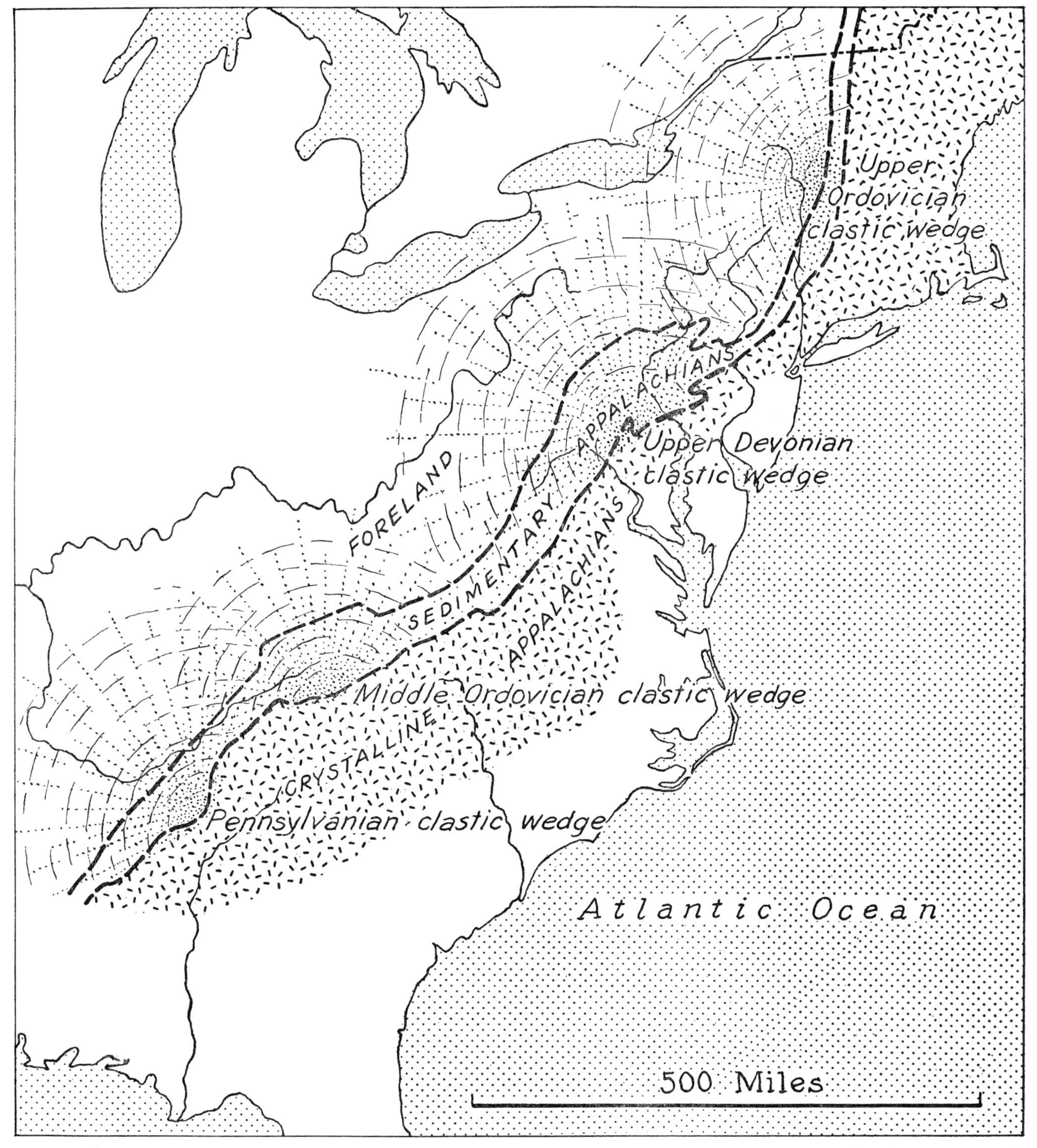

FIG. 34. Sketch map of eastern United States showing extent of the principal clastic wedge deposits of different ages in the sedimentary Appalachians (miogeosynclinal area) and the foreland of the Appalachian system.

more ancient. But we now know that a large part of the metamorphic rocks of that area were formed surficially during Paleozoic time, and that the region was only made into a crystalline area by deformation, metamorphism, and plutonism during that era.

Once this was understood, it became necessary to shift the position of Appalachia farther and farther off the present land area and out to sea, but here a new difficulty became apparent. Geophysical work has demonstrated that sialic continental crust gives place to simatic oceanic crust a little beyond the edge of the continental shelf; no land could have foundered beneath the Atlantic Ocean basin unless there were some mechanism whereby it could be converted from sialic to simatic crustal material. Thus, if Appalachia ever existed, it was hemmed in both on its landward and seaward sides. As inconvenient facts have accumulated, it has been necessary continually to reduce its area.

Twenty-five years ago the theory of Appalachia and other borderlands was firmly entrenched in American geological theory, but it has given way since to more complex and sophisticated explanations. Before interring the idea, let us say to its credit that it was a blurred and somewhat naive means of expressing the known mobility of the continental borders. We now believe that some of the features ascribed to the borderlands existed, but that others did not. Instead of expressing the known features by an obsolete terminology, it is better to speak of them in other terms.

Specifically, therefore, rather than attribute the clastic wedges of the miogeosyncline and foreland of the Central and Southern Appalachians to the erosion of a borderland of Appalachia, it is more reasonable to relate them to crustal activity farther back in the geosyncline itself, as has already been done for similar clastic wedges in the Northern Appalachians.

APPALACHIAN EUGEOSYNCLINE. We are now ready to explore the area "farther back" in the Appalachian geosyncline, or southeast of the miogeosynclinal area and away from the continent. This is the eugeosynclinal area, from whose rocks the crystalline Appalachians—the New England Upland and Piedmont Plateau—were formed.

Let us recall that rocks of the crystalline Appalachians are thoroughly metamorphosed and full of plutonic rocks today, so that their record is difficult to decipher. Original sedimentary or volcanic structures, fossils, and even whole sequences have been more or less obliterated; much reconstruction is necessary to interpret their original character. But these reconstructions can be verified in places where the rocks are less altered. Metamorphic rank thus decreases northeastward along the strike from New England into the Maritime Provinces; original structures and sequences become plainer and fossils are more abundant. By contrast, in the crystalline belt of the Central and Southern Appalachians rocks of extensive areas have been strongly altered, and fossil occurrences are widely separated; even here, the unfossiliferous deposits of the Carolina slate belt on the southeastern side preserve much of their original character.

Such reconstructions indicate that the eugeosynclinal rocks were originally surface sediments and volcanics, much of which were accumulating at the same time as the rocks of the miogeosyncline; the remainder were similar surface rocks formed in the eugeosyncline during later Precambrian time.

The sediments are very different from those of the miogeosyncline. They include little of its carbonates and quartzites; most are graywackes, or poorly sorted and rapidly deposited sandstones, and silty or sandy argillaceous rocks. In many areas these are interbedded with chert which probably originated from silica contributed to sea water during times of volcanism.

Associated with the sediments are volcanic rocks in varying quantities—lava flows of different compositions as well as breccias, tuffs, and other pyroclastic products. Great parts of these were spread out under water; parts which were land-laid probably formed when the accumulations were built up into volcanic islands. The volcanics are a new geosynclinal component, rarely if at all found in the miogeosynclinal deposits.

The eugeosynclinal deposits attain great thickness, although this is commonly difficult to estimate because of the complex structure and alteration of the rocks. A sequence in eastern Vermont, believed to be of Cambrian to Devonian age, is at least 50,000 feet thick, even allowing for possible duplication. Great thicknesses imply great subsidence; moreover, such subsidence appears not to have been compensated entirely by sedimentation, and parts of the eugeosynclinal deposits were probably laid down in water much deeper than were any of the deposits of the miogeosyncline.

Seldom is the base of a eugeosynclinal sequence visible. Toward the west side of the belt sediments and volcanics lie in places on earlier granitic rocks; elsewhere, any rocks beneath the Paleozoic sequences which come to the surface are geosynclinal deposits formed in later Precambrian time. On the side toward the continent the eugeosynclinal deposits were thus laid down in part on a sialic crust, but much of the remainder might have been spread over simatic ocean bottom. Could they, then, have been built over that part of the ocean floor along the edge of the continent during the geosynclinal stage, and have been made part of the continent during the orogenic stage?

RELATION BETWEEN EUGEOSYNCLINE AND MIOGEOSYNCLINE. How do the eugeosynclinal deposits connect with the very different yet contemporaneous miogeosynclinal deposits? Were eugeosyncline and miogeosyncline each synclinal in fact—that is, separate troughs? Here, as in many other phases of the geosynclinal problem, we encounter trouble in our reconstructions because of the considerable deformation and alteration to which the rocks have been subjected.

Throughout much of the Northern, Central, and Southern Appalachians the sedimentary and crystalline parts with their miogeosynclinal and eugeosynclinal rocks are separated by the Green Mountain and Blue Ridge uplifts which bring up basement rocks in the midst of the orogenic belt (Fig. 30). At some places, as in the Taconic Mountains of eastern New York State, the two facies are juxtaposed but lie in different thrust blocks and are still structurally separated. In only a few places, as in northern Vermont, are transitional rocks preserved between those of the eugeosyncline and miogeosyncline.

Some reconstructions of the geosynclinal cross-section (for example, that of Kay, reproduced as our Figure 35) suggest an existence of barriers or partial barriers between the miogeosyncline and eugeosyncline at about the present positions of the Green Mountains and Blue Ridge. Such separations are hypothetical. So far as we know, these uplifts are merely anticlinoria imposed on the rocks by later deformation, and were not inherited from barriers that existed while the rocks were being deposited. Deposits of the miogeosyncline and eugeosyncline more likely were continuous originally, one grading into the other through a transition zone that has mostly been obliterated by later deformation and erosion.

TECTONIC LANDS. Deposits accumulated to much greater thickness in the eugeosyncline than in the miogeosyncline; what was their source? Parts of these deposits were volcanic flows, breccias, and tuffs brought up from within the crust and spread rapidly over the floor of the geosyncline. But other parts were clastic sediments derived from erosion of nonvolcanic lands. By earlier interpretations these should have been derived from erosion of the borderland of Appalachia, but the actual story is probably more complex.

All indications—the volcanism, the rapid subsidence, and the inferred deep water deposits—suggest that the eugeosyncline was a region of greater crustal mobility than the miogeosyncline. In such an environment local structural ridges probably were forced up from time to time whose rapid erosion supplied a greater volume of sediments than one would expect from their small extent; such features Kay has termed *tectonic lands.* Unconformities are known in places in the eugeosynclinal deposits, suggesting the existence of such tectonic lands; traces of other tectonic lands may now have been lost by later deformation, metamorphism, and erosion. These tectonic lands were not necessarily made up of pregeosynclinal basement rocks; many of them no doubt

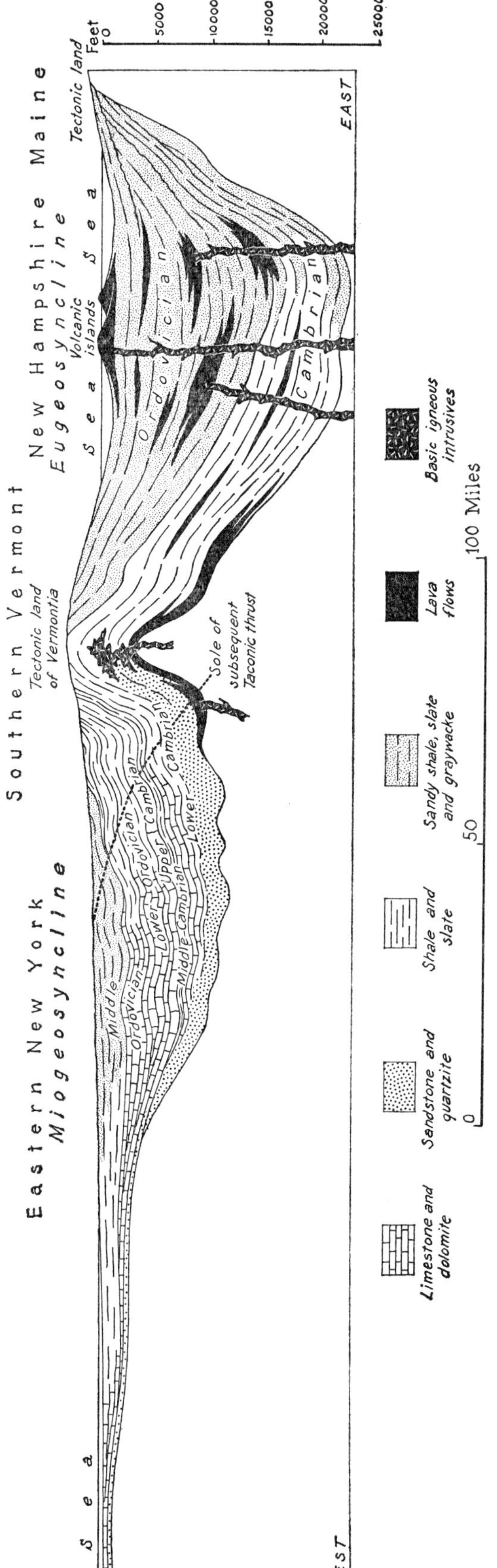

FIG. 35. The Appalachian miogeosyncline and eugeosyncline in the segment across New York, Vermont, and New Hampshire as they existed at the close of Ordovician time, as reconstructed by Kay, 1951. Most geologists would agree with the broader relations represented. One must remember, however, that any reconstruction is based on piecing together of sequences in different areas, connections between which are now partly destroyed by folding and faulting; no two geologists would therefore make the same interpretation of the details.

consisted of deformed and upraised earlier geosynclinal deposits themselves. Erosion of these earlier geosynclinal materials provided sediments for new geosynclinal troughs; thus, the eugeosnycline grew by a process of *cannibalism.*

During the geosynclinal stage one can thus picture the continent as bordered not by massive lands such as Appalachia, but by a eugeosyncline made up of variously subsiding troughs interspersed with chains of volcanic islands and narrow structural ridges—conditions that were probably similar to those in modern island arcs and deep-sea trenches (see Chapter V, Section 3).

In New England and the Maritime Provinces where the record of the Appalachian eugeosyncline is best known, Cambrian and Ordovician deposits form thick, continuous sequences; tectonic lands then were probably widely spaced. Silurian and Devonian deposits are preserved only in more discrete areas between which the rocks differ so much that land barriers were probably more extensive—a manifestation of mounting crustal mobility of the eugeosynclinal region that led to an orogenic climax in mid-Paleozoic time. Mississippian and later deposits of the region are post-orogenic, and accumulated in local basins in a region that was otherwise a highland of deformed earlier rocks.

Igneous and plutonic history of the eugeosynclinal area. Let us now explore another facet of the mobility of the eugeosynclinal area—the igneous and plutonic activity which did much to transform it into a region of crystalline rocks. Different kinds of igneous and plutonic rocks were emplaced during successive stages, but with considerable overlap from one phase to the next.

(1) Volcanic activity characterized the eugeosynclinal area from its beginning; *flows and pyroclastic sediments* were spread over the area of deposition, mainly under water, at various times and from different centers, and were intermingled with the nonvolcanic sediments.

In the Vermont-New Hampshire segment the Cambrian and earlier Ordovician slates and graywackes are interbedded with greenstones or altered volcanics, but volcanism reached a climax in the later Ordovician Orfordville and Ammonoosuc formations which are made up largely of rhyolitic, andesitic, and basaltic lavas. The thin Silurian deposits are nonvolcanic and include beds of limestone and quartzite reminiscent of the miogeosyncline, but volcanism was resumed in the Devonian whose Littleton formation contains rhyolite and trachyte flows.

Elsewhere, the Silurian, too, contains volcanic components. On the south coast of Gaspé it is 12,000 feet thick, with fossiliferous marine sediments below and basalt flows above. Near Eastport, Maine another thick Silurian section likewise consists of interbedded marine sediments, tuffs, and lavas.

(2) *Ultrabasic (ultramafic) rocks* were injected into the eugeosynclinal rocks in small bodies which lie in swarms along several zones of structural weakness. The ultrabasics—the peridotites and their allies—are thought by many geologists to have been derived from a deep crustal layer or "mafe" discussed in Chapter I, Section 1, and to have been forced into much higher levels of the crust by some mechanism associated with marked downfolding of the eugeosynclinal area. The most prominent swarms are on the northwest side of the presumed axis of greatest downfolding, although others occur farther southeast. The age of most of the ultrabasic intrusives is difficult to determine, but they are generally believed to be an early phase in the plutonic cycle—perhaps as early as Ordovician or as late as Devonian. In any event, they were intruded while at least the later part of the eugeosynclinal sediments were still accumulating.

(3) Ultrabasic intrusions were followed by a long succession of *acidic plutonic rocks*—granites and their allies—which form a *granite series* ranging from concordant to through-breaking bodies.

The earliest members of the series are *concordant, foliated plutonic rocks,* which spread widely through the eugeosynclinal materials, mainly while deformation was in progress. Origin of granites and their allies has been much debated among geologists, and widely divergent views have been expressed. My own belief is that these earlier granitics were formed by a *transformation of the eugeosynclinal materials themselves,* when they had been depressed so deeply into the crust that they were subjected to great heat and pressure.

At first, such transformed material was static and not very fluid, but as the plutonic and orogenic cycle advanced, the material became more mobile and spread farther from its source. This group of concordant foliated plutonic rocks is therefore gradational into the next.

(4) *Massive, through-breaking plutonic rocks* were emplaced mostly after the orogeny and ascended high into the deformed country rocks. They themselves were clearly injected as magmas, although their ultimate source may well have been granitized material from some lower level.

When it was thought that the country rocks of the crystalline area were a Precambrian complex like that of the Canadian Shield, it was believed that a large part of the acidic plutonics belonged to this complex. The remainder, however—the through-breaking bodies—were interpreted as having been injected after the climax of the orogeny, those in the Northern Appalachians in late Devonian time, after the Acadian orogeny, those in the Southern Appalachians in late Permian time, after the Appalachian Revolution.

But such inferred ages would mean that part of the granite series was younger by a quarter of a billion years or more than the remainder, and was unrelated to it. Such assumptions have not been substantiated by radiometric determinations, which show that all the acidic plutonic rocks of the crystalline area are of various Paleozoic ages except those raised to the surface in the welt of basement rocks which forms the Blue Ridge and Green Mountain axes.

In New England radiometric determinations on the acidic plutonic rocks from concordant to massive varieties indicate ages of 390 to 180 million years, or roughly from Ordovician to Permian time. The Highlandcroft magma series of New Hampshire, overlain unconformably by the Silurian, has an age of 360 to 390 million years and is probably Ordovician. The Oliverian and New Hampshire magma series of the same state have ages of 290 to 310 million years, and are probably middle Devonian. Alkalic granites of Massachusetts and Rhode Island have ages of 230 to 260 million years and are probably Mississippian and Pennsylvanian. The White Mountain magma series of New Hampshire has an age of 180 to 190 million years and is probably late Permian. In the Southern Appalachians many dates of between 400 and 280 million years have been determined on acidic plutonic rocks in North and South Carolinia; so far, no late Paleozoic dates like those in New England have been recorded. Ironically, the supposed "Permian granites" of the Southern Appalachians have turned out to be much older than hitherto supposed, and some of the supposed "Devonian granites" of the Northern Appalachians have turned out to be much later.

Development of the acidic plutonic rocks therefore occupied a long span of early, middle, and even later Paleozoic time, the earlier phases occurring during times of intense orogenic activity which interrupted and succeeded the accumulation of the eugeosynclinal sediments.

Orogeny in the eugeosynclinal area. Our review of the eugeosynclinal area indicates that it was a region of much crustal mobility, even from its inception, and that mobility increased to an orogenic climax in mid-Paleozoic time, by which time its rocks had been thoroughly deformed, metamorphosed, and plutonized. In the Northern Appalachians considerable stratigraphic information is available, and many known details of these movements can be matched with times of emplacement of the acidic plutonic rocks. Here movements have been assigned to a Taconian and an Acadian orogeny. In the Southern Appalachians less is known of the movements themselves, but the time span of the plutonic rocks is similar and suggests a like deformational history.

The clastic wedges of the miogeosyncline and foreland of both the Northern and Southern Appalachians thus appear to have had a like source—deformed lands that were raised in the eugeosynclinal area. In each segment there were one or more deformational climaxes between Ordovician and Mississippian time, the times varying from place to place along the chain and causing the overlapping, scale-like arrangement of the clastic wedges.

After mid-Paleozoic time mobility gradually diminished in the former eugeosynclinal area. Post-orgenic deposits of later Paleozoic time in the Maritime Provinces are little deformed in places and somewhat affected by compressional deformation in others. The much younger Late Triassic deposits are disturbed by tilting and tensional faulting.

Orogeny in miogeosynclinal area and foreland. What, then, of the deformation of the sedimentary Appalachians with their miogeosynclinal rocks, and of the foreland beyond them? Date of their deformation cannot be proved; it merely can be inferred from ages of the youngest rocks involved.

Rocks of the sedimentary Appalachians and foreland form a nearly unbroken sequence extending from Cambrian to Pennsylvanian in many places, and to as high as lower Permian in a few. The sequence contains occasional disconformities indicative of failures of deposition or moderate erosion, but there are no angular unconformities indicative of times of deformation. Even the clastic wedges which are products of deformation elsewhere are conformable with the deposits on which they lie.

Mississippian rocks are preserved in many remnants in the strongly deformed sedimentary Appalachians, some as far southeast as the front of the Blue Ridge. Pennsylvanian rocks are also preserved in places in the strongly deformed belt as near Birmingham, Alabama, and in the Broadtop and Anthracite coal basins of Pennsylvania. Early Permian rocks are more restricted and lie mostly well out in the foreland.

Deformation was thus certainly later than Mississippian, Pennsylvanian, or early Permian time, depending on locality. Much or all of it may, in fact, have been younger than early Permian. But many parts of the sedimentary Appalachians show a succession of superposed structures; for example, thrust sheets have been emplaced, then folded, then offset by transverse faults. Such superposed structures did not develop during any single deformational event but during a considerable span of time—perhaps as long as a geological period or two. Very possibly, structures in the southeast part of the sedimentary Appalachians began to form earlier than those in its northwestern part.

Be that as it may, this deformation of rocks in the sedimentary Appalachians is the so-called *Appalachian Revolution.* But was it so revolutionary? As we have shown, the whole Appalachian region was not transformed from geosyncline to mountain belt in a single cataclysm. Most of its interior on the southeast had been thoroughly deformed and consolidated by middle Paleozoic time (Fig. 36 B). Marginal deformation of the sedimentary Appalachians and foreland was merely a later, and in part a concluding phase (Fig. 36 C).

To avoid false implications, it would be best not to refer to these structures as having formed in *the* Appalachian Revolution; the word "Appalachian" would apply better to the whole prolonged series of movements. It would be more expressive if we spoke of the deformation of the miogeosyncline and foreland as having taken place during an *Allegheny orogeny,* with the implication that this was of the same magnitude as the earlier Acadian and Taconian orogenies.

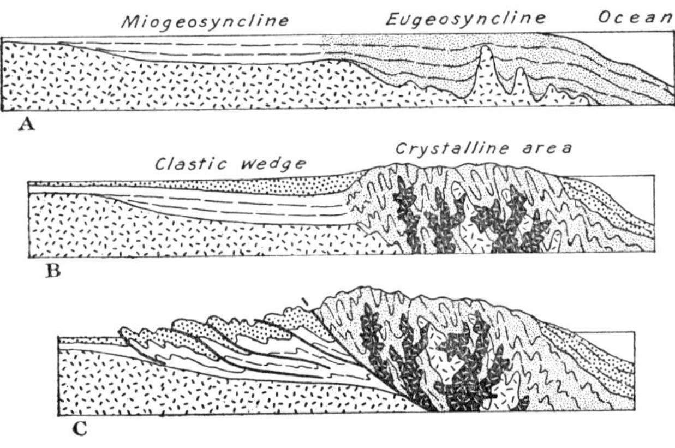

Fig. 36. Sections across Appalachian system at different times in the Paleozoic, showing its inferred development from geosyncline to deformed belt. (A) Geosynclinal stage in early Paleozoic time. (B) After deformation of eugeosynclinal area and deposition of clastic wedges to northwest in middle Paleozoic time. (C) After final deformation of miogeosynclinal area in late Paleozoic time.

Conclusion. We have now carried our story of this great mountain system far enough for present purposes. What followed was pale aftermath—prolonged erosion of the system with burial or submergence of some of the parts, as set forth in Section 1 of this chapter. The lesson of the whole complex story is that the Appalachians were long in creation, and that their building occupied much of Paleozoic time.

Possibly it may be felt that I have labored the story too much, but I have done so in order to present a modern philosophy of mountain growth, using as an example a mountain system where the features are unusually well laid out. We will encounter all the same features and problems again when we take up the Cordillera of the west (Chapter VI, et seq.)—the miogeosyncline, the eugeosyncline, the tectonic lands, the zone of metamorphic and plutonic rocks, and the rest—but these will be overlaid by many more complications. Having been through the story once it will be unnecessary to repeat it in a single piece again, although we will deal frequently with its many parts.

6. PALEOZOIC STRUCTURES WEST OF THE MISSISSIPPI RIVER

We now leave the Appalachians and all the country east of the Mississippi River. But recall that the Appalachian system is only part of an originally more extensive system of mountain structures that once extended much farther west (Plate I). These emerge again from the cover of younger rocks in small areas beyond the Mississippi River, in Arkansas, Oklahoma, and Texas. Branches of the system also extended even farther into the Cordilleran region, where they greatly influenced the structures formed there later, as we will see (Chapter VII, Section 5).

In Alabama the whole Appalachian deformed belt passes southwestward beneath deposits of the Gulf Coastal Plain. At the place where the belt goes underground it shows no change in its usual trend or structure nor any diminution in intensity. Clearly, it must continue farther, but where? We will return to this question later; in the meantime let us jump ahead a little.

Marathon region of Texas. I entered my study of Paleozoic structures on the southeast and south sides of the continent through the back door, so to speak. I first encountered them in one of their westernmost areas of exposure, the Marathon region of western Texas. Instead of launching directly into generalizations on the Paleozoic structures west of the Mississippi, it might therefore be more interesting to approach the subject indirectly, as did geologists of my generation, and see how the puzzles were solved step by step and what problems still remain.

If the reader has traveled by rail on the line of the Southern Pacific from El Paso to New Orleans, or if he has driven over U.S. Highway 90, he has passed through the town of Marathon (Fig. 20). He would recall it as a little ranch settlement several hundred miles southeast of El Paso, set down in a wide plain that is rimmed on the west by the first range of western mountains and elsewhere by limestone mesas and escarpments.

This plain is the *Marathon Basin*, a topographic feature 35 miles across, produced by erosion of the crest of the *Marathon dome*. The dome was formed by arching of the Cretaceous limestones that cover much of the semi-arid southwest Texas country. Many of the mesas and scarps which surround the basin are carved from these Cretaceous limestones; the basin itself is cut on Paleozoic rocks which underlie them (Fig. 37).

From the floor of the Marathon Basin project low, northeast-trending strike ridges formed of the more resistant layers in the deformed Paleozoic rocks; they resemble on a somewhat smaller scale those of the Valley and Ridge province in the Appalachians. The folding and faulting that turned up the Paleozoic rocks in the basin are of pre-Cretaceous age, as shown along its edges where a great angular unconformity can be observed between the Paleozoic and the Cretaceous (Fig. 38).

There is, however, an additional structural element in the Marathon dome. The northern side is bordered by the Glass Mountains which are made up of tilted marine Permian rocks (Fig. 37). Like the earlier Paleozoic rocks, the Permian is truncated by the Cretaceous, but like the Cretaceous, it lies unconformably on the earlier Paleozoic. The Permian rocks are part of the southern margin of the West Texas basin (Chapter III, Section 6); they are foreland deposits, but they differ from the Appalachian foreland deposits in that they lie unconformably on earlier rocks of the adjacent deformed belt.

The Marathon region thus consists of three parts: (1) Deformed earlier Paleozoic rocks which emerge on the crest of the dome. (2) Permian rocks of the Glass Mountains on the north flank, which lie unconformably on the earlier Paleozoic rocks and are tilted northward toward the West Texas basin. (3) Cretaceous rocks which lie unconformably on the truncated edges of both the Permian and earlier Paleozoic strata, but which were later raised into a broad dome.

Let us now say more about the deformed pre-Permian Paleozoic rocks. Their structure is much like that in the sedimentary Appalachians—a series of northeast-trending folds broken by thrust faults that dip southeast, with some low-angle thrusts that have carried the rocks above them several miles northwestward (Fig. 37, section A A′). The date of their deformation is well-defined, unlike the inferred dating of the deformation in the sedimentary Appalachians. Its beginnings can be traced in conglomerates in the Pennsylvanian sequence, and its climax was toward the end of that period, as Pennsylvanian rocks are deformed and Permian rocks lie unconformably on them.

The pre-Permian rocks themselves resemble in many respects those of the sedimentary Appalachians. They occupy about the same age span from Cambrian through Pennsylvanian; one difference is that the oldest beds which come to the surface at Marathon are Upper Cambrian, and it has not been determined whether earlier Cambrian lies beneath. As in the Appalachians the upper beds are clastic wedge deposits—fine-grained below, coarser above. These begin with the late Mississippian and extend through the Pennsylvanian.

But the rocks beneath the clastic wedge deposits are of a considerably different facies from those of the miogeosynclinal deposits of the sedimentary Appalachians—a sequence of Ordovician graptolite-bearing shales, thin limestones, sandstones, and bedded cherts topped by a unit of massive white chert or *novaculite* of Devonian age. These rocks are not of a facies usual in a miogeosyncline, but are almost of eugeosynclinal character; perhaps they represent a facies transitional from the miogeosyncline to the eugeosyncline.

The whole Marathon Paleozoic section beneath the Permian—clastic wedges above and near-eugeosynclinal deposits beneath—is unique for the west Texas area. Little disturbed rocks exposed in mountains nearer El Paso to the northwest and rocks penetrated in wells beneath the West Texas basin to the north are of a quite different facies, from Ordovician through Pennsylvanian—mostly limestones and dolomites with only occasional intercalations of shale and sandstone.

Regional relations of Ouachita system. Taking a broader view, this contrast between pre-Permian Paleozoic facies of the Marathon area and of areas to the northwest and north turns out to be general through Texas and Oklahoma.

The carbonate facies is exposed in the Llano uplift of central Texas and the Ozark uplift of Missouri and nearby states, where the rocks are little deformed. It also appears in the Wichita and Arbuckle Mountains of southwestern Oklahoma, where the rocks are thrown into the folds of the Wichita system; this will be discussed later in this section.

But pre-Permian rocks with a facies and structure nearly identical to that of the Marathon area reappear 500 miles to the northeast in the Ouachita Mountains of southeastern Oklahoma and southern Arkansas. Here again are folds and faults like those in the sedimentary Appalachians; here again are the same graptolite shales, cherts, and novaculites below, and

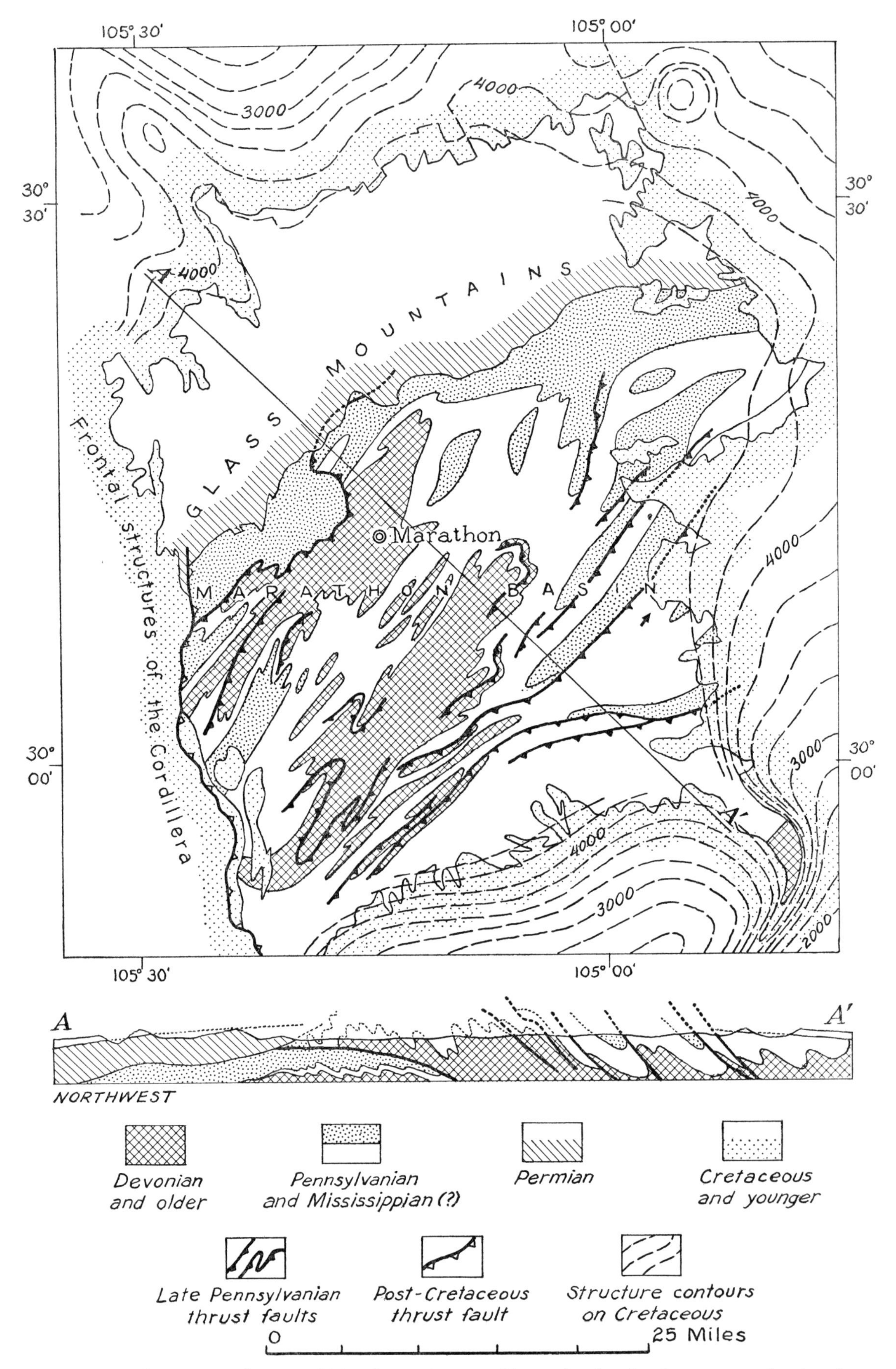

FIG. 37. Geologic map and section of Marathon region, west Texas, showing the three structural units: the domed Cretaceous, the tilted Permian on the north flank, and the strongly deformed earlier Paleozoic of the Marathon Basin. Generalized from King, 1937.

FIG. 38. Angular unconformity between tilted Pennsylvanian sandstones and flat-lying Cretaceous limestones on east rim of Marathon Basin, west Texas. View is at point marked by arrow on FIG. 37. Redrawn from King, 1937.

Mississippian and Pennsylvanian clastic wedges above (Fig. 39).

We thus have some pieces of a jigsaw puzzle, but with many parts missing due to the wide separation of the pre-Permian outcrops. Great intervening areas are covered by younger deposits, partly Mesozoic, partly later Paleozoic (upper Pennsylvanian and Permian), which conceal much of the earlier structures (Plate I).

Little progress on the puzzle could ever have been made without the aid of drilling, which has probed below the younger cover and determined the pattern of the earlier rocks and structures. From drilling we have learned much, although problems yet remain to be solved.

Briefly stated, these data indicate that the deformed rocks exposed in the Marathon and Ouachita areas are parts of a single orogenic belt which may be termed the *Ouachita system*. As noted earlier, the two exposed parts probably lie on apices of sharply curved salients in the belt; a recess lies in a buried segment between, where it curves southeast around the foreland buttress of the Llano uplift (Plate I). This orogenic belt developed from a geosyncline whose growth was about as prolonged as that of the Appalachian geosyncline, but which lacked a well-developed miogeosyncline; its rocks in both the Marathon and Ouachita areas are of near-eugeosynclinal character. This geosyncline was transformed into an orogenic belt by a series of deformations which culminated in the latter part of Pennsylvanian time.

Northwest of the geosyncline from El Paso to Oklahoma the earlier Paleozoic is of carbonate facies, most of which remains a little disturbed foreland, although part has been thrown into the folds of the Wichita system. Unlike the Appalachians, very little is known of the interior of the Ouachita system; it has been peculiarly susceptible to collapse and subsidence, and is now deeply buried beneath deposits of the Gulf Coastal Plain.

RELATIONS OF OUACHITA AND APPALACHIAN SYSTEMS. Having gone this far in our reconstructions, we return again to the greatest puzzle—Where does the Appalachian deformed belt extend southwestward, and how do the Ouachita structures relate to those of the Appalachians?

Complete answers are not yet available, although some progress has been made in the last few decades. Oil has long been known in the western part of the Gulf Coastal Plain in Texas, Louisiana, and Arkansas. Later on, exploration proceeded eastward; now Mississippi has become an oil-producing state and Alabama is being prospected. During this exploration some drill holes have passed through the Coastal Plain deposits into the Paleozoic rocks beneath.

Records of these drill holes indicate that the Ouachita deformed belt bends southeastward beyond its last exposure in the Ouachita Mountains of Arkansas and extends under the Mississippi Embayment and eastern Gulf Coastal Plain across the state of Mississippi (Plate I). Other drilling in Alabama indicates that the Appalachian structures maintain their southwestward course under the Coastal Plain deposits. The two sets of structures have been traced within fifty miles of each other, with every indication of an impending collision somewhere near Meridian, Mississippi. But just as the data become interesting, both sets of structures pass to great depths and are so thickly covered by an overburden of Coastal Plain deposits that it is no longer feasible to reach them by drilling.

The problem is thus like an old-time silent movie serial—two railroad trains are racing for each other headon, yet just as we have braced ourselves in our seats for the inevitable crash, the installment ends with the cheerless announcement, "Continued next week."

But in the same region there is another intersection between two deformed belts where the sequel to our movie serial is known. On the western side of the Ouachita salient in southern Oklahoma the Ouachita structures trend southwest, and those of the Wichita system in the Arbuckle Mountains trend southeast; each disappears beneath Coastal Plain deposits before they intersect. However, the Coastal Plain deposits of this area are relatively thin and have been drilled through in many places. This drilling indicates that structures of the Ouachita system override those of the Wichita system (Fig. 40).

Returning again to our problem of relation of the Ouachita and Appalachian systems, recall that Paleozoic rocks of the sedimentary Appalachians in Alabama are of miogeosynclinal facies with thick carbonates below and clastic wedge deposits above. Re-

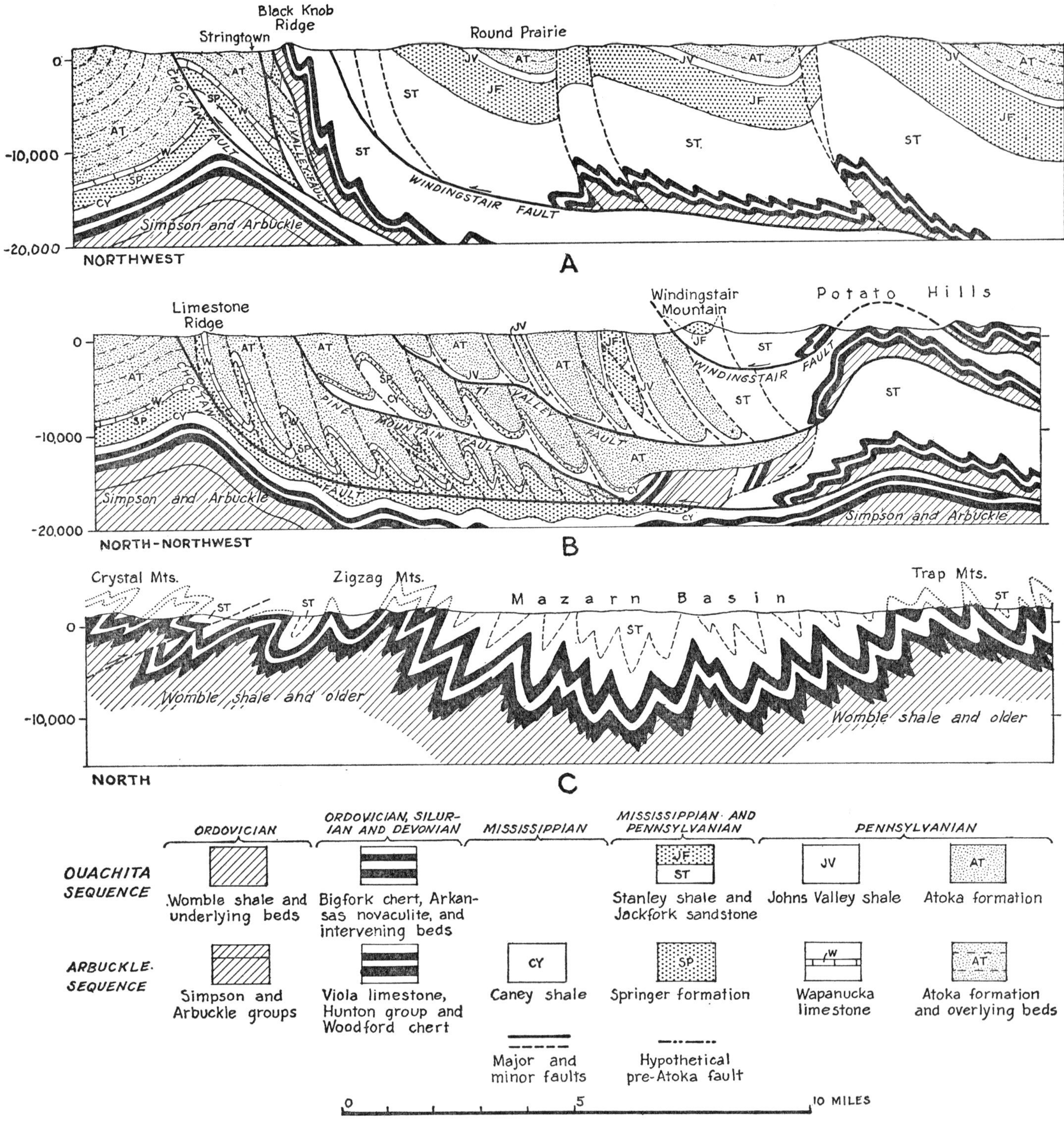

Fig. 39. Sections in Ouachita Mountains of Oklahoma and Arkansas. (A and B) Sections in western part of mountains in Oklahoma where faulting dominates; note changes in sequence from one fault block to another, and in B, the window at the Potato Hills. (C) Section in eastern part of mountains near Hot Springs, Arkansas where folding dominates. After Hendricks and others, 1947; and Purdue and Miser, 1923.

call also that the known Ouachita sequence is not of miogeosynclinal character, but more nearly of eugeosynclinal character. It is therefore possible that passage from the Appalachian to the Ouachita deformed belt was accomplished by an overriding of the interior or eugeosynclinal zone of the Appalachian system across the outer or miogeosynclinal zone. Possibly the miogeosynclinal zone of the Ouachita orogenic belt persists under a cover of the overriding eugeosynclinal zone, disappearing on the southeast in the Appalach-

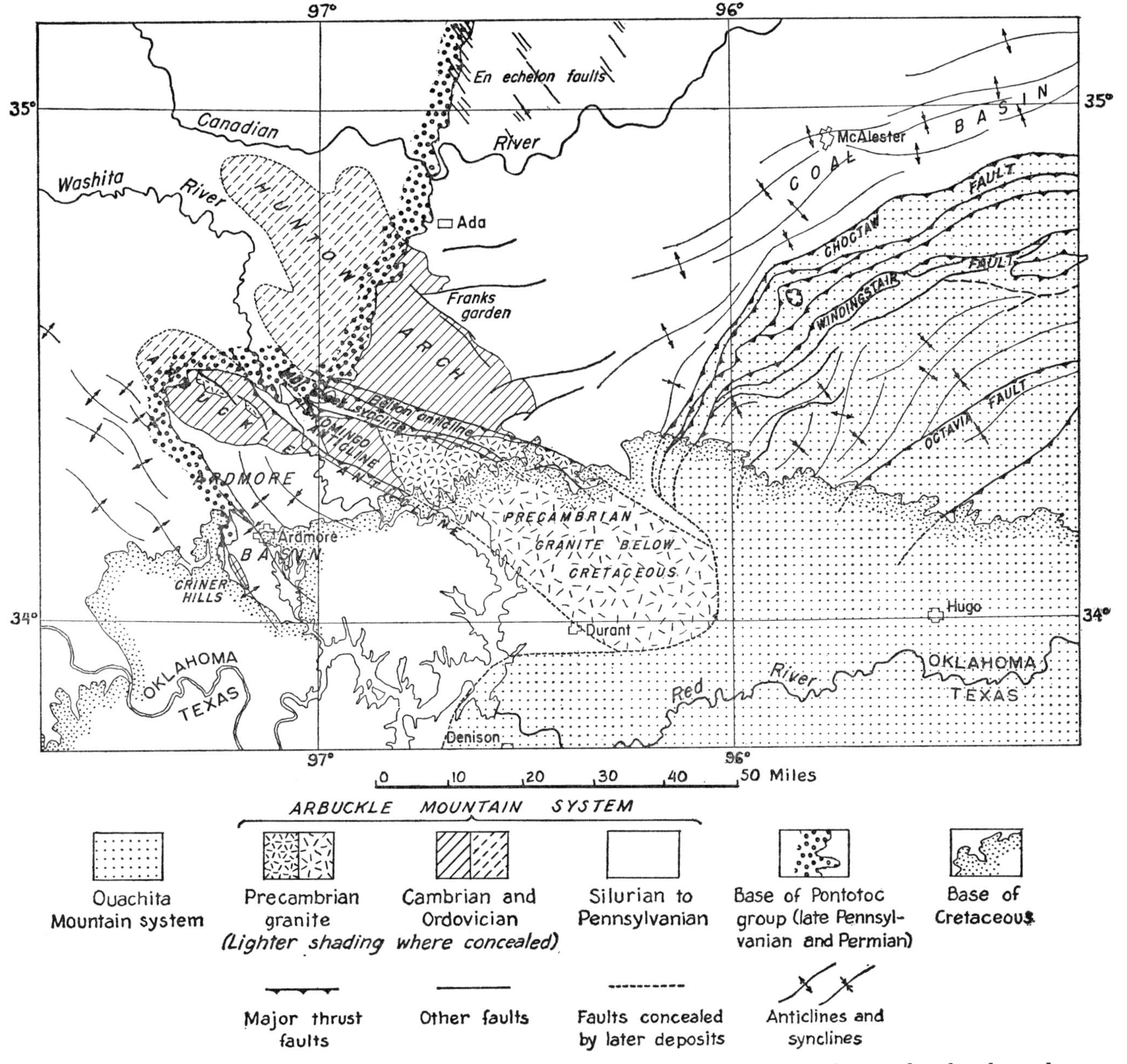

Fig. 40. Map of southern Oklahoma showing relations of Ouachita system to Arbuckle Mountains of Wichita system. The junction of the two systems is concealed on the south by overlapping Cretaceous deposits but has been determined by drilling. Compiled from Geologic Map of Oklahoma, 1954; and other sources.

ians of Alabama, and reappearing again in the Wichita system of Oklahoma. To the latter we will now turn.

Wichita system. In southwestern Oklahoma folded Paleozoic rocks of a character quite different from that in the Ouachita Mountains are exposed in several small areas—the Arbuckle Mountains and Criner Hills on the east (Fig. 41) and the Wichita Mountains on the west. As with other Paleozoic structures of the region, these are greatly obscured by younger deposits—here red beds and various continental sediments of Permian and later Pennsylvanian age. As elsewhere, much has been learned by drilling through these later deposits (Fig. 42).

Earlier Paleozoic rocks of the Wichita system differ greatly from those of the Ouachita system. Whereas in the Ouachita area the base of the section is nowhere exposed and the sequence might extend well down into the Cambrian, that of the Wichita system lies unconformably on Precambrian basement with thin Upper Cambrian sandstones at the base. Succeeding Cambrian and Ordovician rocks are largely carbonates (the Arbuckle limestones and dolomites,

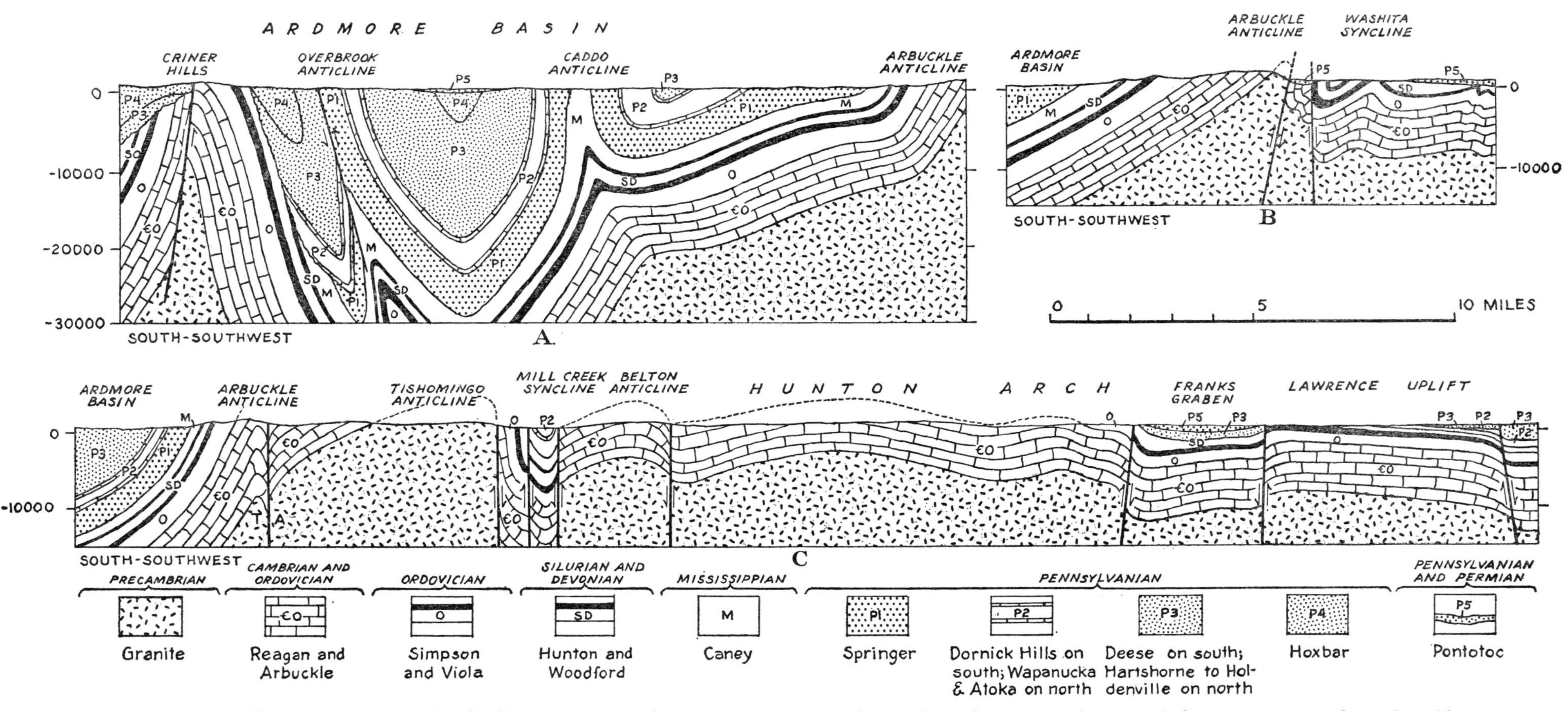

FIG. 41. Sections illustrating structure of Arbuckle Mountains, Ardmore Basin, and Criner Hills. (A) Section across Ardmore Basin from Criner Hills to Arbuckle Mountains. (B) Section across western part of Arbuckle Mountains, nearly continuous with last but slightly offset to west. (C) Section across eastern part of Arbuckle Mountains. The greatest unconformity is at the base of the Pontotoc group (P 5) and expresses a late Pennsylvanian orogeny. Beneath it in the Ardmore Basin (section A) the sequence is nearly conformable from Pennsylvanian down to the base of the Cambrian. In the Criner Hills (section A) and the Lawrence uplift (section C) are other unconformities lower in the Pennsylvanian which express an early Pennsylvanian orogeny. Compiled from Tomlinson, 1929; Dott, 1934; and Ham and McKinley, 1955.

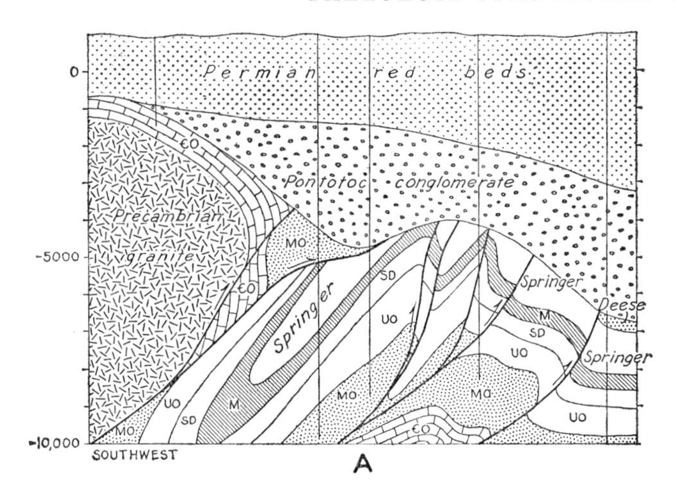

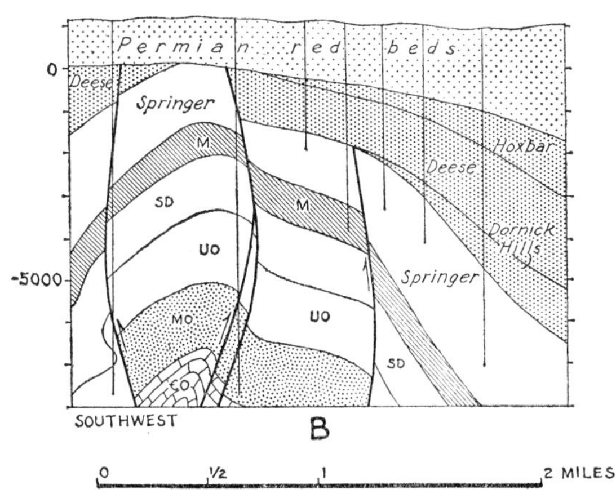

FIG. 42. Complex structures of Wichita system in Oklahoma, revealed by drilling. (A) Eola pool, Garvin County, immediately west of Arbuckle Mountains. (B) Velma pool, Stephens County, about 12 miles west-southwest of preceding. Drill holes indicated by vertical lines. The same unconformities shown in FIG. 41 are also present here—one beneath the Pontotoc conglomerate and Permian red beds indicating a late Pennsylvanian orogeny; another beneath the Dornick Hills and Deese indicating an early Pennsylvanian orogeny. After Swesnik and Green, 1950; and Selk, 1951.

Explanation of symbols. CO Upper Cambrian and Lower Ordovician (Arbuckle limestone, with Reagan sandstone at base). *MO* Middle Ordovician (Simpson group). *UO* Upper Ordovician (Viola limestone and Sylvan shale). *SD* Silurian and Devonian (Hunton group and Woodford chert). *M* Mississippian (Sycamore limestone and Caney shale).

and the Simpson limestones and sandstones) rather than cherts, shales, and sandstones. They thus resemble those of nearby parts of the continental interior, except that the carbonates above the basal sandstones are many times thicker, and more like those in the sedimentary Appalachians. It would thus appear that the area of the Wichita system remained part of the continental platform until Late Cambrian time, after which the crust beneath it gave way, and the area acquired a miogeosynclinal character.

That the Wichita area thereby became a truly mobile belt is indicated by its later history. Beginning in late Mississippian time and continuing through the Pennsylvanian it was broken into a series of west-northwest-trending mountain ridges and intervening troughs. Their history was complex; as the uplifts rose, troughs subsided between them in which great thicknesses of Pennsylvanian deposits accumulated. These were themselves deformed during progress of the orogeny, so that sequences in the troughs are broken by many unconformities.

One of the largest troughs is the *Anadarko basin* which flanks the north side of the Wichita system in Oklahoma (Fig. 43), where Pennsylvanian and Permian sediments were piled to a thickness of more than 20,000 feet. The basin is asymmetrical with its deepest part close to the Wichita chain on the south and its floor rising gradually northward toward the interior region of Kansas. Farther southeast is the similar but narrower and smaller *Ardmore basin* crowded between the uplifts of the Arbuckle Mountains and the Criner Hills (Fig. 41, Section A).

The foreland basins of the Appalachians which we have considered previously (Chapter III, Section 5) were covered by clastic wedge deposits derived from erosion of the inner parts of the mountain belt but lying conformably over the earlier deposits. By contrast, sources of the clastic wedges in the basins of the Wichita system are in the uplifts immediately adjacent, and the deposits themselves are broken by successive unconformities; uplifts and troughs are thus parts of the same structural complex.

After the deformations of Mississippian and Pennsylvanian time the Wichita structures were nearly or wholly buried by Permian deposits. These spread indiscriminately over older rocks of every age, although they thicken somewhat into the earlier basins. Unlike the Pennsylvanian deposits which were at least partly marine, all those of the Permian are continental—conglomeratic near uplifted earlier sediments, arkosic near granitic basement rocks, and passing into red beds away from the uplifts.

BURIED PARTS OF WICHITA CHAIN. Drilling has greatly expanded our knowledge of the Wichita structures and has shown that many more mountain ridges were formed than one would suspect from the small areas which happen to project to the surface.

One set of uplifts (the Red River uplift) which nowhere comes to the surface lies south of the exposed ridges in Oklahoma and extends westward across northern Texas under the Great Plains, finally dying out in eastern New Mexico.

The main Wichita chain passes beneath the surface at the west end of the Wichita Mountains, but continues beyond into the Amarillo district of the

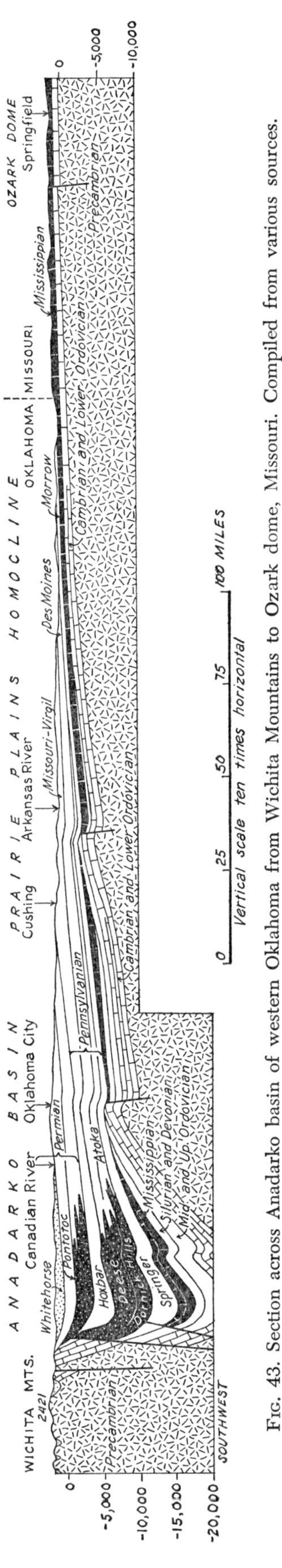

FIG. 43. Section across Anadarko basin of western Oklahoma from Wichita Mountains to Ozark dome, Missouri. Compiled from various sources.

Texas Panhandle where its structures have been referred to as the *Amarillo Mountains.* These are, of course, mountains in a geological sense only; they are completely buried by Permian, Mesozoic, and Tertiary deposits, and the surface of the Amarillo district is flat as only the Great Plains can be flat. Permian deposits that were spread from the old mountain ridges form wedges of arkosic detritus—the so-called *granite wash.* In them, great quantities of natural gas have accumulated, and the district was one of the great natural gas fields of the world; this has now been partly depleted, although it is still producing.

The Wichita chain continues beyond the Amarillo district through various other buried uplifts but comes to the surface once more in the Rocky Mountains of Colorado (Plate I) where its structures have been greatly obscured by those imposed on the region in Late Cretaceous and Tertiary time; these Paleozoic structures form the *Colorado system,* which we will discuss later (Chapter VII, Section 2).

We have now arrived in the Cordilleran system of the west, but before we discuss it, let us retrace our steps for a while and search for modern analogues of geosynclines.

REFERENCES

2. *Geological investigations*

Rodgers, John, 1949, Evolution of thought on structure of middle and southern Appalachians: *Am. Assoc. Petrol. Geol. Bull.*, v. 33, p. 1643-1654.

3. *Appalachian cross-section*

Billings, M. P., 1945, Mechanics of igneous intrusion in New Hampshire: *Am. Jour. Sci.*, v. 243 A, p. 40-48.

———, 1956, The geology of New Hampshire, part II, Bedrock geology: *New Hampshire Planning and Development Commission*, 203 p., map.

Butts, Charles, 1940, Geology of the Appalachian Valley of Virginia: *Virginia Geol. Survey Bull.* 52, pt. 1, 568 p.

Cloos, Ernst, 1947, Oolite deformation in the South Mountain fold, Maryland: *Geol. Soc. America Bull.*, v. 58, p. 943-918.

Ewing, Maurice, Crary, A. P., Rutherford, H. M., and Miller, B. L., 1937, Geophysical investigations in the emerged and submerged Atlantic Coastal Plain, pt. 1, Methods and results; pt. 2, Geological significance of the geophysical data: *Geol. Soc. America Bull.*, v. 48, p. 753-812.
(Some eight additional papers of this series, by Ewing and his collaborators, have appeared in subsequent years.)

Keith, Arthur, 1923, Outlines of Appalachian structure: *Geol. Soc. America Bull.*, v. 34, p. 309-380.

King, P. B., 1950, Tectonic framework of southeastern United States: *Am. Assoc. Petrol. Geol. Bull.*, v. 34, p. 635-671.

Rich, J. L., 1934, Mechanics of low-angle overthrust faulting as illustrated by the Cumberland Mountain thrust block, Virginia, Kentucky, and Tennessee: *Am. Assoc. Petrol. Geol. Bull.*, v. 18, p. 1584-1596.

Rodgers, John, 1953, The folds and faults of the Appalachian Valley and Ridge province: *Kentucky Geol. Survey Spec. Publ. 1*, p. 150-166.

4. *Geosynclines*

Dunbar, C. O., and Rodgers, John, 1957, *Principles of Stratigraphy*: New York, p. 309-315.

Glaessner, M. F., and Teichert, Curt, 1947, Geosynclines, a fundamental concept in geology: *Am. Jour. Sci.*, v. 245, p. 465-482, 571-591.

Kay, Marshall, 1947, Geosynclinal nomenclature and the craton: *Am. Assoc. Petrol. Geol. Bull.*, v. 31, p. 1289-1293.

———, 1951, North American geosynclines: *Geol. Soc. America Mem. 48*, 143 p.

Knopf, Adolph, 1948, The geosynclinal theory: *Geol. Soc. America Bull.*, v. 59, p. 649-670.

Schuchert, Charles, 1923, Sites and nature of the North American geosynclines: *Geol. Soc. America Bull.*, v. 34, p. 151-229.

5. *Growth of the Appalachians*

Hess, H. H., 1939, Island arcs, gravity anomalies, and serpentine intrusions, a contribution to the ophiolite problem: *17th Int. Geol. Cong. Rept.*, U.S.S.R., 1937, v. 2, p. 263-283.

Kay, Marshall, 1951, North American geosynclines: *Geol. Soc. America Mem. 48*, p. 48-60.

Lyons, J. B., Jaffe, H. W. Gottfried, D., and Waring, C. L., 1957, Lead-alpha ages of some New Hampshire granites: *Am. Jour. Sci.*, v. 255, p. 527-546.

Read, H. H., 1957, *The Granite Controversy*: Interscience Publishers, New York, 430 p.; especially Chap. 7, Granite series in mobile belts, p. 374-398.

Woodward, H. P., 1957, Chronology of Appalachian folding: *Am. Assoc. Petrol. Geol. Bull.*, v. 41, p. 2312-2327.

6. *Paleozoic structures west of the Mississippi River*

Dott, R. H., 1934, Overthrusting in Arbuckle Mountains, Oklahoma: *Am. Assoc. Petrol. Geol. Bull.*, v. 18, p. 567-602.

Hendricks, T. A., Gardner, L. S., and Knechtel, M. M., 1947, Geology of the western part of the Ouachita Mountains of Oklahoma: *U. S. Geol. Survey Oil and Gas Inves.*, map 66, 3 sheets.

King, P. B., Geology of the Marathon region, Texas: *U.S. Geol. Survey Prof. Paper 187*, 148 p.

Miser, H. D., 1929, Structure of the Ouachita Mountains of Oklahoma and Arkansas: *Oklahoma Geol. Survey Bull.* 50, 30 p.

———, 1943, Quartz veins in the Ouachita Mountains of Arkansas and Oklahoma; their relation to structure, metamorphism, and metalliferous deposits: *Econ. Geology*, v. 38, p. 91-118.

Tomlinson, C. W., 1929, The Pennsylvanian system in the Ardmore basin: *Oklahoma Geol. Survey Bull.* 46, 79 p.

Van der Gracht, W. A. J. M., 1931, The Permo-Carboniferous orogeny in the south-central United States: *K. Akad. Wetensch. Amsterdam Verh., Afd. Natuurk.*, Deel 27, no. 3, 170 p.

CHAPTER V

LANDS AND SEAS SOUTH OF THE CONTINENT; MODERN ANALOGUES OF GEOSYNCLINES

1. GEOSYNCLINES AND UNIFORMITARIANISM

We have described Paleozoic structures along the southeast and south sides of North America and how they grew from a geosyncline into a deformed belt whose mountainous character was afterwards greatly modified by erosion and burial. Before taking up the similar Cordilleran structures along the west side of the continent, it will be profitable to study the lands and seas to the south along the Gulf Coast and in the West Indies.

Much of the human history, scenery, geography, geology, and geologic history of these regions is of interest, but our treatment of such matters will be incidental to another theme—Are geosynclines forming today, and do these regions furnish us with possible modern analogues of geosynclines?

One of the basic tenets of geology is that of *uniformitarianism*—the present supplies the key to the past. We thus believe that various earth processes—rock weathering, erosion, sedimentation, glaciation, volcanism, crustal movements, and the rest—have been of the same kind, varying only in degree, whether we observe them at work today or read their record in rocks that formed in earlier times. Hence, ancient features such as geosynclines should have modern counterparts, and if we could observe a geosyncline now undergoing formation, it would give us greater understanding of geosynclines that formed earlier.

But the problem of identifying a modern analogue of a geosyncline is complicated. We see the ancient ones after they have gone through the full cycle from sedimentary trough to deformed mountain belt. Their strata have been turned up, and we are thereby able to study their sequence and facies from base to top; yet the very upturning and erosion has produced enough gaps across the feature so that it is not always easy to deduce the geography amidst which the geosyncline formed. If geosynclines were now in process of formation, we could see their geography and their surfaces of sedimentation, but we would not be able to observe directly the masses of sediment which had already accumulated, nor the floor beneath the geosyncline. Even if we were able to probe the substructure of a modern geosyncline by drilling or geophysical means, we would still be unable to look into the future and predict whether this particular sedimentary body is of the sort which inevitably would be deformed and built into mountains at some future time.

Our inquiry will thus not provide us with certain answers. Nevertheless, some thought-provoking comparisons are possible between ancient geosynclines and areas where sedimentation is in active progress today, or was active recently, and where the crust is unstable and undergoing rapid subsidence.

2. GULF COAST AREA

Modern sedimentation along Gulf Coast. Some of the conditions set forth above exist along the Gulf Coast of the United States.

On the map note the extensive areas underlain by Tertiary and Quaternary rocks and the many large rivers which flow into the Gulf of Mexico from the interior region. The largest of these rivers is the Mississippi which, with its many tributaries, drains most of the Interior Lowlands of the United States and enters the Gulf in Louisiana at about mid-length along the coast. But to the west are the Sabine, the Trinity, the Brazos, the Colorado, and the Rio Grande, and to the east are the Tombigbee and the Chattahoochee as well as many other rivers which drain smaller areas than the Mississippi, all of which are bringing to the coast sediments that were derived from erosion of the land.

The Mississippi is estimated to deliver to its mouth 730 billion tons of solid and dissolved material each year. The solid material is laid down as sediments off the mouth of the river; it is estimated that these amount to .068 cubic miles per year, or that a cubic mile of sediments is added about every fifteen years. The projection of land into the Gulf in Louisiana was built by sediment brought down by the river and is the *Mississippi Delta.* If the river were not there, the coast would have been much farther inland, more nearly in line with the coast farther east and west. Deposits of the Mississippi form a low embankment which has been built across the continental shelf nearly to its edge.

One complication makes modern sedimentation on all continental shelves different from that in the past. At several times during the immediately preceding Pleistocene ice age, sea level was lowered some hundreds of feet when ocean water was withdrawn to form the continental ice caps. Through parts of Pleistocene time, therefore, the continental shelves were above water and were being eroded. Sediments which accumulate on the continental shelves today are thus being laid over this eroded surface, or surface of unconformity (Fig. 44).

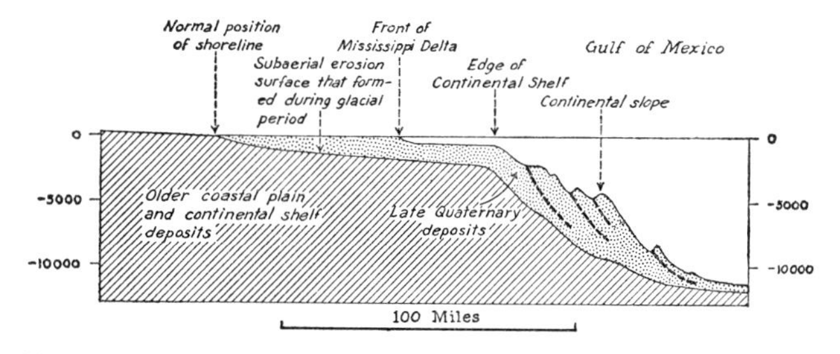

FIG. 44. Generalized section across the Mississippi Delta from the land to the continental shelf and slope, showing the different environments in which sediments are now accumulating. Vertical scale greatly exaggerated. Compiled from Fisk and McFarlan, 1955; and other sources.

Modern sedimentary processes around the Mississippi Delta are thus building the surface up to sea level from the Pleistocene erosion surface. Louisiana geologists assure us, however, that the Pleistocene erosion surface has been downwarped beneath the delta area so that actual volume of post-Pleistocene sediments is much greater than would have resulted from upbuilding alone. This raises again the oft-debated question as to whether or not a surface of sedimentation can subside because it is loaded, but we will pursue this no further here.

Even though a large volume of the sediment discharged by the Mississippi and other rivers has been laid down on the delta and continental shelf, an even greater volume is lost to delta and shelf. Waves and currents gradually shift the sediments out to sea, carrying them beyond the forward edge of the shelf and down the continental slope into the deeper waters of the Gulf. The Mississippi whose delta has nearly reached the shelf edge must especially contribute great volumes of sediment in this manner to deeper water.

Oceanographic work shows that the continental slope off the Mississippi River has a complex, hummocky topography scoured by peculiar channels and many little scarps (Fig. 44). These submarine topographic forms seem to have been produced by slumping and landsliding of the unconsolidated, freshly deposited sediments. The hummocky topography is probably the result of landslides; the channels were formed where the sediments moved as turbid flows, and the scarps were produced where the sediments faulted as they slipped and settled.

In places the continental slope is marked by a few longer, higher, and steeper scarps which were apparently formed on faults that extend beneath the sediments into the basement (Plate I). These seem to result from some more fundamental cause than mere sedimentary loading, and are dropping the Gulf floor relative to the surface of the shelf and the land. Whatever their cause, they accentuate the lesser processes of slipping, flowage, and settling of the continental slope sediments.

In short, modern sedimentation is continually building forward the continental shelf and slope into deeper water, and is thereby extending the continental margin.

Beyond the continental slope is the deeper water of the main basin of the Gulf of Mexico lying between Texas and Yucatan, which reaches a depth of about 13,000 feet in the Sigsbee Deep on its south side (Fig. 50). Beyond this, sea bottom rises again into another continental shelf off the coast of Yucatan, where few rivers enter, and where the surface is receiving mainly limestone deposits. Apparently the deepest part of the main basin of the Gulf is placed away from the Texas-Louisiana side and toward the Yucatan side because of an imbalance in the amount of sedimentation from north to south.

SURFACE FEATURES OF GULF COASTAL PLAIN. So much for modern conditions. Now let us extend our inquiry into the past, and discover what went on along the Gulf Coast during Tertiary and later Mesozoic time.

The reader will recall the *Mississippi Embayment* which we mentioned earlier—the sag containing Cretaceous and Tertiary deposits which extends inland from the coast as far as southern Illinois (Fig. 49, Section A 1). This began to form during the Cretaceous period, since which time it has constituted a sort of funnel toward which all drainage of the interior region has been drawn. From well back in Tertiary time, anyway, rivers ancestral to the Mississippi have flowed into and down the sag, carrying much of the erosional waste of the interior region to the Gulf Coast. Sedimentation like that going on along the coast today must therefore have gone on in much the same manner for the last sixty million years or so of geologic time.

Turning to the geologic map, we can view the Gulf Coastal Plain as a whole (Plate I). Note that it has a width from its inland edge to the coast of 150 to 300 miles. But beyond the coast the same structure extends beneath the surface of the continental shelf, which is only a few miles wide where the Mississippi has built its delta nearly across it, and as much as 150 miles wide in other places.

On the geologic map note the belted outcrops of the formations of the Coastal Plain, with the Cretaceous farthest inland followed successively by the Paleocene, Eocene, Oligocene, and Miocene, and by Pliocene and Pleistocene along the coast. All the strata of these ages dip gently seaward, probably at angles no steeper than inclinations of the Paleozoic strata of the Interior Lowlands which we discussed earlier. Let us make a rough calculation. If we assume a width of the Coastal Plain of 300 miles and an average dip of fifty feet to the mile, the thickness of Coastal Plain sediments at the coast line from Cretaceous to Recent should be about three miles (Fig. 45). We will see presently that for various reasons they are actually very much thicker.

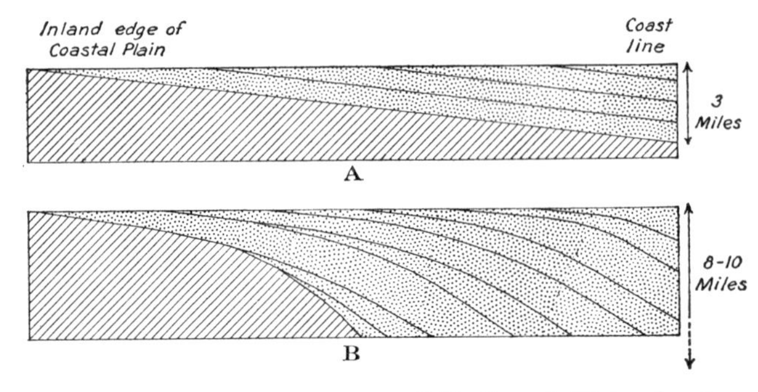

FIG. 45. Sketch sections across Gulf Coastal Plain showing: (A) Structure which would be inferred from surface features alone on the assumption of a uniform seaward dip of all the units. (B) The actual structure as known from subsurface data.

SUBSURFACE FEATURES OF GULF COASTAL PLAIN. The Gulf Coastal Plain, as we know, is one of the great oil provinces of the United States. In the search for its oil resources it has been penetrated by many closely spaced wells to depths of three miles or more. Stratigraphy and structure of the Coastal Plain are thus known far beneath its relatively featureless surface. Geophysical work has extended our knowledge even farther, beyond depths reached by the drill.

Considering first the Mesozoic rocks which are all Cretaceous at the surface, we know from drilling that the Cretaceous of the outcrop thickens greatly down dip and seaward, and that some distance down dip and beneath the surface still older rocks wedge in which are nowhere exposed (Fig. 46). Some of these

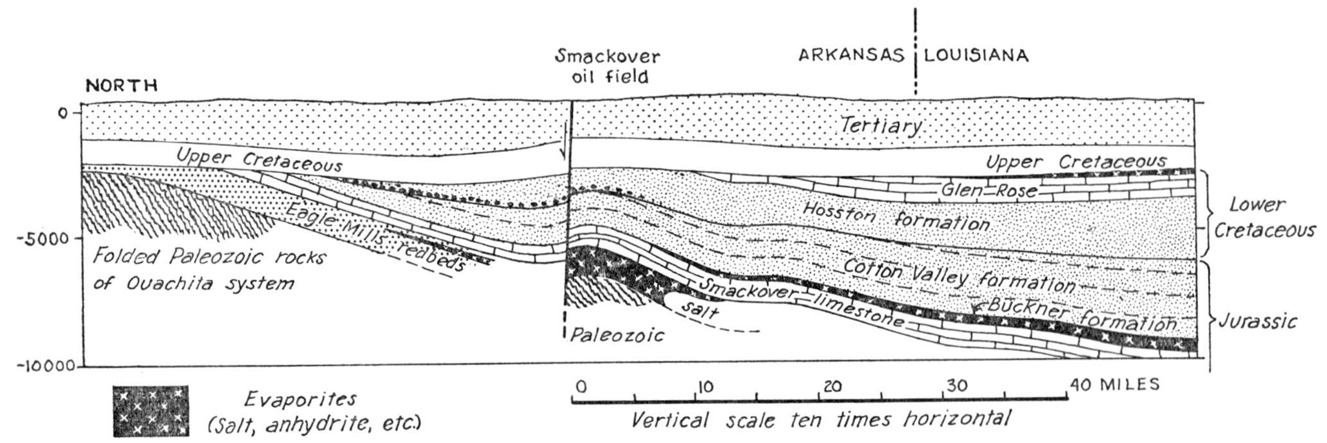

FIG. 46. Section in southern Arkansas and northern Louisiana to show manner in which earlier Mesozoic strata, not exposed at the surface, wedge in down the dip. The Upper Cretaceous lies on the older Mesozoic rocks with a major unconformity. A minor unconformity separates the Lower Cretaceous (Hosston formation) from the Jurassic. Based on Imlay, 1940; and other sources.

are of early Cretaceous age, but others have been proved by contained fossils (recovered from drill cores) to be of later Jurassic age.

Beneath the fossiliferous Jurassic an even older series wedges in. It is largely rock salt which was evidently laid down as an evaporite deposit on the Coastal Plain on the landward side of the advancing initial Mesozoic seas. This salt bed is probably the source of the salt domes which were pushed through the younger sediments of the Coastal Plain (Fig. 47). The salt must extend southward as far as the coast under all the younger deposits, as salt domes occur from the inner part of the Coastal Plain down to the shore or even beyond. Geology of the salt domes is a fascinating subject in itself, with which we cannot deal further here.

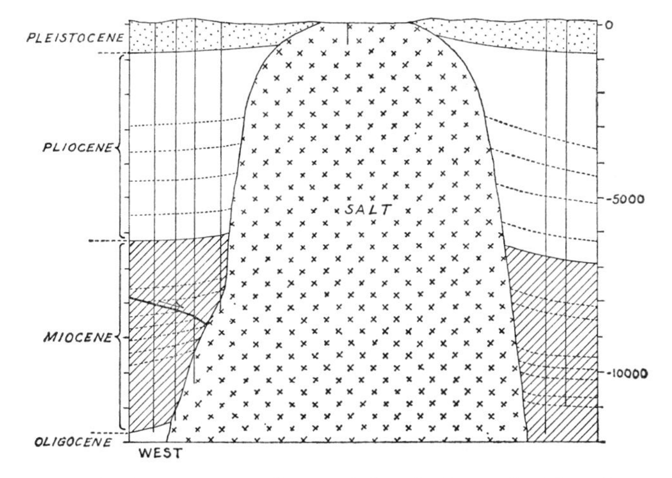

FIG. 47. Cross-section of a typical salt dome that has been outlined by deep drilling; Avery Island, south-central Louisiana. Drill holes shown by vertical lines; horizontal scale same as vertical. After Carsey, 1950.

The lesson of the Mesozoic rocks is that their thickness in their seaward parts is much greater than we would have inferred from their surface outcrops alone, the greater thickness resulting in part from thickening of each layer down the dip and in part from wedging-in of formations down the dip which never extend to the surface (Fig. 45 B).

Many of the same relations hold for the Tertiary deposits. On the outcrop and toward the land these are sands and clays, in part containing beds of coal and various plant remains which indicate a continental origin, in part containing oysters and other shells indicating that they were laid down in salt water along a shore or in lagoons of brackish water behind the shore (Fig. 48). But if any layer is followed down its dip and beneath the surface, it is found to change into marine clays and shales whose contained Foraminifera indicate that they were deposited in progressively deeper water the farther they are from the outcrop and toward the Gulf.

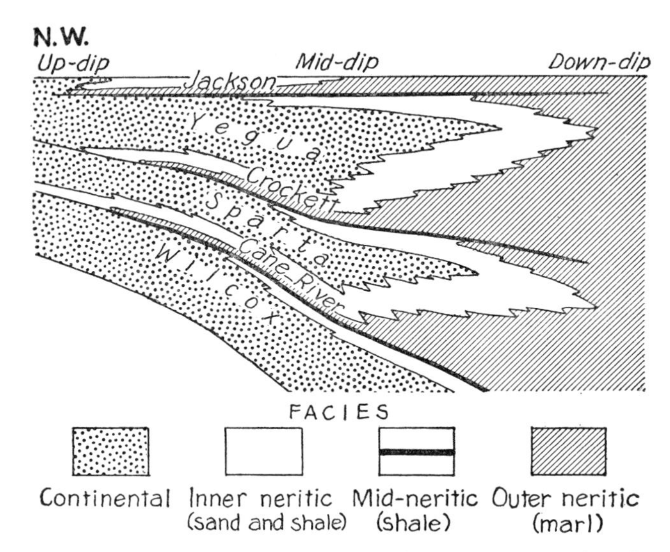

FIG. 48. Diagram illustrating typical arrangement of sedimentary facies in the Eocene deposits of the Gulf Coastal Plain, east Texas and Louisiana. After Lowman, 1949.

Moreover, when any layer in the Tertiary is traced down dip toward the coast, it thickens and steepens. At a certain point in each layer thickening and steepening reach a maximum, and it plunges so rapidly Gulfward that it can no longer be reached by drilling. Obviously the deposits will not thicken and steepen indefinitely—otherwise they would "go right on to China." Further changes in the deposits must take place at depth and farther out, although what these are remain speculative.

Compare these relations with those of the modern sediments. In the modern sediments the same changes take place toward the Gulf—continental deposits, shore deposits, marine deposits of the continental shelf laid down in deepening water outward, and continental slope deposits which thicken and steepen abruptly (Fig. 44). Could not the abruptly thickened and steepened deposits in the Tertiary likewise be continental slope deposits? The zones of maximum steepening and thickening of the various Tertiary deposits each lie farther inland the older they are. If they represent former positions of the continental slope, the slope has moved progressively toward the Gulf through Tertiary time by a distance of several hundred miles.

Let us now consider the whole mass of sediments laid down in the Gulf Coastal Plain (Fig. 49). We can determine this partly from surface dips, although, as we have seen, these give a much smaller total than the true thickness. We can supplement surface information by drill data and by geophysical data on

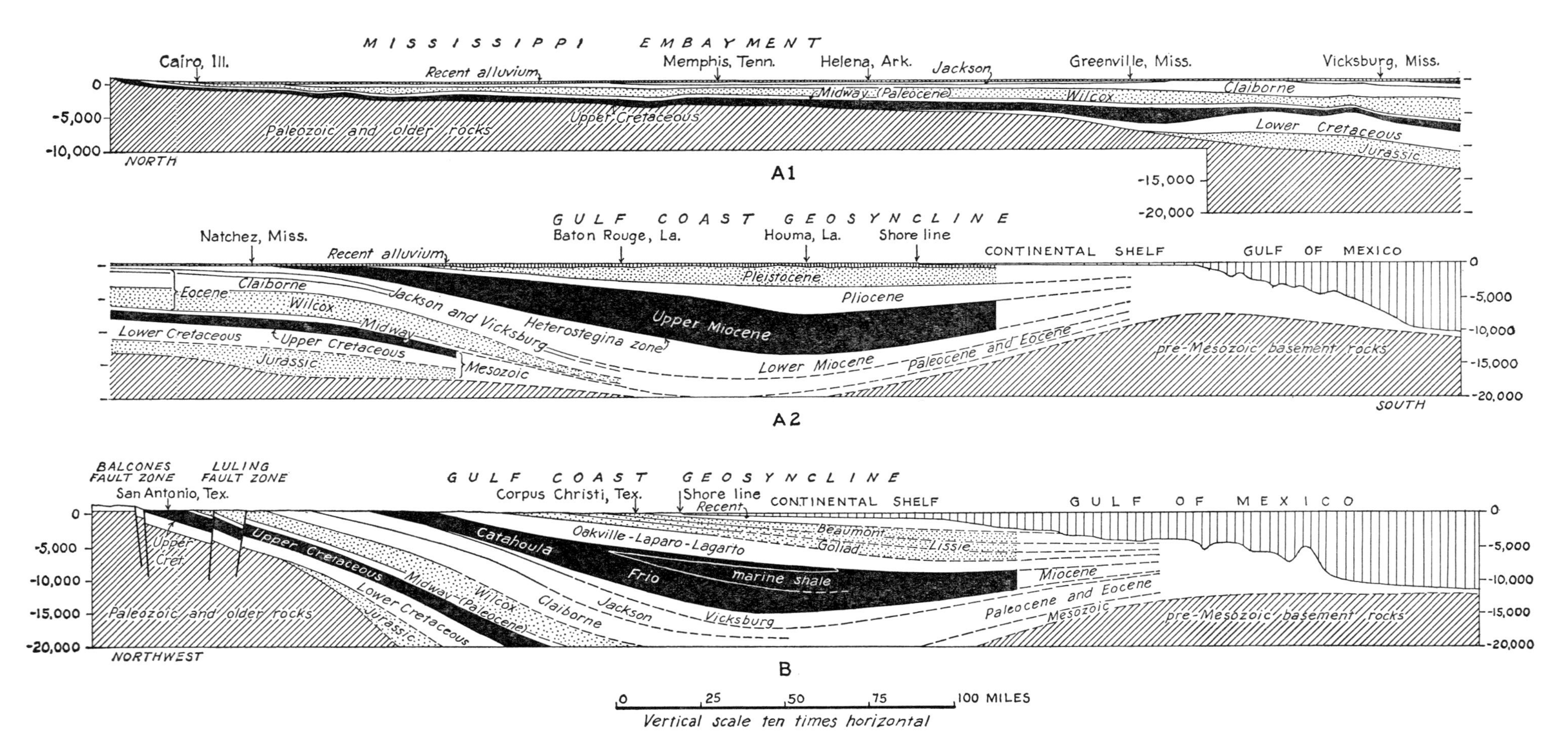

FIG. 49. Sections across Gulf Coastal Plain. (A 1 and A 2) From head of Mississippi Embayment in southern Illinois to edge of continental shelf off Louisiana coast. (B) In Texas from San Antonio to edge of continental shelf. Patterns shown are arbitrary, and no attempt has been made to indicate the complex variations in lithology of the sediments. After Fisk, 1944; Storm, 1945; and other sources.

the still deeper layers. The deepest well near the Louisiana coast, which is currently the deepest producing well in the world, reached a total depth of 22,570 feet without passing through the Miocene; Miocene, Pliocene, and later deposits in this vicinity are at least 25,000 or 30,000 feet thick. When we add the older Tertiary and Mesozoic deposits which undoubtedly underlie them, total thickness of sediments near the Louisiana coast must amount to 40,000 to 50,000 feet or eight to ten miles.

The floor beneath the Mesozoic and Tertiary deposits must therefore descend from the inner edge of the Coastal Plain where it emerges, to the coast; here the great thickness of sediments just mentioned is attained. As none of the sediments were deposited in more than a few hundred or thousand feet of water, present descent of the floor is a result of the progressive subsidence and downwarping of the edge of the continent. Relations are less certain on the side toward the deeper basin of the Gulf of Mexico (Fig. 49). However, recent geophysical work indicates that this basin is underlain by a body of sediments which is not as thick as the one along the coast; the floor beneath the deposits must rise again in this direction. Deposits of the Gulf Coast area therefore lie in a trough whose axis is near to and approximately parallel with the coast (Fig. 50).

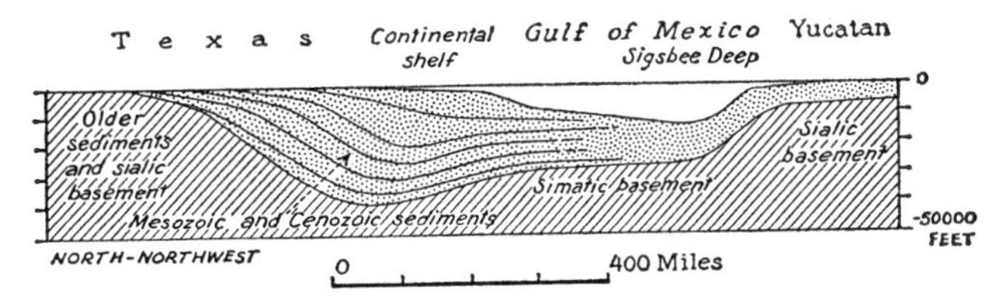

FIG. 50. Sketch section across Gulf of Mexico from Texas to Yucatan, showing inferred variations in thickness of Mesozoic and Cenozoic sediments and their zone of maximum accumulation near the Texas coast.

On the landward side the crust below the Coastal Plain sediments is of continental material or sial. Many geologists have assumed that the basin of the Gulf of Mexico is itself a sag in the continental plate. Recent geophysical work indicates that the crust beneath the sedimentary cover in the main basin of the Gulf is not sial but sima; somewhere beneath the Gulf Coast sediments the continental crust must thin out leaving only oceanic crust beyond. Sedimentation along the Gulf Coast therefore causes the continental area to expand progressively over the oceanic area.

COMPARISONS WITH OLDER GEOSYNCLINES. We have gone far enough to see that there are many resemblances between the Gulf Coastal Plain and the continental shelf, with their Mesozoic, Tertiary, and Quaternary sediments, and the Appalachian geosyncline discussed in the preceding chapter:

(1) Thickness of sediments amounts to as much as 40,000 or 50,000 feet, or close to the maximum thickness of sediments observed in any part of the Appalachian geosyncline.

(2) Floors of the troughs in which the Gulf Coast and Appalachian sediments accumulated were both downwarped into a synclinal form.

(3) The landward half of the sediments of the Gulf Coastal Plain was laid down in relatively shallow water like that of the Appalachian miogeosyncline; the depths to which part of these sediments have been depressed was accomplished by a gradual subsidence.

(4) The seaward half of the sediments of the Gulf Coastal Plain was built forward from the sialic continental crust over the simatic oceanic crust, thus extending the continental area in the same manner as we have suggested for the eugeosynclinal deposits of the Appalachians.

But we can also observe differences between deposits of the Gulf Coastal area and those of the Appalachians:

(A) There is no clearly defined miogeosyncline or eugeosyncline. Deposits on the landward side, at least from Tertiary time onward, include very little limestone and there are no volcanic or tectonic lands farther out.

(B) Limestone is being deposited on parts of the continental shelves around the Gulf of Mexico, but not according to any miogeosynclinal pattern. Most extensive areas of limestone deposition are in those parts where few rivers enter the Gulf, notably off Yucatan and Florida.

(C) Some volcanism occurred in the Gulf Coastal area during Cretaceous and early Tertiary time but not according to the eugeosynclinal pattern. It was along the inner side near a hinge line probably related to downwarping of the trough along the coast, rather than near the edge of the continental shelf.

(D) Main source of sediments is from the landward side, not from tectonic lands offshore; this condition is related to the existence of numerous rivers which are bringing sediments down from the land. This has prevented an accumulation of limestone deposits along the shore. These relations are nearly the opposite of those in the Appalachian geosyncline.

(E) The Gulf Coastal Plain is not especially mobile, the only crustal activity being a slow subsidence. It is not a seismic area and there are no significant earthquakes.

(F) No mountains have developed from the geosyncline—if it is one. Possibly mountains will form here during later stages of the geosynclinal cycle, but no present evidence affirms that they ever will.

Whether such differences are so fundamental that they debar the Gulf Coastal area from comparison with the Appalachian or other geosynclines is a subject much debated among geologists.

I myself am inclined to believe that differences between the Gulf Coastal deposits and those of the Appalachian geosyncline are not very fundamental, and that they are mainly the result of differences between internal geography of the continent in Paleozoic and Cenozoic time. In the first half of Paleozoic time the interior of the continent was low; streams which drained it brought to the coast mainly fine clastics or material in solution. During Cenozoic time the interior of the continent had a much greater relief, so that a much larger volume of clastic material was delivered to the shore by streams.

It is true that crustal unrest and volcanic activity in the Appalachian geosyncline was much greater than along the Gulf Coast, but this is a matter of degree and reflects variations in mobility from one geosyncline to another, as we have noted in our list of "geosynclinal attributes" (Chapter IV, Section 4). Presently we will consider another possible modern analogue of geosynclines in which there was more crustal mobility.

WHERE ARE THE ANCIENT COASTAL PLAIN DEPOSITS? Before leaving the subject of the Gulf Coastal Plain, we should consider a broader question which has troubled the minds of geologists—What has become of the coastal plain deposits of ancient geologic times?

The Atlantic and Gulf Coastal Plains of southeastern North America are perhaps better developed than the rest, yet there are many narrower and shorter coastal plains in other continents—in parts of Europe, Africa, and South America—but all known coastal plain deposits of the world are relatively young, mainly of Quaternary, Tertiary, and Cretaceous age. In a few, as in the Gulf Coastal Plain, the initial deposits are Jurassic, but none are Triassic or earlier.

The continents obviously must have had edges before Cretaceous or Jurassic time. We would suppose by comparing present conditions with the past that before Jurassic time these edges would have received deposits brought down from the land by rivers, and that these would have been built up as coastal plains. Nevertheless, nothing that can certainly be identified as such has been recognized.

I believe that one of the difficulties geologists have had in thinking about earlier coastal plain deposits has been too great a preoccupation with the belief that "borderlands" such as Appalachia once lay along the present continental margins. As indicated in Chapter V, Section 5, no compelling reason exists for believing that the Appalachian geosyncline or the others did not extend directly to the edges of the ocean basins, just as deposits of the Gulf Coastal Plain extend directly to the edge of the Gulf of Mexico today.

So far in this chapter we have posed two problems: Where are the modern geosynclines, and where are the ancient coastal plains? Is it possible that the two problems are opposite sides of the same coin, and that each provides an answer to the other—that the ancient coastal plain structures are simply what we have called geosynclines? Our failure to identify ancient coastal plain deposits and structures might thus be because they have been made unrecognizable by conversion into mountain belts which were then incorporated into the continents.

For the present we need not attempt a final solution of these problems, but we can leave the answers we have suggested as intriguing possibilities.

3. THE WEST INDIES

In our search for modern analogues of geosynclines we have examined one example—a deep sedimentary trough that formed along the edge of a continent in a coastal plain area by slow subsidence but with little other crustal activity. Let us take up another and somewhat different example.

ISLAND ARCS AND DEEP SEA TRENCHES. In discussing the eugeosyncline of the Appalachians we mentioned that its troughs and intervening tectonic lands and volcanic islands might be comparable to modern island arcs and deep sea trenches. This comparison we will now consider further.

If we examine a map of the world we will find around the edges of the Pacific Ocean remarkable festoons of islands which form a series of *arcs* convex toward the ocean. In the northwest Pacific some of these such as the Aleutian, Japanese, and Indonesian islands, lie close to the continents. When the Aleutian arc is traced eastward, in fact, it passes from a chain of islands onto the mainland of Alaska where it merges with the Cordilleran mountains along the western border of North America (Plate I). The latter continue thence along the Pacific to the south tip of South America and no island chains lie offshore; perhaps the Cordillera is the equivalent on the land of the island arcs at sea. But when the island arcs are traced in the opposite direction from Japan into the southwestern Pacific, they run far out into the ocean, forming such chains as the Marianas, New Zealand, and the Tonga Islands.

Alongside the island arcs on their oceanward sides are equally remarkable *trenches*—long and narrow like the rows of islands, with steeply sloping sides which descend far below any normal level of the ocean bottom (Fig. 51). The trenches contain some of the greatest deeps of the world; the deepest extend to 35,000 feet below the surface of the ocean, or farther below sea level than Mount Everest rises above it. They are, in fact, mountain ranges in reverse. Such deep trenches lie next to the island arcs that are close to the continents in the northwest Pacific, next to the Cordilleras of North and South America in the eastern Pacific, and next to island arcs at sea in the southwestern Pacific. Obviously island arcs and deep sea trenches are parts of the same earth structure; we can designate the whole as an *island arc-trench system.*

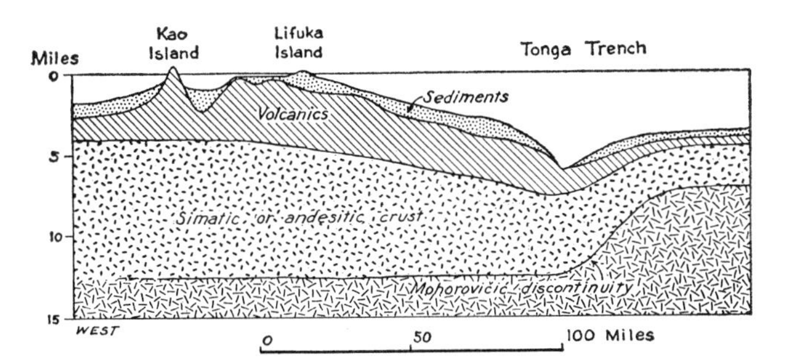

FIG. 51. Section across Tonga Trench, southwestern Pacific Ocean, to illustrate island arc-trench structure. Positions of different crustal layers are derived from geophysical observations. Note exaggerated vertical scale. Modified from Raitt and others, 1955.

The features so far set forth are merely topographic; origin of island arc-trench systems must be sought in their geological and geophysical features. They coincide with some of the most seismic and volcanic belts of the world; the latter has been called the "Pacific circle of fire." When bedded rocks are present in the islands, they are mainly of later Tertiary age, yet are strongly deformed. On some of the islands, as well, coral reefs have been uplifted many hundreds of feet above the level of the present shores. These features and others indicate that something very unusual must have happened and may be still happening to the crust of the earth in the island arc-trench systems; they must have been shaped by strong movements late in geologic time.

THE WEST INDIES AS AN EXAMPLE. Fortunately, we have a sample of an island arc-trench system at the back door of North America—not in the Pacific which has no island arcs on the American side, but in the West Indies and Caribbean Sea southeast of the Gulf of Mexico which we have just discussed. We are doubly fortunate, moreover, because the structures of this region are the southeastward continuation of the North American Cordillera, with which we will soon deal.

In the West Indies, as in the Pacific island arc-trench systems, we find the same arcuate island chains —chains of volcanoes, deep trenches, earthquake belts, and indications of deformation in later geologic time. Unfortunately for the example, some features are not as well displayed as in the systems of the Pacific, but these deficiencies appear to be of degree rather than kind, so that we can fit our example into the general class with which our inquiry is concerned. Let us, then, review the geographic and geologic features of the area.

CENTRAL AMERICA. The land connection between North and South America, as we can see from the map, is through the isthmus of Central America. From the map alone one would be tempted to connect the Cordillera of North America with the Cordillera of South America (or Andes Mountains) through this land bridge, but examination of the geologic features indicates that the linkage is more complex.

The middle part of Central America centering in Nicaragua is primarily a chain of volcanoes that was built up rather recently to join deformed Tertiary rocks on the south with older structures on the north. While the volcanic chain must follow some young line of weakness in the crust, this would seem to be less fundamental than the structures which we will now discuss.

Northern Central America, by contrast, in Guatemala and Honduras consists of folded rocks of Paleozoic and Mesozoic ages. Northwestward in Mexico these folds are continuous with the North American Cordillera and are thus the southeastern extensions of our own mountains of the western states. But instead of trending down the Central American isthmus, they trend across it in an eastward direction and run out beneath the Caribbean Sea (Plate I).

GREATER ANTILLES. The structural zone of northern Central America is not lost in the Caribbean, however, as submarine topography shows that it continues eastward into the Greater Antilles or northern massive islands of the West Indies—Cuba, Jamaica, Hispaniola (Haiti and Santo Domingo), and Puerto Rico. Structures of their southern part in Jamaica and southern Hispaniola are in direct line with those of northern Central America; structures of their northern part in northern Hispaniola and Cuba seem to die out westward toward the flat-lying rocks of Yucatan.

Rocks and structures of the Greater Antilles are much like those of such mature mountain belts as the Appalachians and western Cordillera. They are composed of a thick body of Mesozoic and Tertiary rocks formed in a geosyncline of considerable mobility. Their Cretaceous rocks include large volcanic components, as do the Paleozoic rocks of the eugeosynclinal belt of the Appalachians. They were strongly deformed toward the end of the Mesozoic time, as well as several times earlier and later. Some of the older rocks have been metamorphosed and plutonized; they contain bodies of ultrabasic, basic, and intermediate intrusives, much like those of the crystalline part of the Appalachians. But there are no young volcanic fields and only shallow seismic activity.

The Greater Antilles are thus the emerged parts of a mountain system which has gone through the full cycle of deformation and has now attained relative stability.

FORELAND OF THE BAHAMAS AND FLORIDA. Northeast of the Greater Antilles are the broad, partly submerged platforms of Florida and the Bahama Islands. They are covered by thick, flat-lying Tertiary and Cretaceous deposits—largely limestones. Deep wells in southern Florida and on Andros Island in the Bahamas have penetrated 10,000 to 15,000 feet of Tertiary and Cretaceous rocks with Jurassic beneath. These thick limestone deposits are comparable to those of the miogeosyncline and foreland in the Appalachians. The limestone was probably laid down in part on a sialic plate, a small continental platform that lay in the triangle between the Antillean mountain belt, the Appalachian Mountain belt, and the Atlantic Ocean basin.

LESSER ANTILLES. Beyond Puerto Rico much more of the structural belt is submerged, but parts project above water in the islands of the Lesser Antilles. These extend southward in a great arc which again joins the continent in Trinidad and Venezuela.

The arc of the Lesser Antilles is double, the islands forming two concentric chains of unlike topographic and geologic character. The inner chain of the arc is a line of volcanoes, some dormant and deeply eroded, some active, in part violently so. The outer chain is formed of limestones and other sediments of Tertiary age which lie on a basement of deformed older rocks, perhaps an extension of the mountain belt of the Greater Antilles. Its islands are spaced more widely than those of the inner chain and extend around only the north half of the arc. In the south half farther out are the islands of Barbados and Tobago which are emerged parts of another structure; this has special characters which we will examine later. Although the crustal forces which produced the Greater Antilles have now largely come to rest, those of the Lesser Antilles are still active, as shown by the modern volcanic and seismic activity of the area.

TRINIDAD AND VENEZUELA. South of the arc of the Lesser Antilles the structural belt which we have been tracing is again mainly emergent. Its northern border, it is true, appears only in a chain of islands off the coast of Venezuela, but the main body forms the island of Trinidad and continues thence into the ranges of northern Venezuela, which curve westward to join the main Cordillera of the Andes along the Pacific side of South America (Plate I). From Trinidad westward these ranges are again a mature mountain structure like those of the Greater Antilles and northern Central America.

FAULT ZONES NORTH AND SOUTH OF ANTILLEAN ARC. Another item must be added to complete the picture of the surface geologic structure—great zones of transcurrent faults on each side of the Antillean arc, trending generally eastward through Trinidad and Venezuela on the south and the Greater Antilles on the north.

Thus, the North Range of Trinidad and the Cordillera de la Costa in Venezuela together form a line of narrow ridges along the Caribbean Sea, composed of metamorphosed Mesozoic rocks unlike the unmetamorphosed Mesozoic and Tertiary rocks south of them. The metamorphic rocks have reached their present position by movements along a fault zone on the south side of the narrow ridge—a sideward shift to the east relative to the rocks on the south—that is, a *right-lateral strike-slip displacement*. Other faults with the same sense of movement occur farther west in Venezuela (Plate I).

On the opposite side of the Caribbean Sea the *Bartlett Trench* extends diagonally through the Greater Antilles between Cuba and Jamaica. Its north and south sides descend in steep submarine escarpments to depths as great as 22,800 feet. Land features on the south coast of Cuba and submarine features elsewhere indicate that the block which forms the trench has been depressed by faulting. It was, however, probably not merely dropped by gravity and extension of the crust; rather, pattern of the adjacent islands and their structure suggests that movements were primarily a sideward shift of the southern side toward the east relative to the side on the north—a *left-lateral strike-slip displacement*, or movement in the opposite sense from that on the faults of Venezuela and Trinidad.

The two zones of faulting—in Venezuela and Trinidad and in the Bartlett Trench—apparently separate stable parts of the crust north and south of the Caribbean Sea (with mature mountain structures) from a more mobile intervening block of the crust which is shifting eastward toward the arc of the Lesser Antilles, an immature mountain structure.

Caribbean Sea floor. So much for the geology of land areas in the West Indies. But, as we have seen, West Indies structures are only partly emergent; the remainder lie beneath the sea. To complete the picture submarine features must be added.

The Caribbean Sea lying between the Greater Antilles and Venezuela and inside the arc of the Lesser Antilles is floored by a fairly level plain 12,000 to 17,000 feet deep. This is diversified only by ridges concentric to the arc of the Lesser Antilles, which may be folds or volcanic belts related to the same deformation. Recent geophysical work indicates that the crust beneath the floor is of a sort intermediate between the usual simatic and sialic crusts of the oceanic and continental areas. It is thicker than the crust beneath the oceans and has a lower seismic velocity, suggesting that it has a composition more like andesite than basalt.

This special andesitic crust occurs only in association with island arc-trench systems. In the western Pacific, as in the West Indies, it floors the seas behind the island arcs—for example, the Philippine Sea between the Mariana Islands and Asia. In the Pacific the boundary between the island arcs and the central simatic ocean floor is termed the *andesite line.* Some geologists have speculated that the areas of andesitic sea floor behind the island arcs are foundered edges of the continental plates. Modern knowledge suggests, however, that they were originally areas of oceanic crust which are being converted by magmatic and sedimentary processes into less dense crustal material approaching that of the continental area.

Puerto Rico Trench. On the opposite side of the Antillean arc from the Caribbean Sea along the edge of the Atlantic Ocean basin is the Puerto Rico Trench. This originates on the west between Hispaniola and the Bahama Islands, passes north of Puerto Rico, and curves southeast concentric to the island arc, fading out at about mid-length in the arc (Plate I). Its deepest part north of Puerto Rico lies 27,000 feet beneath the surface of the sea, and is the deepest point in the Atlantic Ocean.

The Puerto Trench seems to be a different kind of structure from the Bartlett Trench previously discussed. Instead of cutting through the island arc and across the structural grain, it extends concentrically with the arc and outside it on its oceanward side. It is, moreover, marked by a strong negative gravity anomaly, whereas the Bartlett Trench exhibits a positive anomaly; the significance of this we will see presently. The Puerto Rico Trench is of the same structural class as the deep sea trenches which border the island arcs of the Pacific on their oceanward sides.

The Puerto Rico Trench has a flat floor some miles wide, which geophysical work indicates is underlain by a thick body of sediments. The rock floor beneath the sediments must have a more nearly V-shaped cross-section which was probably shaped by downwarping or downfaulting. In initial stages of trench formation the rock surface may have been nearly bare; sediments were brought to it later, probably by turbid flows derived from adjacent lands or from shallower parts of the ocean bottom. Various stages of this sequence are illustrated in the trenches of the Pacific, some of which are V-shaped and rock-floored to their bottoms (as in Fig. 51), whereas others are flat-floored and sediment-filled like the Puerto Rico Trench.

Processes of trench formation and destruction are illustrated on the island of Barbados, which lies southeast of and outside the Lesser Antilles arc, and is believed to be on the structural prolongation of the Puerto Rico Trench. Here the following stratigraphic sequence is exposed:

STRATIGRAPHIC SEQUENCE ON ISLAND OF BARBADOS

Coral rock (Pleistocene). Forms a thin mantle over wide areas of the island. Deposited after the island had been uplifted in much its present form.

UNCONFORMITY

Bissex Hill marl (upper Oligocene-lower Miocene). Globigerina marl and foraminiferal limestone. A marine deposit seemingly laid down at moderate depth .... 100 feet

UNCONFORMITY

Oceanic formation (upper Eocene-lower Oligocene). Radiolarian earth, Globigerina marl, and red clay. A deep sea marine deposit .... 1,800 feet

GRADATION IN MOST PLACES

Joes River beds (upper Eocene). Silts of chaotic structure containing small to large fragments of clay, sandstone, and limestone from Scotland formation. A mudflow deposit formed by submarine landslides while deformation was in progress .... 0-1,300 feet

ANGULAR UNCONFORMITY

Scotland formation (middle Eocene). Silty and sandy shale, sand, grit, and fine conglomerate strongly folded and faulted. A shallow-water marine deposit. Exposed thickness about .... 5,000 feet

This sequence suggests that after Scotland time a sea floor of moderate depth was sharply deformed and downbent into a deep sea trench, the Joes River mudflows representing submarine landslides which took place during its creation. The trench was then filled more slowly by the deep sea deposits of the

Oceanic formation (Fig. 52), and the area was finally elevated once more, probably by further compression, to form the present island and the ridge on which it stands. The structure of Barbados is thus in a more advanced stage of development than the Puerto Rico Trench which lies along its continuation to the north.

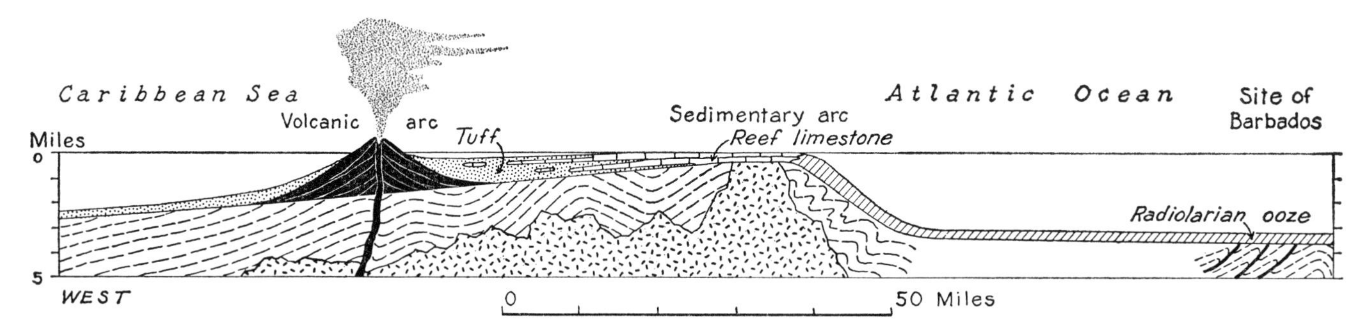

FIG. 52. Section from Caribbean Sea to Atlantic Ocean in latitude of Barbados, showing inferred geography and structure at the time of deposition of the Oceanic formation (late Eocene and early Oligocene). After Senn, 1940.

NEGATIVE GRAVITY ANOMALIES. We mentioned briefly a remarkable feature of the Puerto Rico Trench and others like it—their relation to belts of strong negative gravity anomaly. We will now consider this further.

Geophysicists have found that over the earth the value of gravity varies by slight amounts. The famous apple which struck Sir Isaac Newton would not strike with the same force everywhere. Neither we nor Sir Isaac would be able to detect these variations with our own senses; they must be measured by delicate instruments.

Most of these variations can be accounted for by well-known physical laws, and result from latitude, altitude, and topographic irregularities. One would expect that when observed readings were corrected for these factors, the value of gravity at all points would be equal. Nevertheless, variations still remain called *gravity anomalies* which result from the interposition of geological factors such as differences in density of rocks of the crust from place to place.

Thus, land areas yield chiefly negative anomalies, and oceanic areas chiefly positive anomalies. These are caused by the contrasting densities of the sialic continental plates and the simatic ocean floors. Negative anomalies are largest in mountain areas, showing that a greater volume of lighter rocks is there present. The crust is thus in *isostatic balance*; high areas stand high because they are underlain by rocks less dense than those of lower areas.

These principles were established mainly by study of gravity on the land; up to twenty-five years ago little could be determined about gravity conditions at sea. Observations of gravity are measured by means of the rate of swing of a pendulum; the instrument is set on a solid base and a series of measurements is made, sometimes consuming many days. Such observations could not be made on shipboard because of the tossing of the vessel. However, the Dutch geophysicist *F. A. Vening Meinesz* conceived a method of making gravity measurements at sea in a submarine, submerged far enough below the surface to be free of oscillation of the water. With the help of the Dutch Navy he made a series of such observations in the region of island arcs and trenches in Indonesia, which at that time was Dutch territory.

These observations yielded startling results. Near the trenches it was found that there was a narrow belt of strong negative gravity anomalies. Trench and anomaly clearly had a close relation, although the axes of the two did not necessarily coincide; sometimes the axis of the anomaly lay on one side of the trench, sometimes on another; in places it crossed an island which lay in its course.

Work done afterwards in other island arc-trench systems has shown that most of them have similar gravity properties. Thus, in the West Indies a belt of negative gravity anomalies lies near the axis of the Puerto Rico Trench on the convex side of the Antillean arc. It swings southward in front of the arc, continues beyond the point where the trench fades out, and passes through the islands of Barbados, Tobago, and Trinidad. Other less continuous negative anomaly belts lie in the Caribbean Sea north of the coast of Venezuela.

These negative anomalies cannot be accounted for by ordinary geological factors such as isostatic compensation. Oceanic areas generally yield positive anomalies, and one would expect these to be greatest in the trench areas. In fact, the Bartlett Trench, apparently with a different structure from the other trenches, does show a positive anomaly. The negative anomaly belts near the trenches seem to be caused by underlying masses of light rocks which are being held down by crustal forces, out of isostatic balance.

The belts of negative anomalies associated with the island arc-trench systems prove that some structure even more profound than the remarkable surface features lies in the earth beneath them. They also show that the trenches are the primary surface features, and the island arcs are merely accessory.

DEEP-FOCUS EARTHQUAKES. Another set of features, earthquake foci, also indicate great depth of disturbance beneath the island arc-trench systems. Regrettably, the earthquake picture is not as complete in the West Indies as in the island arc-trench systems of the Pacific; although the structure here is like that in the Pacific, the area is apparently somewhat less mobile. Even so, what has been learned of earthquake foci is worth describing here because of its bearing on the general problem of island arc-trench systems.

By modern methods not only can the geographic position of an earthquake epicenter be located, but also the depth at which the shock took place. It has thus been observed that shallow earthquake foci occur along the deep-sea trenches, and that successively deeper foci extend under the adjacent island arcs and for several hundred miles inside them. These foci define a plane which descends into the earth, dipping from the trench beneath the island arc at an angle of about 35° to a depth of 300 kilometers, then at an angle of about 60° to a depth of 700 kilometers. Deepest recorded earthquake foci lie at the astonishing depth of 700 kilometers or 435 miles.

Recall that the Mohorovičić discontinuity, mentioned earlier as forming the base of the crust of the earth, lies at a depth of only four to twenty-three miles. We thus see that disturbances which create the earthquakes near the island arc-trench systems extend into the earth a score or more times as far as the mere crust with which we are mainly dealing.

EXPLANATIONS OF ISLAND ARC-TRENCH STRUCTURES. So much is fact or reasonable inference; now we enter the realm of speculation. What is the origin of the island arc-trench structures?

Vening Meinesz and his associates, in order to explain the negative gravity anomaly belt which they had discovered, suggested that the deficiency of mass implied by it was caused by a great downbuckle of the crust termed a *tectogene*, which carried light sialic rocks far below their normal level. Surface expression of the downfold, at least in its early stages, was the deep-sea trench; the island arcs formed on upbulges along its sides. It was suggested that as downbuckling continued, sediments deposited in the trench were themselves crushed and folded, and that after relaxation of the compression, isostatic compensation asserted itself and the deformed mass of light rocks was raised into a mountain belt.

It is now known, however, that most deep-sea trenches lie in areas of simatic or at least andesitic crust, so that not much light crustal material is available to form such a downbuckle. Moreover, geophysical profiles across the trenches indicate no such deep downfolding of the base of the crust as would be required to form a tectogene. One set of calculations for the Puerto Rico Trench implies that the crust beneath it is not bent down at all, but most calculations for the trenches imply at least moderate downbending (Fig. 51). The deep-focus earthquakes indicate a brittleness of the rocks of the earth far below expected levels, which would suggest that they yielded not only by buckling but by fracturing.

I myself feel that the observed geologic facts require that there be some sort of crustal downbending near the deep sea trenches and negative anomaly belts. Nevertheless, the substructure of the island arc-trench systems is currently being given a new evaluation, so that it would be hazardous to predict what the final solution might be.

COMPARISON OF WEST INDIES WITH NORTH AMERICAN GEOSYNCLINES. We have now carried our discussion far enough to hazard some judgments on our original question—Are the West Indies a modern analogue of a geosyncline?

In the West Indies there is a mountain belt still in a state of growth—parts emergent, parts still beneath the sea. The two ends show a mature structure comparable to that in the Appalachian and Cordilleran systems of North America; the middle is still mobile and immature. These structures are traceable northward and southward into the Cordilleras of North and South America, so that all are members of the same system of deformation.

The West Indies are an island arc-trench system comparable to systems which border the Pacific Ocean. Island arc-trench systems are the most active tectonic, seismic, and volcanic areas in the world today, yet they seem to lie on continuations of more mature mountain belts on the land. It is unlikely that they represent structural features wholly novel and unrelated to more mature mountain belts; rather, one is tempted to think that they are made up of the same sorts of structures now in the process of growth.

In the ancient examples eugeosynclinal rocks and the structures which developed in them seem to lie in the most mobile parts of the geosynclinal belts. The record of the rocks indicates that growth of the eugeosynclinal areas was attended by tectonic lands formed of uplifted earlier sediments and by chains of volcanic islands; these may be analogous to the various concentric chains in modern island arcs. The record indicates that such tectonic and volcanic lands were separated by rapidly subsiding depositional troughs. In the modern examples deep sea trenches alongside the island arcs are the fundamental structures and lie near the axis of the negative gravity anomaly. In the ancient examples it is difficult to determine which one of the troughs was such a fundamental feature and whether any of them attained excessive depths; most of the sedimentary features by which this could be determined have been obliterated by the great deformation and metamorphism to which the rocks have since been subjected.

Downfolding of the crust implied in the modern examples would provide a means by which the eugeosynclinal rocks could have been carried to depths great enough to have been subjected to metamorphism and plutonism, and a means by which they could have been deformed earlier than the rocks of adjacent areas.

The miogeosynclinal areas, or those parts of the ancient geosynclines with which we are most familiar, might correspond to depositional basins on the concave sides of modern island arcs. In the modern as well as the ancient examples these are areas of slow subsidence covered by waters of shallow to intermediate depth which received sediments derived principally from the more mobile central parts of the system. In the modern examples as well as the ancient, one would expect that these sediments would be deformed only after the central part of the system had been consolidated.

Be this as it may, perhaps the principal lesson which we can draw from our comparison between ancient geosynclines and mountain belts, and modern island arc-trench systems is that neither are mere superficial features produced by processes of sedimentation and erosion, or of loading and unloading of the crust. Instead, they appear to be products of an activity within the crust of the earth or even far beneath the crust. The geosynclinal and mountain structures which we can observe do not express the whole feature, but are merely surface manifestations of much more deep-seated structures.

REFERENCES

2. *Gulf Coast area*

Bornhauser, Max, 1958, Gulf Coast tectonics: *Am. Assoc. Petrol. Geol. Bull.*, v. 42, p. 339-370.

Ewing, Maurice, and others, 1955, Geophysical and geological investigations in the Gulf of Mexico: *Geophysics*, v. 20, p. 1-18.

Fisk, H. N., and McFarlan, E., Jr., 1955, Late Quaternary deltaic deposits of the Mississippi River: *Geol. Soc. America Spec. Paper* 62, p. 279-302.

Gealy, B. L., 1955, Topography of the continental slope in northwest Gulf of Mexico: *Geol. Soc. America Bull.*, v. 66, p. 203-228.

Murray, G. E., ed., 1952, Sedimentary volumes in Gulf Coastal Plain of United States and Mexico: *Geol. Soc. America Bull.*, v. 63, p. 1157-1228.

Russell, R. J., 1940, Quaternary history of Louisiana: *Geol. Soc. America Bull.*, v. 51, p. 1199-1234.

Storm, L. W., 1945, Résumé of facts and opinions on sedimentation in Gulf Coast region of Texas and Louisiana: *Am. Assoc. Petrol. Geol. Bull.*, v. 29, p. 1304-1335.

Weaver, Paul, 1955, Gulf of Mexico: *Geol. Soc. America Spec. Paper* 62, p. 269-278.

3. *West Indies*

Benioff, Hugo, 1954, Orogenesis and deep crustal structure; additional evidence from seismology: *Geol. Soc. America Bull.*, v. 65, p. 385-400.

Bucher, W. H., 1952, Geologic structure and orogenic history of Venezuela: *Geol. Soc. America Mem. 49*, 113 p.

Ewing, J. E., and others, 1957, Geophysical investigations in the eastern Caribbean; Trinidad shelf, Tobago trough, Barbodos Ridge, Atlantic Ocean: *Geol. Soc. America Bull.*, v. 68, p. 897-912.

Ewing, Maurice, and Heezen, B. C., 1955, Puerto Rico Trench; topographic and geophysical data: *Geol. Soc. America Spec. Paper* 62, p. 255-268.

Fisher, R. L., and Revelle, Roger, 1955, The trenches of the Pacific: *Scientific American*, v. 193, no. 5, p. 36-41.

Hess, H. H., 1938, Gravity anomalies and island arc structure with particular reference to the West Indies: *Am. Philos. Soc. Proc.*, v. 79, p. 71-96.

Officer, C. B., and others, 1957, Geophysical investigations in the eastern Caribbean; Venezuelan basin, Antillean arc, and Puerto Rico Trench: *Geol. Soc. America Bull.*, v. 68, p. 359-378.

Roberts, R. J., and Irving, E. M., 1957, Mineral deposits of Central America: *U.S. Geol. Survey Bull.* 1034, 205 p., map.

Rod, Emile, 1956, Strike-slip faults of northern Venezuela: *Am. Assoc. Petrol. Geol. Bull.*, v. 40, p. 457-476.

Schuchert, Charles, 1935, *Historical Geology of the Antillean-Caribbean Region*: New York, 811 p.

Senn, Alfred, 1940, Paleogene of Barbados and its bearing on history and structure of Antillean-Caribbean region: *Am. Assoc. Petrol. Geol. Bull.*, v. 24, p. 1548-1610.

Woodring, W. P., 1954, Caribbean land and sea through the ages: *Geol. Soc. America Bull.*, v. 65, p. 719-732.

CHAPTER VI

THE MOUNTAIN BELT OF WESTERN NORTH AMERICA; INTRODUCTION TO THE CORDILLERAN SYSTEM

1. GEOGRAPHIC FEATURES AND THE PROCESSES WHICH SHAPED THEM

We take up now the mountain system of western North America, to which we will devote the remainder of this book. It is a belt of ranges, in part massive and continuous, in part disconnected and isolated, and of intermontane valleys, basins, and plateaus, which extends 500 to 1,000 miles inland from the Pacific Coast along the entire length of North America (Plate I).

Many of the varied features of the mountain system will be treated in this book, yet it will not be possible to do justice to all of them in view of the size and complexity of the system. Chief emphasis will therefore be given to the central segment lying mainly in the United States. It is thought that exposition of one segment of the Cordillera will explain much of the remainder in Canada, Alaska, and Mexico, although some different and novel features do occur in those regions.

THE NAME CORDILLERA. The western mountains are commonly termed the *North American Cordillera* or *Cordilleran system.* Why use this rather odd name? It is because some general term is needed, and because none of the more familiar names such as Rocky Mountains, Sierra Nevada, Coast Ranges, or Sierra Madre adequately cover the whole.

The name Cordillera is derived from the Spanish word for "cord" or "rope," and is used in that language as we in English would use "chain" in the sense of a mountain chain. In Spanish the name is pronounced Cor-di-*yerra,* but in North America it is frequently Anglicised to Cor-*dill*-era.

In the Western Hemisphere the term Cordillera was first applied to the Cordillera de los Andes or Andes Mountains, which form a compact and continuous bundle of ranges along the western side of South America. Later, the term was extended by the German geographer *Alexander von Humboldt* to the more complex and less compact system of ranges in the continent to the north.

BROADER RELIEF FEATURES. As with the mountain belt southeast of the Interior Lowlands we can introduce the one on the west by examining the maps—first physiographic maps, next geologic maps—and then pass on to features not evident from maps alone.

Recalling the Appalachian Mountains southeast of the Interior Lowlands (Chapter IV, Section 1), it is at once evident that the mountains of the Cordillera on the west are both wider and higher. In mid-section in the United States between San Francisco and Denver the mountain belt is about 1,000 miles wide, and includes the highest points of the continental United States—Mount Whitney in the Sierra Nevada of California projecting to 14,495 feet, and Mount Elbert in the Rocky Mountains of Colorado projecting to 14,431 feet. Northward and southward from this mid-section the Cordilleran ranges are relatively lower and narrower, but even compared with these parts the Appalachian Mountains are insignificant.

PHYSIOGRAPHIC SUBDIVISIONS. In the segment between San Francisco and Denver we cross a number of contrasting regions (Plate I). Along the Pacific Coast on either side of San Francisco Bay are the *Coast Ranges.* Inland follows the *Great Valley* of California drained by the Sacramento and San Joaquin

Rivers, beyond which rises the lofty *Sierra Nevada*. East of the Sierra is the lower and much more broken country of the *Great Basin*, whose surface does not drain to the sea but into interior basins. Beyond the Great Basin is the high tableland of the *Colorado Plateau* at whose farther edge stand the *Rocky Mountains*. Passing through the defiles of these ranges we come abruptly to the *Great Plains* on the edge of the Interior Lowlands which, by contrast with the broken country we have left behind, appear as level and featureless as the ocean.

But these units are not constant along the strike of the Cordillera. Northward in Oregon, Washington, and Idaho the extension of part of the features just mentioned is masked by a great sheet of volcanic rocks, expressed partly by the *Cascade Range* with its line of high, young volcanic cones, and partly by the *Columbia Plateaus* to the east, mantled by gently dipping lavas.

Still farther north in Canada the Cordillera draws together into a narrower, more continuous bundle of ranges that intervene between the Pacific border and the Interior Lowlands. These continue into Alaska where they partly run out to sea in the Aleutian Islands and partly cross Bering Strait into Asia.

Southward, also, the high Sierra Nevada, Colorado Plateau, and Rocky Mountains die away, so that near the Mexican border the Cordillera is merely a succession of little mountains separated by desert basins, a fact utilized in laying out the line of one of the transcontinental railroads—the Southern Pacific between Texas and Los Angeles.

In Mexico beyond, the mountains rise again to form two groups of ranges: the *Sierra Madre Occidental* and the *Sierra Madre Oriental*. Here, relations of the Cordillera differ significantly from those farther north; instead of fronting eastward on the Interior Lowlands and Central Stable Region, it now faces directly on the Gulf Coastal Plain and Gulf of Mexico. Still farther south in northern Central America (Guatemala and Honduras) the continuation of the Cordillera turns eastward and passes thence into the West Indies (Chapter V, Section 3).

GEOGRAPHICAL VERSUS GEOLOGICAL "MOUNTAINS." We have made an unflattering comparison between the Appalachians and the Cordillera with respect to the breadth and height of their present mountains; certainly all the surface features of the Appalachians are much less impressive. But we should recall the distinction we made earlier between geographical and geological "mountains" (Chapter I, Section 2).

The Appalachians were not always so low and unimpressive. During the last half of Paleozoic time when their construction was nearly completed, the Appalachians and kindred ranges must have been a high, massive mountain system along the eastern and southeastern sides of the North American continent—raised partly by deformation of the rocks themselves, partly by processes following the deformation. Subsequently, crustal forces became quiescent so that the system was worn down and partly buried or submerged. The present ranges are erosion remnants lying well inland from the former southeastern edge of the highland.

The Cordilleran system is higher and wider than the Appalachians mainly because it is younger. Main orogeny took place in the last half of Mesozoic time and early in Tertiary time. As in the Appalachians, initial features shaped by this orogeny have been deeply eroded and partly buried, but height and width have been maintained by a continuation of crustal mobility through much of Tertiary and Quaternary time. The eastern half of the Cordillera is now largely stable, but modern earthquake shocks indicate that in the western half from Nevada to the Pacific Coast crustal processes are still at work.

Thus, even though the Cordilleran system began to grow later than the Appalachians, their present form is not caused directly by folding, thrust faulting, metamorphism, and plutonism—the combination of orogenic structures which we have described in the Appalachians. Such structures occur in the Cordillera as well, but they were formed mainly during an early chapter in its history. As in the Appalachians other events took place later which had a greater effect in shaping the present surface. These later events and structures were more extensive than in the Appalachians and have greatly confused the initial pattern, as may be seen on both the physiographic and geologic maps.

PROCESSES WHICH MODIFIED THE FUNDAMENTAL CORDILLERAN STRUCTURE. Among the processes which operated during Tertiary and Quaternary time to modify the original orogenic structure of the Cordillera we can list the following:

(1) Volcanic activity and accompanying shallow intrusions. Such volcanism occurred during Tertiary and Quaternary time at one place or another over nearly the whole Cordilleran area, although there were several conspicuous volcanic centers, including those in the Cascade Range and Columbia Plateau of the northwestern United States, and the Sierra Madre Occidental of western Mexico.

(2) Formation of basins between the mountain uplifts soon after the main orogeny in which Tertiary and especially Eocene sediments were deposited. Such basins are extensive in the Central and Southern

Rocky Mountains of Wyoming, Colorado, and adjacent states.

(3) Breakup of the deformed terrane by block faulting resulting from crustal tension or other causes to produce a succession of block mountains and intervening basins. Block faulting has largely shaped the surface forms of the Great Basin between the Colorado Plateau and Sierra Nevada, and the Sierra is itself a tilted mountain block larger than the rest. Block faulting also took place along the Rio Grande in New Mexico and at many other places.

(4) Regional uplift of broad areas without much folding or faulting in later Tertiary and Quaternary time. Such uplifts late in the history of a mountain system have also taken place in the Appalachians and other mountain belts, probably because of isostatic and other crustal adjustments. Regional uplift is thus responsible for much of the present height of the Central and Southern Rocky Mountains and Colorado Plateau. Streams invigorated by the uplift have etched out the mountain ranges of the first two areas and have carved the myriad canyons of the third.

(5) Continuing sedimentation, deformation, and orogeny along the Pacific Coast, notably in the Coast Ranges of California, where the Tertiary and even parts of the Quaternary strata have been greatly folded and faulted, and where seismic records indicate that the crust is still active.

(6) In the Coast Ranges of California and elsewhere, development of great faults such as the San Andreas, along which movements were not upward and downward but sidewise—that is, transcurrent faults with strike-slip displacement.

SIGNIFICANCE OF GEOGRAPHIC UNITS. Most of the present landscape of the Cordillera has resulted from the processes which operated in Tertiary and Quaternary time. Present geographic units thus have only indirect meaning in terms of fundamental orogenic structure.

This is most evident in the volcanic fields of the northwestern United States, where lavas and other eruptive products effectively bury the underlying structure.

Also, the Sierra Nevada is made up of rocks with a complex fundamental structure, yet its present shape has little relation to that structure. The fundamental structures are much more extensive than the present range, and the Sierra is a block that was raised within them by later faulting and tilting.

The most confusing geographic name used for any part of the Cordillera is "Rocky Mountains." It came into use by explorers, trappers, and pioneers a century and a half ago to denote the first mountain chain encountered after crossing the Great Plains. This might mean any of the geologically diverse frontal structures of the Cordillera—the uplifted masses of basement rock in Colorado, or the folded and faulted geosynclinal strata of Montana and Alberta; outlying peaks of intrusive igneous rock in the Great Plains of Montana have been called the "Little Rocky Mountains." Geographers have perpetuated the name because high summits of these diverse rocks and structures extend from Canada southeastward into Colorado and New Mexico as though they were parts of a single system. But in order to give the term any geological meaning it will be necessary in this account to break the "Rocky Mountains" into several parts—the *Northern, Central,* and *Southern Rocky Mountains.*

2. THE FUNDAMENTAL OROGENIC STRUCTURE

Where, in the midst of all this confusion of surface features, are features like those of the Appalachians which we had come to believe were characteristic of mountain belts in general—the continental platform, miogeosyncline, and eugeosyncline of the formative stage, and the foreland, sedimentary belt, and crystalline belt of the orogenic stage?

In the Cordillera, as in the Appalachians, our problem is to clear away the battering and modification which the system has undergone after the climactic orogeny. Here, however, we must not only restore what has been worn down, buried, or submerged, but we must unravel those features which were imposed on the fundamental structures later—the uplifts and downwarps, folds and faults, and igneous features which have been responsible for most of the surface forms.

OROGENIC FEATURES IN CANADA. Our analysis will be simplest if we begin with the Cordillera in Canada, where resemblances are strongest between the mountains on the west and those on the southeast of the Central Stable Region.

In Canada the "Rocky Mountains" of local terminology front the Great Plains eastward and are made up of miogeosynclinal strata, mainly Paleozoic but with Triassic and Jurassic above, topped by clastic wedges of Cretaceous age. These have been thrown into long folds and thrust blocks like those in the Valley and Ridge province of the Appalachians. The interior ranges west of the Rocky Mountains consist of eugeosynclinal rocks, likewise of Paleozoic, Triassic, and Jurassic age, which have been heavily deformed, metamorphosed, and plutonized.

Here, as in the Appalachians, climactic orogeny was mainly earlier on the side toward the ocean in the eugeosynclinal area than it was on the side toward the continent in the miogeosynclinal area. Much of the eugeosynclinal area was deformed and plutonized during the latter half of Mesozoic time, whereas the miogeosynclinal area was not deformed until near the end of the Mesozoic or a little later. According to terminology used in the United States, the western part was deformed during the *Nevadan orogeny* and the eastern part during the *Laramide orogeny*.

Some geologists have pictured these two orogenies as having produced two mountain systems: an earlier one on the west and a later one formed alongside it on the east. Concepts which we have developed in the Appalachians suggest, instead, that the whole Cordillera is a unit which was formed by a continuing process of deformation.

EASTERN RANGES AND PLATEAUS OF THE UNITED STATES. On proceeding southward from Canada into the United States, relations became more complex, and additional structural elements appear (Plate I).

If we define the eastern edge of the miogeosyncline as a line along which Paleozoic rocks thicken abruptly with Lower Cambrian wedging in at their base, that line would not be at the eastern edge of the Rocky Mountains near Denver, but at the western edge of the Colorado Plateau in Utah. At least a third of the Cordilleran system in mid-section in the United States—comprising the Southern Rocky Mountains and Colorado Plateau—thus lies outside the miogeosynclinal belt and consists of structural elements unlike any in the Appalachians. We will call this part of the Cordillera the *Eastern Ranges and Plateaus.*

The Colorado Plateau is an old stable block, a part of the continental platform when this was outlined in Cambrian time (Chapter III, Section 4), which later was detached from the main platform by development of structures in the Rocky Mountains to the east. Its strata still lie nearly flat over wide areas, but elsewhere they have been thrown into broad folds or depressed into basins that were filled with Tertiary sediments or covered by lavas.

The Southern Rocky Mountains on the east were likewise originally part of the continental platform, but during later Paleozoic time they grew into a mountain belt that was the northwestern extension of the Wichita system. Its folds were rejuvenated and accentuated during the *Laramide orogeny* toward the end of Mesozoic time. The region is now a succession of great anticlinal uplifts, many raised so high that their cores of basement rock are brought to the surface, with deep intervening downfolds containing sediments that are much deformed.

MIOGEOSYNCLINAL AND EUGEOSYNCLINAL BELTS IN THE UNITED STATES. We can now return to the main belt of the Cordillera, which we find passes southward from Canada behind and to the west of the Eastern Ranges and Plateaus.

Miogeosynclinal rocks make their appearance west of the edge of the Colorado Plateau and continue thence across western Utah into Nevada. Here there is a great thickness of Paleozoic carbonate rocks with Triassic and Jurassic strata at the top. Orogenic deformation of the miogeosynclinal rocks must have occurred before the end of Mesozoic time, or before the climax of Laramide orogeny in the Rocky Mountains to the east. Clastic wedges in the Cretaceous rocks thicken westward toward the Great Basin, along the edge of which they contain coarse conglomerates that lie unconformably on earlier deformed strata.

Farther west in the Great Basin, in the western half of Nevada, Paleozoic and earlier Mesozoic miogeosynclinal rocks give place to eugeosynclinal rocks of the same ages, which continue with greater development into the Sierra Nevada and Klamath Mountains beyond. In the Sierra Nevada and Klamath Mountains they have not only been deformed, but have been metamorphosed and invaded by acidic plutonic rocks. As in Canada, orogeny reached its climax earlier in the eugeosynclinal area than in areas to the east. Important Paleozoic movements took place along the eastern edge of the eugeosyncline in Nevada, but the main deformation here and westward took place during the *Nevadan orogeny* which extended, from one place to another, from early in the Jurassic until well into Cretaceous time. On the west sides of the Sierra Nevada and Klamath Mountains Cretaceous sediments dip gently off the heavily deformed earlier rocks; neither the Cretaceous nor the later Jurassic of the Coast Ranges to the west are regionally metamorphosed.

In this segment of the Cordillera, as farther north, its structures are frequently interpreted as consisting of two systems of different ages: one in the Rocky Mountains formed during the Laramide orogeny, another in the Sierra Nevada formed during the Nevadan orogeny. In this segment contrast between the two has been heightened because the two mountain ranges are now separated by a wide tract of diverse country, the Colorado Plateau and Great Basin, whose place in the Cordilleran pattern has not been well understood.

We now know, however, that Nevadan orogeny was itself prolonged, and that deformation at a time intermediate between the Nevadan and Laramide orogenies took place in the Great Basin (witness the Cretaceous conglomerates and clastic wedges near it), so that crustal activity seems to have progressed eastward across the Cordillera through the latter half of Mesozoic time. There is thus no sharp boundary between any "Nevadan" and "Laramide" mountain systems, and it is best to consider the whole mountain belt as a single system formed by a continuing process.

OROGENIC FEATURES IN MEXICO. South of the complex segment in the United States the Cordilleran system narrows again, mainly by the ending of the Eastern Ranges and Plateaus. In Mexico Paleozoic rocks are only scantily exposed, so that little is known here of the earlier history of the system. The available record is mainly of Mesozoic and later time.

To the east is a broad Mesozoic miogeosyncline in which great thicknesses of Jurassic and Cretaceous limestones accumulated; these are now thrown into the series of parallel folds of the Sierra Madre Oriental which face the Gulf Coastal Plain and in places the Gulf of Mexico itself.

Fundamental structures farther west are much obscured by a thick pile of Tertiary lavas which forms the great dissected plateaus of the Sierra Madre Occidental in western Mexico; the boundary between the miogeosynclinal and eugeosynclinal rocks is thus not plain. However, the basement of the peninsula of Baja California along the Pacific Ocean is much like that of the Sierra Nevada farther north—a body of metamorphosed Mesozoic and earlier rocks invaded by basic and acidic plutonic rocks. The basement rocks of Baja California probably originated in the eugeosynclinal part of the Cordilleran system.

3. GEOLOGICAL EXPLORATION OF THE CORDILLERAN SYSTEM

Before describing details of the Cordilleran rocks and structures it will be well to set forth how some of our knowledge of them was acquired. Our knowledge was not "handed to us on a silver platter," nor was it delivered in one package. It was learned through the labor of many geologists working over a wide region, mostly within the span of the last hundred years.

WESTERN HISTORY IN GENERAL. The reader is doubtless familiar with the broader facts of the opening of the West:

Of acquisition by the United States of the northwestern part of its domain by division of the Oregon Country with Great Britain along the 49th parallel into Oregon Territory (later to become the states of Oregon, Washington, and Idaho) on the south, and into British Columbia on the north. Of acquisition of the southwestern part by cession from Mexico as an aftermath of the Mexican War of 1846 and 1847.

Of the discoveries of mineral riches in the new domain: discovery of gold at Sutter's Mill at Coloma in the foothills of the Sierra Nevada, and the great Gold Rush to California which followed in 1849; the discovery of silver in the Comstock Lode of Nevada a decade later, in time to provide an economic boost to San Francisco when prosperity of Gold Rush days was declining; and other discoveries a little later, such as those in the Rocky Mountains of Colorado which for many years made Denver a mining capital.

Of the filling up of the region by settlers from the East: the great trek to the Oregon Country, beginning even before United States sovereignty had been established; the migration of the Mormons to Utah, with Brigham Young going ahead of the main party to find "the place" which was to become Salt Lake City; the rapid settlement of California, first in the north, later in the south; and many others.

Exploration went on hand in hand with the rest of the human drama. First exploration was by the fur trappers and "mountain men" who ranged over the whole country as early as the 1820's. But while they found out much about the land, they did little accurate surveying and left few written records. First scientific exploration was largely geographical. Early scientific explorers such as *John Charles Frémont* devoted their efforts to finding routes of travel and location of mountains, rivers, deserts, and spots favorable for agriculture.

Later it became evident that something more must be learned about what this new land was made of—as a guide to mineral development and to aid the settlement of people who were arriving from the East in increasing numbers.

PACIFIC RAILROAD SURVEYS (1853-1854). First serious geological exploration of the West came in the 1850's immediately before the Civil War, when there were set up a series of "explorations and surveys to ascertain the most practicable and economical route for a railroad from the Mississippi River to the Pacific Ocean," so that the west could be joined with the settled eastern part of the country by communications better than the Prairie Schooner or Pony Express.

These explorations were made by Army Engineers under direction of the Secretary of War, Jefferson Davis, who was later to attain greater fame as President of the Confederate States of America. In these explorations one sees the names of many men who came more into public eye during the Civil War. One of the surveys, for example, was under Captain John Pope, later to become the General Pope who led the Union forces to their great defeat at the Second Battle of Bull Run in Virginia in 1862. Pope must have been a better engineer than he was a leader of men, or at least a leader of great armies into battle.

All these exploring parties were accompanied by geologists whose observations gave the first idea of the geology of many parts of the West. Or perhaps we can call them "geologists," for geological observations were part of their assigned duties. Many of the geologists of the expeditions had been educated in medicine, and did double duty as physicians for the party and as scientific explorers.

GEOLOGICAL SURVEYS AFTER THE CIVIL WAR. Geological study of the West, rather than mere exploration, did not begin until after the Civil War had ended. Opening of the West now began apace. Many young men who were mustered out of service found ordinary civilian existence dull and searched for new fields and new careers. Some of these young men found an outlet for their energies in scientific exploration of the West.

It seems sad today that these men remain merely as names in musty books, to be memorized by students, when once they were full of life, enthusiasm, and a spirit of adventure. They thought and felt much as young men of our own day who have been mustered out of service after our later wars, but at that time the great west was still open before them.

Some of these young men with more vision and genius for organization than the rest worked up and presented to Congress proposals for geological explorations and surveys of the western country.

U.S. GEOLOGICAL AND GEOGRAPHICAL SURVEY OF THE TERRITORIES (HAYDEN SURVEY). Among these men was *Ferdinand Vandiveer Hayden*, who had served as physician in the Medical Corps during the Civil War. He first received a grant from Congress of a few thousand dollars a year for geological and geographical surveys of the country immediately west of the Missouri River. Appropriations were gradually expanded until they reached the munificent sum (for those days) of $75,000 a year. The field of inquiry was also expanded, and between 1870 and 1880 large crews of his geologists and surveyors were being sent each year for surveys and studies in Colorado, Wyoming, and adjacent territories. Geologists of the Hayden Survey were among the first white men to see the geysers and other wonders at the head of the Yellowstone River, and their reports were influential in setting aside this area as the first National Park.

GEOLOGICAL EXPLORATION OF THE FORTIETH PARALLEL (KING SURVEY). Another of the young men was *Clarence Rivers King*, who proposed to Congress that the resources along the line of the new transcontinental railroad (the Union Pacific and Central Pacific) should be explored and studied. This survey was intended to cover a belt of country on each side of the railroad, approximately along the Fortieth Parallel from Wyoming to California.

King's organization was the aristocrat of the Surveys. It was staffed largely by men trained at Yale or Harvard, many of whom had gone on for advanced study in German universities. It is related that, later on, when all the Surveys had been combined into a single U.S. Geological Survey, the men of the former King Survey would not speak to those of the former Powell and Hayden Surveys if they met them on the street—these were too plebeian to be noticed.

King himself, in his day, must have been a very dynamic and charming man. His enthusiasm for geology and his way of presenting the facts of science to laymen made him the friend of important people in Washington—Senators, Congressmen, and many others.

King's most spectacular achievement was really a side issue of the Fortieth Parallel Survey itself, and had to do with the "great diamond swindle."*

In 1872 two weatherbeaten prospectors, Philip Arnold and John Slack, came out of the mountains and presented themselves to William Ralston, president of the Bank of California in San Francisco. They showed him a bag of diamonds which they had collected at some remote spot in the West. They came at a psychological time, for a speculative madness had taken over San Francisco after the opening of the Comstock Lode. Every day brought disclosures of new mineral deposits in the West—of gold, silver, and the baser metals. The Kimberley diamond fields had been opened in South Africa only a few years before. Could there not be similar deposits in western North America? Ralston, a great plunger and speculator who had made a fortune in the Comstock mines, was greatly excited, and made up his mind to gain control of the new diamond deposit.

After some persuasion the prospectors consented to have their find examined on the spot, provided the inspectors were brought to it blindfolded. The inspectors came back even more impressed than Ralston—diamonds were all over the place, as well as rubies, sapphires, and emeralds—on the surface of the ground, in ant hills, in crevices in the rocks. A sample of the stones collected by them was submitted to Tiffany's in New York and Mr. Tiffany personally valued the sample alone as worth $150,000. The prospectors reluctantly sold their claim for $360,000, with a stock interest which they sold in turn for $300,000, a return of $660,000 in all.

The San Francisco group then organized the San Francisco & New York Mining & Commercial Co. to develop the deposit, capitalized at $10,000,000. All the stock could easily have been sold to the avid public

* Many accounts of this famous incident have been published, each varying somewhat in details. In order to make a consistent story I have had to pick and choose from several sources.

in San Francisco, but instead it was offered to twenty-five of the outstanding business and financial leaders of the city.

It was at this point that Clarence King entered the picture. Returning from field work in Nevada in the fall of 1872, he learned of the new discovery which rumors placed at such widely separated points as Arizona, New Mexico, and Utah. He wondered whether the deposit lay within the area covered by his Fortieth Parallel Survey, and whether it had somehow been overlooked during investigations of the survey.

"Feeling that so marvelous a deposit as the diamond fields must not exist within the official limits of the Fortieth Parallel Survey, unknown and unstudied, I availed myself of the intimate knowledge possessed by the gentlemen of my corps, not only of Colorado and Wyoming, but of the trail of every party traveled there, and was enabled to find the spot without difficulty."

This turned out to be in the northern foothills of the Uinta Mountains, only a short distance south of the Union Pacific Railroad. Going to the place, King observed that it lay in a region of tilted sedimentary rocks without any evidence of igneous intrusion or mineralization; besides, the association of diamonds, rubies, sapphires, and emeralds seemed incongruous. He found that most of the stones lay on the surface; none were embedded in the rocks themselves, and some even bore marks resulting from the work of jewelers! Tiffany's appraisal to the contrary, the stones were very inferior—mere jewelers' waste that had been bought up and planted for purpose of fraud. Bit by bit, contrary to his expectations, King was forced to conclude that Ralston and his friends had fallen into a trap.

Once having made up his mind, King traveled night and day to San Francisco and demanded of Ralston and the Bank of California that sales of stock be stopped at once, announcing that he intended to publish his findings. All Hell must have broken loose over the head of this young government geologist, for he was facing up to some of the richest and most powerful men in California. But he stood his ground and the company collapsed.

That was the beginning of the end for poor Ralston, whose bank failed in the panic of 1875, and he ended as a suicide in the waters of the Golden Gate. The prospectors disappeared, but Arnold was eventually traced to Kentucky where, on threat of lawsuit, he surrendered $150,000 of his ill-gotten gains.

But I have strayed too far into this side issue. Let us return to the geological story.

U.S. Geographical and Geological Survey of the Rocky Mountain Region (Powell Survey). The other survey was led by *John Wesley Powell*—Major Powell who served under Grant in the Civil War and lost an arm at the Battle of Shiloh, and was afterwards a teacher of science at a little college in Illinois. Like the other men of whom I have spoken, he had the urge to get into the West, and organized summer expeditions to the Rocky Mountains under the auspices of the Illinois Academy of Sciences. Once in the West, the country cast its spell, and Illinois saw little more of him.

As he explored beyond the Rocky Mountains he became aware of the great remaining challenge of western exploration. The head streams of the Colorado River, the Green, and the Grand were known to enter the Colorado Plateau on the northeast; the Colorado was known to issue from the plateau on the southeast. Almost nothing was known of the river between. Why not find out? Why not descend the river by boat through the whole plateau country?

To carry out this daring design he enlisted the aid of the Smithsonian Institution, the first Federal support which he had received. On May 24, 1869, his four little boats cast off into the Green River at Green River Station, Wyoming, on the line of the newly built Union Pacific Railroad.

What followed was one of the great adventures of western exploration. We do not need to dwell on the perils and excitements of the trip; they are wonderfully set forth in Powell's own account of his "Exploration of the Colorado River of the West." On August 30 the expedition came out of the lower end of the canyon, and were welcomed by Mormon settlers of the Virgin River valley in Nevada.

But of the nine men who began the trip, three were missing. In the last discouraging week before the end, with provisions nearly gone and formidable rapids still ahead, some of the party concluded that there was no hope of completing the river journey, and they begged Powell to give up and scale the canyon walls on foot. After much debate the party divided, six continuing by boat, three setting off to climb their way out. Unfortunately, the three on foot were waylaid and killed by Indians before they reached the settlements.

But Powell was more than an adventurer. He was a scientist at heart, something of a crusader for his beliefs, and probably also an ambitious man. In a sense, his thrilling adventure was designed to "put him on the map" in western geological studies. On the strength of his exploration still another organization was set up by Congress, the U.S. Geographical and Geological Survey of the Rocky Mountain Region (part of the time also called U.S. Geological and Geographical Survey of the Territories, Second Division). The name was a misnomer, as the work of the survey was largely in the Colorado Plateau rather than the Rocky Mountains. It was never as large as the other surveys, did not receive as much money from Congress, and mapped less country. But its contributions were as significant, or more so, than the rest—on the theories of the West and the theories of geology.

For one thing, Powell had a strong social conscience and strong convictions about the needs of the West. He was concerned over the fate of the Indians and the fate of the white settlers too, and was one of the first to realize that western resources were exhaustible—that the timber, soil, water, and mineral resources had limits of exploitation, and must be *conserved* if permanent, prosperous communities were to arise in the region. These aspects of Powell's career are important in the history of the West at large. We are more concerned here with his geological achievements and those of his men.

The Colorado Plateau was a wonderful place to develop new principles of geology—much of it is a bare desert region in which the rocks are well exposed and laid out on a vast scale with great simplicity of form. Over wide tracts the strata dip gently, but here and there they are bent abruptly into great arches or flexures, all exposed in cross-section on the walls of the myriad canyons. These structures could be observed and analysed in a manner seldom possible elsewhere.

Then, too, there were problems of river work and erosion. What made the Grand Canyon—or any other canyon or valley? It is well to remember that only twenty or thirty years before, *Charles Darwin*, the great British naturalist, in his account of "The Voyage of the Beagle" around the world, ascribed every great canyon or chasm which he saw to "some great cataclysm of nature," and every great escarpment to erosion by the sea at some former time and level—not to erosion by streams on the land. Darwin was a young man at the time of his voyage, but his ideas were not mere youthful fancies. He was well trained in science, and his conclusions represented the prevailing scientific thinking of his day.

Powell did not invent the theory of *stream erosion*; an unfashionable minority of geologists had realized its importance long before. But his work and that of his men in the Colorado Plateau did much to establish the theory against opposing views. Although the Colorado Plateau was dry and bare, it was proved that the gigantic chasm of the Grand Canyon was cut by the stream now flowing in it. Moreover, this was not the whole story; there was an additional *great denudation* of the surface of the plateau, mainly during a time before the canyon cutting. The rim of the Grand Canyon is formed by the Kaibab limestone of Permian age. Off to the north the Kaibab is succeeded by younger Triassic, Jurassic, and Cretaceous formations, all of which once overlay the Grand Canyon region and have now been stripped away by processes of denudation.

Also, many of the rivers showed a curious disregard for the folds in the plateau. The Green River, for example, flows directly across the great fold of the Uinta Mountains (Fig. 64), whereas, by only a slightly longer course it could have passed around the end of the mountains. Why? For this Powell developed the principle of *antecedent streams*—streams which were flowing across the country before the uplifts began, and were able to continue cutting down as the uplifts were raised across their courses. Some of Powell's supposed antecedent streams are now believed to have originated in other ways, but the principle remains valid.

Powell's co-workers also made important contributions:

Major Clarence Edward Dutton worked for Powell from 1875 to 1891 while on leave from the Ordnance Department of the U.S. Army. Among his other tasks he filled in many of the geological details of the Grand Canyon which had been so boldly sketched by Powell himself. Dutton's publications are notable for containing some of the most vivid writing in American geological reports.

Grove Karl Gilbert studied the *laccoliths* of the Henry Mountains near the center of the Colorado Plateau, which he found had been formed by blister-like masses of igneous rock that had domed the strata above them. In explorations farther west before he joined Powell, he concluded that the peculiar mountain ranges of the Great Basin had been shaped by *block faulting*, the blocks being uplifted on faults along their edges, the faults now largely concealed by outwash from the ranges themselves. Gilbert went on later to study the great vanished lakes of the Great Basin, especially *Lake Bonneville* of which the present Great Salt Lake is the shrunken remnant.

CONSOLIDATION OF THE SURVEYS. As these different surveys progressed they inevitably came into conflict, for areas covered by them began to overlap, with much duplication of effort and much rivalry between men and organizations.

A phase of this duplication was that of the vertebrate paleontologists, *Othniel Charles Marsh* and *Edward Drinker Cope*. Marsh was located at Yale and served with the King and Powell Surveys, whereas Cope was located in Philadelphia and served with the Hayden Survey. They competed for fossil finds. At times one of the men, when he discovered a new bone in the field, would dash off a description on the spot and send it to a scientific journal by telegram from the nearest railroad station so as to obtain publication before his rival. All this was a side issue to the main conflict between the Surveys and seems only amusing now, but it was bitterly felt at the time. The rivalry is significant because both Marsh and Cope were wealthy, influential men, each of whom had a strong following and many friends in Congress.

The real difficulties were deeper and were not merely a struggle between intellectuals and theorists. They were as serious as any duplication between government agencies today, and much of the future of the development of the West depended on how the problem would be resolved.

After Congressional investigations through several years and consideration of various alternate plans, the warring Surveys were consolidated in 1879 into a single United States Geological Survey. The consolidation owed much to the negotiation and scheming of Powell, although he remained behind the scenes and had retired to the Smithsonian Institution to direct his Bureau of American Ethnology. Clarence King was named first director of the united bureau.

But some spark seemed to have gone out of King since the bold adventure of the Fortieth Parallel Survey. He was puzzled as to how to organize the functions of the new organization and was more interested in mining promotion for himself. After a year he resigned, and Powell took his place as director, a post he held for nearly fifteen years. In those years he did much to organize the Geological Survey on its present footing, and also to carry out his ideas on conservation in the West.

Aftermath. After Powell's day the Geological Survey became the "mother" of many other government bureaus, for its different functions expanded until they had to be split from the parent organization—the Bureau of Mines to deal with mining technology, the Bureau of Reclamation to deal with use of water for irrigation, the Forest Service to deal with timber resources, and others. The Geological Survey has retained its essential core—a study and mapping of geological features in the West and elsewhere, and appraisal of the mineral resources which the rocks contain.

There are many sequels to this story—the geologists who continued to investigate the Cordilleran region for the U.S. Geological Survey, the Geological Survey of Canada, and other organizations public and private; the things they discovered, the conclusions they drew, the controversies in which they became involved, and how all this resulted in our present knowledge of the region. But these sequels are too long to set down here; we will mention some of them at appropriate places in later chapters.

REFERENCES

1. *Geographic features*

Fenneman, N. M., 1951, *Physiography of Western United States*: New York, 534 p.

King, P. B., 1958, Evolution of modern surface features of western North America, *in* Zoogeography: *Am. Assoc. Advancement of Science Mem.*, chap. 1. (in press).

2. *Fundamental orogenic structure*

Billingsley, P. R., and Locke, Augustus, 1939, Structure and ore districts in the continental framework: *Am. Inst. Min. Met. Eng.*, 51 p.

Eardley, A. J., 1949, Paleotectonic and paleogeologic maps of central and western United States: *Am. Assoc. Petrol. Geol. Bull.*, v. 33, p. 655-682.

Kay, Marshall, 1951, North American geosynclines: *Geol. Soc. America Mem. 48*, p. 11-14, 35-49.

Longwell, C. R., 1950, Tectonic theory viewed from the Basin Ranges: *Geol. Soc. America Bull.*, v. 61, p. 413-434.

Stille, Hans, 1936, Die entwicklung des Amerikanischen Kordillerensystems in zeit und raum: *Prussischen Akad. Wiss. Sitzungberichten, Phys. Math. Kl.*, v. 15, p. 134-155.

3. *Geological exploration of the Cordilleran system*

Darrah, W. C., 1951, *Powell of the Colorado*: Princeton Univ. Press, 426 p.

Merrill, G. P., 1924, *The First One Hundred Years of American Geology*: Yale Univ. Press, chap. 8, p. 500-552.

Powell, J. W., 1875, *Exploration of the Colorado River of the West and its Tributaries*: Washington, 291 p. (Reprinted as *The Exploration of the Colorado River*: Univ. Chicago Press, 138 p., 1957).

Rabbitt, J. C., and Rabbitt, M. C., 1954, The U.S. Geological Survey; 75 years of service to the nation, 1879-1954: *Science*, v. 119, p. 741-758.

Rickard, T. A., 1932, A history of American mining: *Am. Inst. Min. Eng. series*, 419 p. (especially "The great diamond hoax", p. 380-396).

Stegner, Wallace, 1954, *Beyond the Hundredth Meridian*: Boston, 438 p.

Stone, Irving, 1956, *Men to Match My Mountains; the Opening of the Far West, 1840-1900*: Garden City, New York, 459 p.

Wilkins, Thurman, 1958, *Clarence King; a biography*: New York, 441 p.

CHAPTER VII

THE EASTERN RANGES AND PLATEAUS; A NOVEL STRUCTURAL ELEMENT

1. A NOVEL FEATURE OF AMERICAN GEOLOGY

The first unit of the Cordillera which we will consider will be the Central and Southern Rocky Mountains, the ranges of New Mexico, and the Colorado Plateau—that is, the ranges extending from central Montana southeastward through Wyoming and Colorado into New Mexico, and the plateaus behind them on the west which extend into Utah and Arizona. These, for want of better title, we will refer to as the *Eastern Ranges and Plateaus.*

It is interesting to recall that Major Powell early recognized these great geographic and structural divisions, and that he termed the Southern Rocky Mountains the "Park province," the Colorado Plateau the "Plateau province," and the Great Basin (or "Basin Ranges" of Gilbert) the "Basin province." The term "Park province" is somewhat curious, as it emphasises the high mountain valleys or "parks" which are enclosed within the Rocky Mountains, rather than the mountains themselves.

Powell also emphasised the differences between these western mountains and plateaus and the more familiar eastern mountains such as the Appalachians with which most of the geologists of his generation had grown up. He realized that this was more than a matter of scenery and climate, and that there were fundamental differences in the geology—in the strata, the structure, and the geologic history—that here was something truly novel in American geology which had no counterpart in the Eastern States.

Powell did not realize fully why this was, but our later knowledge has given us appreciation of its meaning. As we have seen in Chapter VI, the Eastern Ranges and Plateaus lie inland from the true miogeosyncline of the Cordillera, and were originally part of the continental platform; it was only later that they became mobile so that their rocks were disordered in the manner we now see.

As in the Appalachians, we will deal with this region in both space and time—of what its forms and structures are and how they developed. Here, however, we will follow an order the reverse of the earlier one. We will first describe the rocks and events which led to the great deformation of later Mesozoic time, then the structures produced by this deformation. Finally, we will treat the events which followed the great deformation and how the region was modified between then and the present.

2. ROCKS AND EVENTS BEFORE THE GREAT DEFORMATION

THE PRECAMBRIAN. Our first concern will be with the Precambrian, or the rocks which form the basement on which all the succeeding rocks and structures have been built.

Here, in contrast to the Interior Lowlands and Appalachians, we can learn much about the Precambrian, because it projects to the surface in many more places and over much wider areas. Its manner of exposure, it is true, is not like that in the Canadian Shield; it does not form the whole surface over vast areas. Rather, it emerges in the cores of the great uplifts in the mountains and is revealed in the canyon bottoms in the plateaus (Fig. 53), forming perhaps

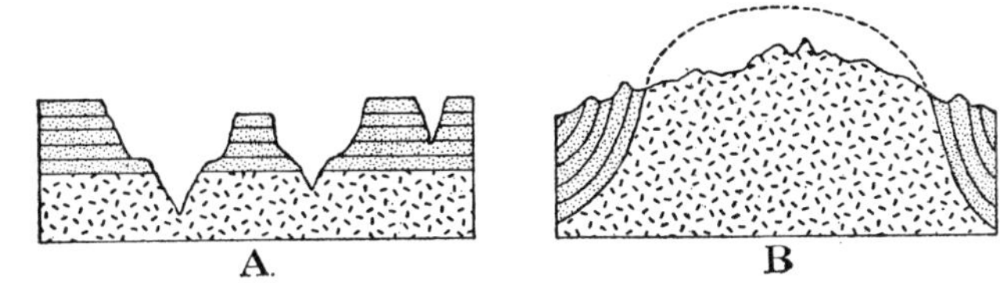

FIG. 53. Sketch sections illustrating manner in which Precambrian rocks are exposed in Eastern Ranges and Plateaus. (A) In plateau areas. (B) In the mountain uplifts.

only five or ten percent of the surface in all. The remaining ninety or ninety-five percent of the surface is covered by younger rocks, downfolded between the ranges in the mountain areas, or spread across the country between the canyons in the plateau areas.

As we have seen (Chapter II, Section 3), even in the Canadian Shield where exposures are nearly continuous it is not easy to make correlations or work out any general Precambrian history. In the Eastern Ranges and Plateaus where exposures are much more interrupted, the problems are greater. We cannot make precise correlations between different parts of the local region, much less establish any convincing correlations with the better known area in the Canadian Shield. Some clues can be obtained, however, from the few available radiometric dates, from lithologic features, and from other evidence, which enable us to make some imperfect generalizations.

EARLIER PRECAMBRIAN ROCKS. Much of the exposed Precambrian of the Eastern Ranges and Plateaus, as in the Canadian Shield, consists of heavily metamorphosed, steeply tilted rocks, originally of surficial origin, in which small to large masses of acidic plutonic rocks are embedded. Such rocks form most of the cores of the ranges in Colorado and Wyoming, and also come to the surface in the outlying Black Hills of South Dakota.

Some of the metamorphic rocks have been so changed that it is difficult to decipher their original character, but in many areas it can be proved that they were once sediments and lavas. Beds of quartzite, originally sandstone, occur in the Medicine Bow Mountains of Wyoming, the Needle Mountains of Colorado, and elsewhere, and several ranges in east-central Wyoming contain iron formations like those in the Huronian of the Canadian Shield. In Colorado the prevailing trend of the structures is southwestward (Fig. 54), and this trend continues into Arizona, suggesting the existence of an ancient mountain structure along the prolongation of those in the southeast part of the Canadian Shield, running nearly at right angles to the much later trend of the Rocky Mountains. In the Black Hills and the ranges of Wyoming, however, prevailing trends are northwestward, more nearly parallel to those of the Rocky Mountains; relation of these structures to the ones farther south is undetermined.

The plutonic rocks include a body of anorthosite east of Laramie, Wyoming, recalling the anorthosites of the Grenville province in the Canadian Shield, but most of them are granites or their allies. One of the most extensive masses of plutonic rocks is the Pikes Peak granite, named for the peak at the eastern edge of the Front Range of Colorado overlooking Colorado Springs. This granite and related intrusives occur in various bodies through most of the Front Range, and its relative, the Sherman granite, extends northward to the vicinity of Laramie, Wyoming (Fig. 54).

The highly altered nature of the surficial rocks suggests, but hardly proves, that they are of early Precambrian age. Early ages are indicated more definitely by radiometric determinations. A series of these in northwestern Wyoming in the Beartooth, Bighorn, and Owl Creek Mountains have yielded surprisingly ancient dates of about 2,500 million years, or as great as the oldest rocks in the Canadian Shield. Younger dates of about 1,600 million years have been reported in the Black Hills to the east, and of about 1,350 million years in Colorado, New Mexico, and Arizona to the south (including the basement rocks of the Grand Canyon). These determinations suggest the existence of provinces or of orogenic belts of different ages, but their patterns are as yet unknown. An age of 1,100 million years has been determined for the Pikes Peak granite, suggesting that it is one of the latest units in the basement rocks; this is also attested by its massive, through-breaking character.

LATER PRECAMBRIAN ROCKS. More revealing of historical record than these highly altered and plutonized rocks, but of less extent and occurring mainly toward the west, are little altered or deformed rocks which are comparable to the last class of surficial rocks which we have listed in the Canadian Shield—"tilted or flat-lying sediments and lavas, unaltered or little metamorphosed."

As with rocks of this class in the Canadian Shield, they are not necessarily young in absolute terms; some may actually be quite ancient. They are at least younger than other Precambrian rocks in their immediate neighborhoods; they have had a much less complex history and in places lie unconformably on them. They appear so fresh, in fact, that they are not unlike Paleozoic strata, yet they are overlain with more or less unconformity by fossiliferous Cambrian

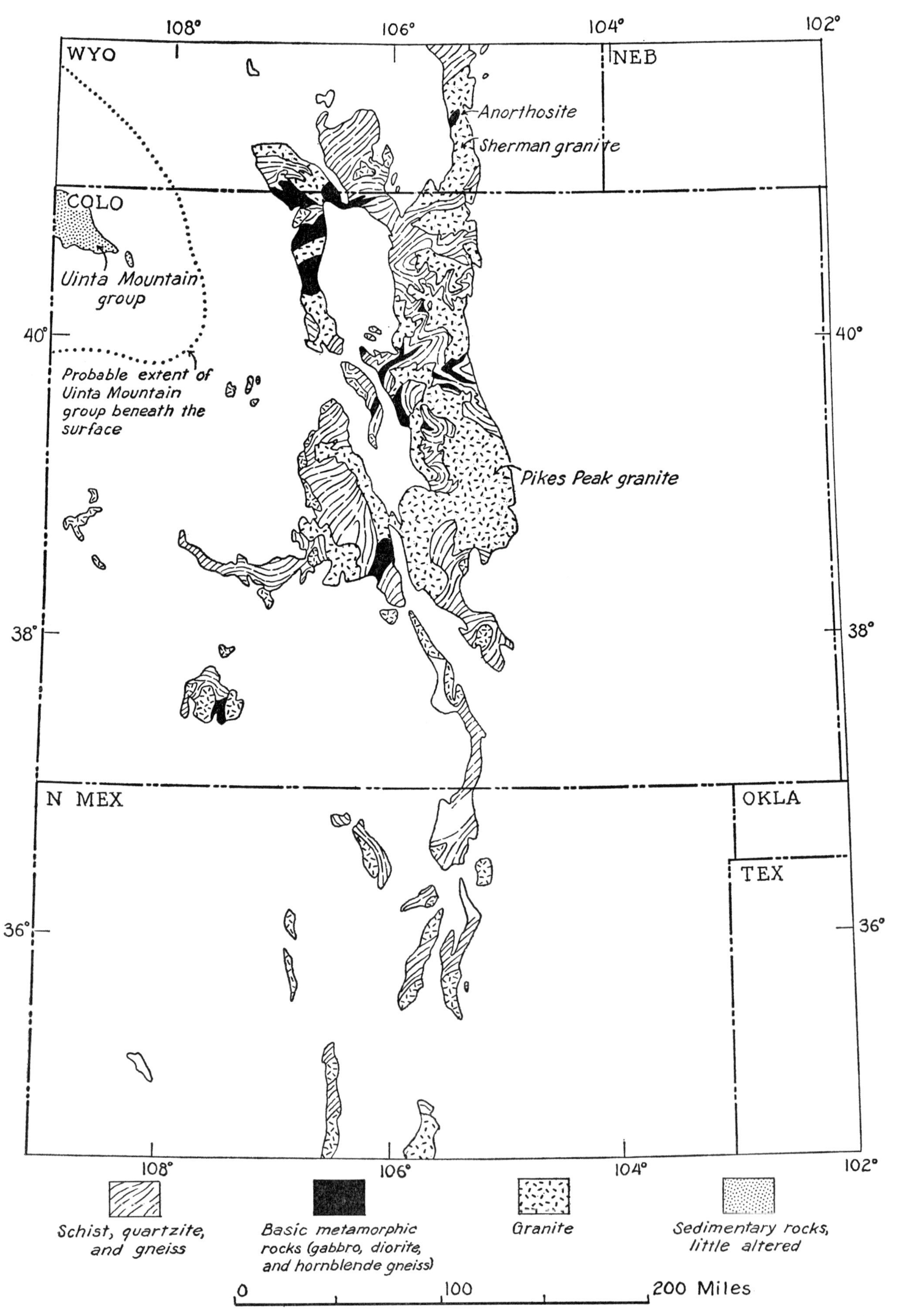

FIG. 54. Part I of a series showing structural development of Southern Rocky Mountains. Map showing exposed areas of Precambrian rocks and their lithology and structure. After Burbank and Lovering, 1933; and other sources.

rocks—although in most places these are not the earliest Cambrian. Some geologists believe firmly that they are of Early Cambrian age, or only a little older at most. Nevertheless, mineralized veins in one of the series in northwestern Montana have yielded radiometric dates of more than 1,200 million years. While the significance of this determination remains to be evaluated, it is possible that formation of these fresh-looking rocks began well before Cambrian time toward the west, and that they include from place to place a wide range of ages down nearly to the beginning of Cambrian time.

The little altered, mainly later Precambrian rocks differ from most of those of the same general class in the Canadian Shield. Unlike the Keweenawan series, for example, they are not volcanic and red clastic rocks that were laid down in a continental environment. Most of them are fine-grained well-bedded sediments such as argillites and cleanly washed sandstones, but with considerable thicknesses of limestone in places and only occasional lavas. Clearly, they were deposited under water and very probably in a marine environment. During later Precambrian time when local basins in the Canadian Shield were receiving continental deposits, perhaps marine deposits were accumulating in the geosyncline along the edges of the continent, thus foreshadowing the growth of the marginal geosynclines of Cambrian and later times.

The similarity in distribution of these little altered Precambrian deposits to those of the Cambrian is illustrated by their occurrence only along the western edge of the Eastern Ranges and Plateaus—that is, at about the place where the thicker and earliest Cambrian deposits set in westward. They also come to the surface in many parts of the miogeosynclinal area west of the region we are now considering.

In the far north these little altered Precambrian rocks are the *Belt series*, named for the Big Belt and Little Belt Mountains of west-central Montana, which are part of our "Eastern Ranges." But the main development of the Belt is in the miogeosynclinal area of the Northern Rocky Mountains farther west and northwest, which we will consider later. It seems best to reserve discussion of the Belt series until that time (Chapter VIII, Section 2).

Farther south in northeastern Utah is the *Uinta Mountain group*, which is raised in the core of the Uinta Mountains (Fig. 54). It is a mass of sandstone and quartzite 10,000 feet or more thick whose base, lying unconformably on older metamorphic rocks, is exposed at Red Creek on the north flank of the range; it is overlain disconformably by Upper Cambrian strata. Like the Belt series the Uinta Mountain group extends westward into the miogeosynclinal area, and its equivalents occur in the Wasatch and nearby mountains of the Great Basin.

But one of the finest displays of these later Precambrian rocks is in northwestern Arizona, where the *Grand Canyon series* is laid bare in the depths of the Grand Canyon.

Rock sequence in the Grand Canyon. Here it is appropriate to digress, and to anticipate some of the later parts of our story by describing the Grand Canyon and all its rocks at one place along with the geologic history which they imply.

The plateau country of northern Arizona is about what one would expect in a high region of little rain—a topography of moderate relief whose lower parts are bare desert, the parts a little higher clad in groves of juniper and pinyon, and the highest covered by spruce forests. Among the highest parts of the region are the Coconino and Kaibab Plateaus, which rise to 6,900 and 8,100 feet, respectively; here in an area about one hundred miles long and twenty-five miles wide rocks of the plateau have been raised higher than elsewhere. According to Powell, Kaibab is an Indian word meaning "mountain lying on its side"—an appropriate description of this great swell in the plateau surface.

The plateaus of northern Arizona are capped partly by highest Paleozoic limestones, partly by lowest Mesozoic redbeds which dip so gently over wide areas that only small thicknesses of strata come to the surface. Little in the plateaus gives much hint of the underlying rocks or structures. For these our recourse in Texas would have been the records of deep drill holes. Here there are few drill holes as yet, but we actually have something better, for the Colorado River has cut across the highest part of the region, laying bare in the Grand Canyon a great cross-section of the lower rocks a mile in depth. All the Paleozoic and later Precambrian rocks are revealed, as well as the basement of earlier Precambrian rocks on which they lie.

More than half the upper slope of the canyon is formed of Paleozoic sedimentary rocks which descend in a series of giant steps (Fig. 55):

Rimrock of the canyon is the *Kaibab limestone** of Permian age, which is underlain in turn by the *Coconino sandstone*, *Hermit shale*, and *Supai redbeds*, also part of the Permian. The Kaibab and Coconino make great lines of cliffs, but the Hermit and Supai

* It is now recognized that the upper limestones of the plateau country represent two distinct cycles of sedimentation to the higher of which the name Kaibab is restricted, the lower unit being termed the Toroweap. (See figs. 55 and 56) These two units are not far apart in age, however, and both are middle Permian. For our purposes the earlier and broader definition of Kaibab is useful.

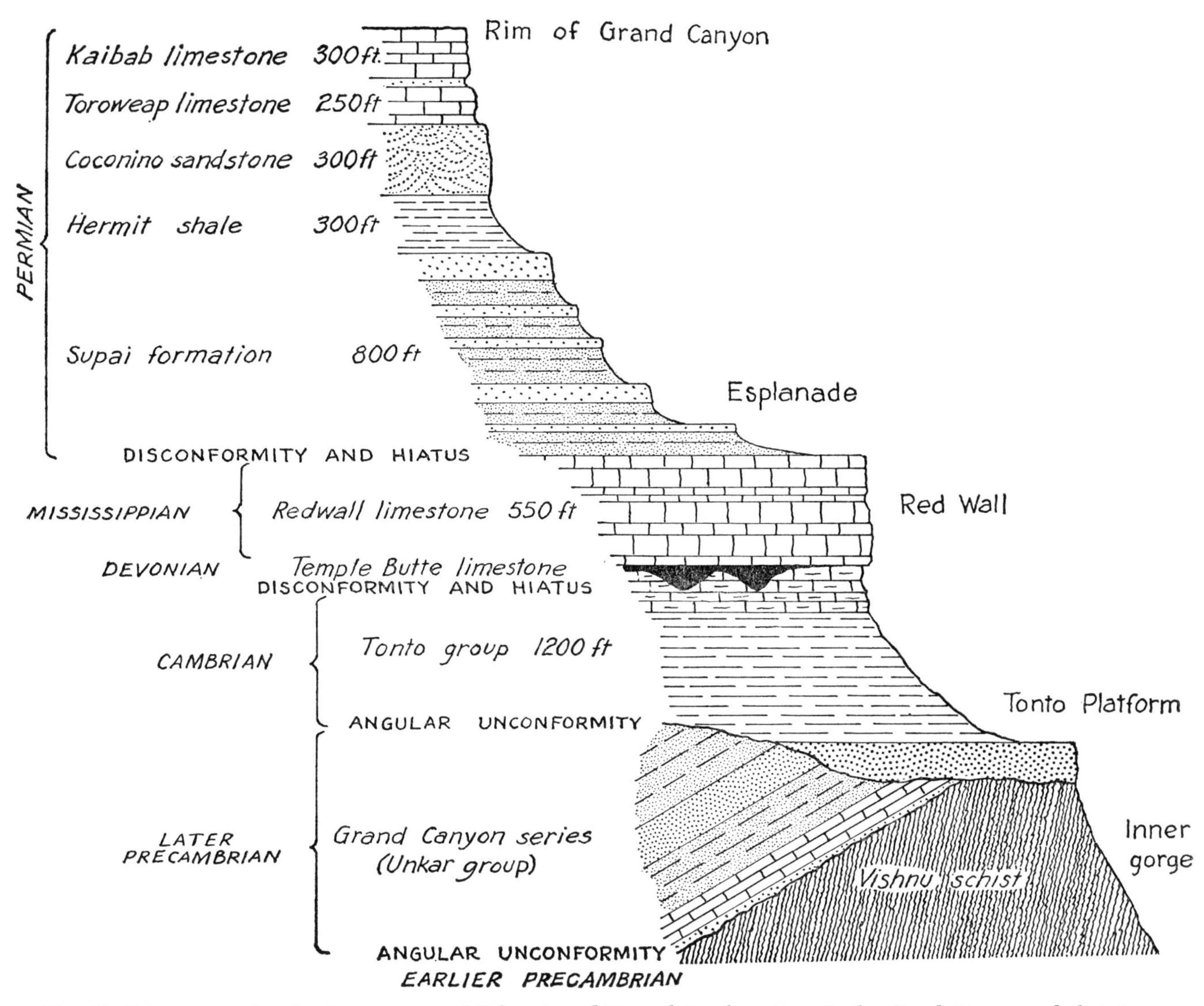

FIG. 55. Columnar section showing sequence of Paleozoic and Precambrian formations in the Grand Canyon and their topographic expression. Compiled from McKee, 1931; and other sources.

form a series of steps, the lower part spreading out in a broad bench or shelf called the Esplanade.

Below the Supai is another great line of cliffs formed by the *Redwall limestone* of Mississippian age. The Redwall is not actually red as its name and appearance implies, but is an ordinary gray marine limestone; its surface has been colored by red silt washed down from the Supai above it.

Beneath the Redwall cliffs is another set of steps formed by the *Tonto group* of Cambrian age, which leads out onto a broad bench well down in the canyon, the Tonto Platform, whose surface is maintained by the basal layer of the group or *Tapeats sandstone.*

The whole span of Paleozoic time from Cambrian to Permian is thus represented in the Grand Canyon sequence, but observe that we have failed to mention the presence of deposits of many of the Paleozoic systems—the Pennsylvanian which should occur between the Supai and Redwall, and the Devonian, Silurian, and Ordovician which should occur between the Redwall and Tonto. All the Paleozoic strata in the canyon lie parallel and are seemingly conformable, yet there are breaks in the sequence, each representing an *hiatus,* or interval of time for which no deposits are preserved; the sedimentary record is thus incomplete. Contacts between the Supai and Redwall, and between the Redwall and Tonto, where these breaks in the record occur, are *disconformities* (Fig. 55).

In some parts of the canyon country the record is partly filled. In places Devonian deposits (*Temple Butte limestone*) lie in pockets and channels on the surface of the Tonto group, and in the western part of the canyon Pennsylvanian limestones wedge in between the Redwall and Supai. But any record in the region of Ordovician and Silurian time, if ever it were present, is now lost.

The Tapeats sandstone of the Tonto group, or initial Cambrian deposit, has another relation to the rocks beneath. In much of the canyon it lies with *angular unconformity* on the eroded edges of steeply tilted, dark, ragged, old-looking metamorphic rocks, the *Vishnu schist*—a basement of early Precambrian rocks upon which all the superstructure of the plateau region has been built (Fig. 56). As noted earlier, intrusives which invade it have been dated at about 1,350 million years.

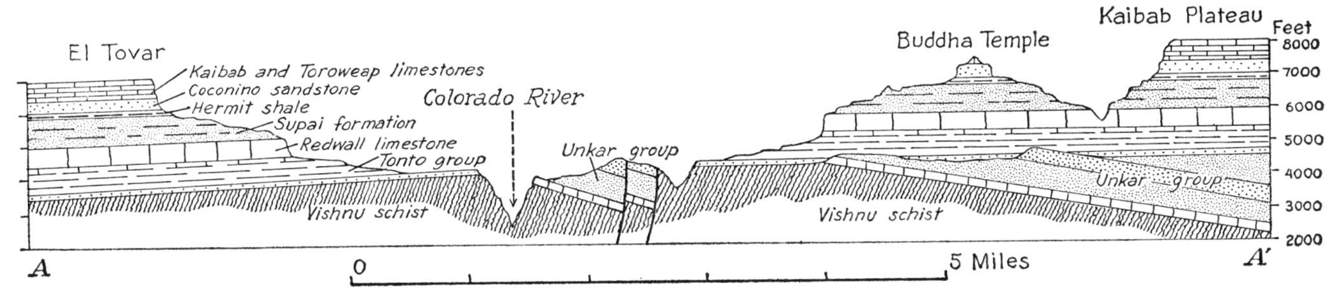

Fig. 56. Section showing structure of the Grand Canyon. Section from El Tovar northward to Kaibab Plateau along line A A′ of Fig. 59. After Darton, 1925.

But in parts of the canyon the Tapeats lies on other rocks—the strata of the *Grand Canyon series.* As with the Vishnu, the Tapeats lies on the Grand Canyon series with angular unconformity, but here one sedimentary series lies on another sedimentary series that had previously been tilted, faulted, and then truncated (Fig. 56). The Grand Canyon series, like the Tapeats, lies with angular unconformity on the Vishnu schist. No doubt the tilting and faulting of the Grand Canyon series produced block mountains like those of later times in the Cordilleran region, but these mountains were planed off by erosion at some time before the Cambrian so that the initial Cambrian deposits lie indiscriminately on either Vishnu or Grand Canyon.

If we will recall our example in Iowa of *layer cake geology* discussed in Chapter III, Section 5, we will recognize in the Grand Canyon another fine example of the same structure—the highest layer made of the Paleozoic formations, the next of the later Precambrian Grand Canyon series, and the lowest of the highly altered earlier Precambrian Vishnu schist. Here, in contrast to Iowa and elsewhere in the Interior Region, it is not necessary to search over a wide area or to have recourse to drill data to prove the relation; it is all exposed in cross-section at the surface in the walls of the Grand Canyon.

The Grand Canyon series consists of two parts: the Unkar and Chuar groups. The lower or *Unkar group,* which is the part seen by most visitors to the canyon, is a sequence of limestone, sandstone, and red shale. The upper or *Chuar group* is preserved above the Unkar only in the eastern part of the canyon, and is dominantly shaly but has some beds of lava in the lower part. When these different parts of the Grand Canyon series are added together, the whole has a thickness of over 12,000 feet, or about three times the thickness of the Paleozoic rocks which overlie it.

Earlier Paleozoic rocks and events. As already noted, the site of the Eastern Ranges and Plateaus was like the remainder of the continental platform during earlier Paleozoic time; its deposits of that time thus have much the same character as those of the Interior Lowlands.

As elsewhere, Cambrian deposits overlapped the continental platform, so that through much of the Eastern Ranges and Plateaus the Upper Cambrian, where present, lies directly on the Precambrian, with earlier Cambrian missing. Some exceptions seem to invalidate the rule; the Tonto group of the Grand Canyon belongs to the Middle rather than the Upper part of the system, and Lower Cambrian beds wedge in near the lower end of the canyon. But although these Cambrian deposits are earlier than most of those which lie on the continental platform, they are much like the later ones and are the edges of deposits which attain a much greater thickness in the miogeosyncline farther west (Fig. 77).

In many parts of the Eastern Ranges and Plateaus the Cambrian is missing entirely, either because it was never deposited, or because it was laid down thinly and afterwards eroded. One such area is that near Albuquerque, New Mexico, where, as we have seen (Chapter III, Section 4), the Pennsylvanian lies directly on the Precambrian (Fig. 11 B). This area is a part of the *continental backbone* or belt of country extending southwestward from Minnesota across the Interior Lowlands, where relatively young rocks lie directly on the basement, and where earlier rocks either were never deposited or have since been eroded.

Of the remaining earlier Paleozoic deposits through the Mississippian we need say little. They are relatively thin and consist mostly of limestone; there are many hiatuses between the units—for example, those below and above the Redwall limestone in the Grand Canyon. Different systems are present in different places recording the extent of successive ephemeral seas, but some of the systems are notable only for their poor representation; no Silurian occurs in Colorado or most of Wyoming, and these areas were probably land throughout the period.

Colorado system of later Paleozoic time. A marked change took place in part of the Eastern Ranges and Plateaus in later Paleozoic time, and its character as a continental platform began to break down. Beginning in the late Mississippian and continuing through Pennsylvanian into Permian time structures of the *Colorado system* were formed in the Southern Rocky Mountains at the same time and along the same trend as in the Wichita system of Oklahoma.

Because of deformation which took place in the Southern Rocky Mountains later on during the Laramide orogeny, these later Paleozoic structures are obscure and hard to identify. Main record of their former existence and their history is in the sediments which were eroded from the uplifted areas and deposited along their flanks. These sediments indicate the following structural pattern and history:

During later Paleozoic time two main uplifts or geanticlines were raised in Colorado and adjacent states (Fig. 57). The *Front Range geanticline* lay to the northeast on about the position of the present Front Range that was raised again during the Laramide orogeny. The *Uncompahgre geanticline* lay to the southwest in southwestern Colorado and is less clearly expressed in surface features. Its Precambrian rocks still emerge in places, as in the Uncompahgre Plateau and Needle Mountains, but most of the geanticline in the Paleozoic and earlier rocks has been concealed by Mesozoic sediments and Tertiary volcanics. Extending southeastward across Colorado between the two geanticlines was a long trough similar to the Anadarko basin of Oklahoma, which received sediments derived from the geanticlines on each side. Pennsylvanian and Permian deposits in the trough are more than 10,000 feet thick.

The two geanticlines were formed at different times, as shown by the record of the sediments derived from them:

The Front Range geanticline was raised in late Mississippian and early Pennsylvanian time. Its existence is indicated by the thick mass of coarse red arkose and sandstone of the Fountain formation of Pennsylvanian age, whose strata are turned up in hogbacks along the eastern edge of the present Front Range where they form such picturesque localities as the Garden of the Gods near Colorado Springs (Fig. 58). Similar deposits of about the same age occur west of the Front Range. Overlying red clastic deposits of Permian age along the edge of the Front Range are finer grained, indicating that by then the geanticline was quiescent and had been worn down to low relief.

The Uncompahgre geanticline was raised in late Pennsylvanian and Permian time. This is suggested toward the northwest end in the present Uncompahgre Plateau, where Triassic rocks lie directly on basement rocks along the crest of the uplift, whereas Permian and Pennsylvanian rocks that had been truncated before Triassic time wedge in along its flanks (Fig. 68). The age of the geanticline is shown more definitely in the Sangre de Cristo Mountains east of its southeast end, where there is a great thickness of coarse bouldery sediments of late Pennsylvanian and Permian age. These sediments were deposited in the trough that lay between the two geanticlines, and were derived from the Uncompahgre geanticline, which then stood not far to the west but is now deeply buried beneath Tertiary sediments and volcanics of the San Luis Valley and San Juan Mountains.

Later Paleozoic rocks. Away from the uplifts of the Colorado system the later Paleozoic deposits are thinner and finer grained. Some of them are still sandstones and redbeds, but they were derived from distant rather than nearby uplifts, and there are some rather widespread marine units. Two of the latter are worth recording—the *Kaibab limestone* (including the Toroweap limestone) which forms the rim of the Grand Canyon and extends over most of the southern part of the Colorado Plateau, and the *Phosphoria formation* of shale, chert, phosphate rock, and limestone which covers a wide area in southeastern Idaho, western Wyoming, and adjacent states. Both are of Permian age, the Phosphoria being somewhat younger than the Kaibab.

At least one large evaporite basin formed in the region, represented by the *Paradox formation* of Pennsylvanian age, which underlies an area in the northeastern part of the Colorado Plateau along the southwest edge of the Uncompahgre geanticline. The Paradox consists of thick beds of salt and gypsum which accumulated in a basin that probably became landlocked by uplift of ridges of the Colorado system. The Paradox comes to the surface mostly in a remarkable series of anticlines that were produced

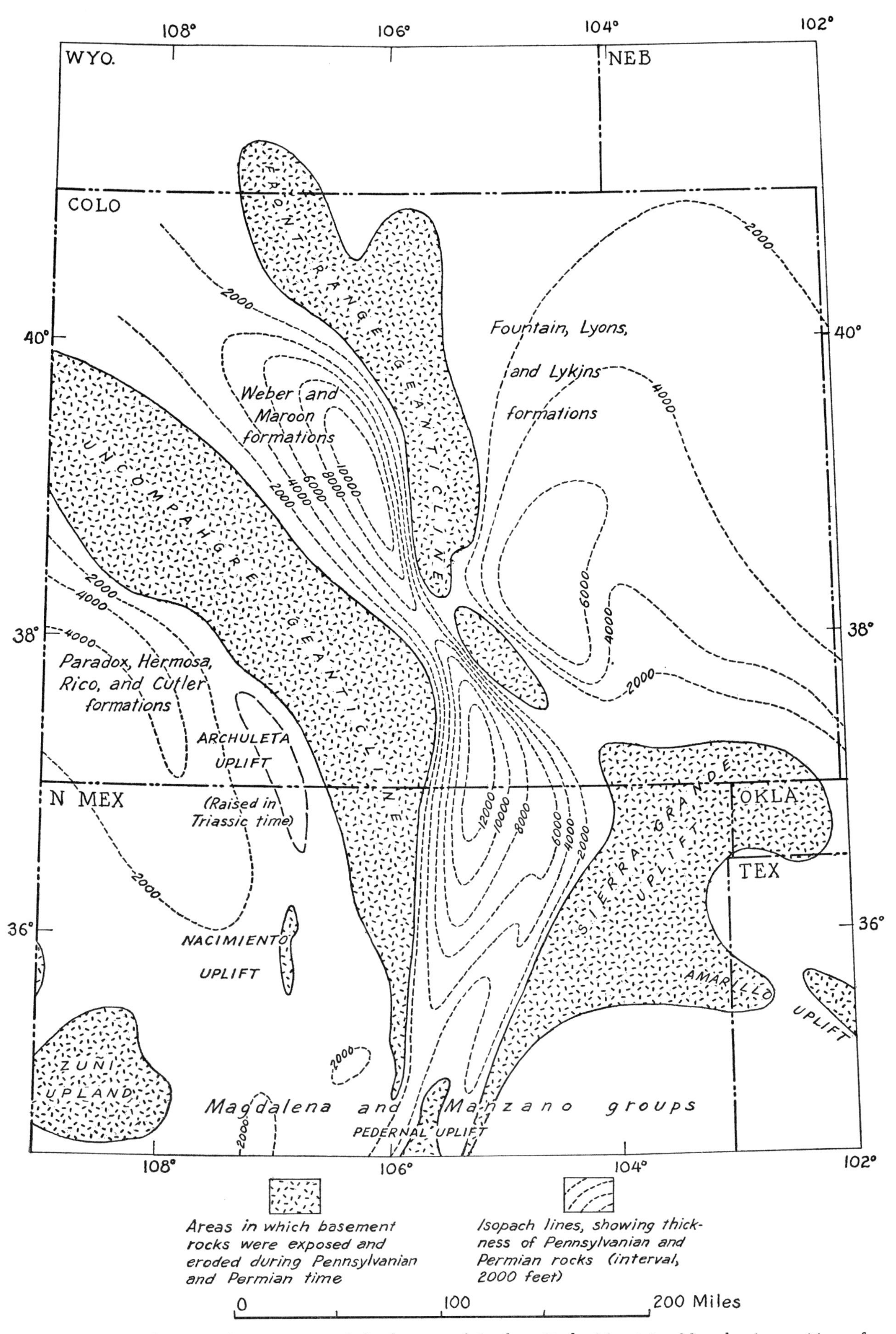

FIG. 57. Part II of a series showing structural development of Southern Rocky Mountains. Map showing positions of late Paleozoic uplifts and basins of the Colorado system. Compiled from Burbank and Lovering, 1933; and other sources.

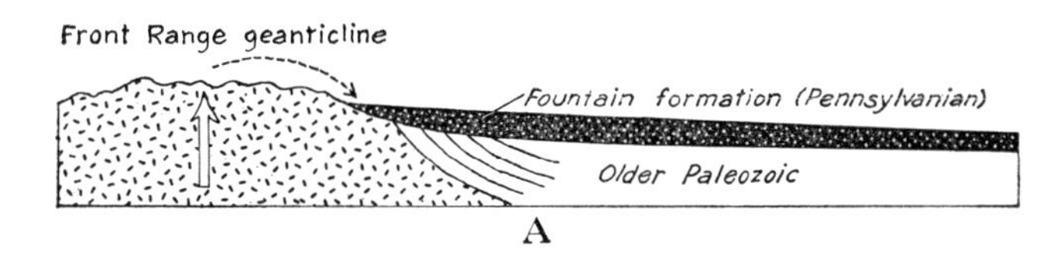

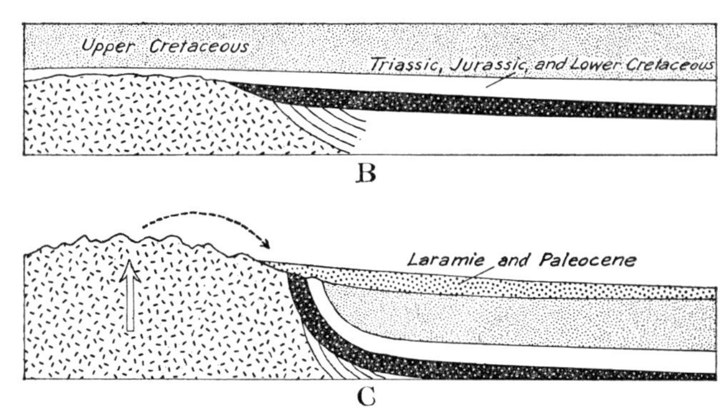

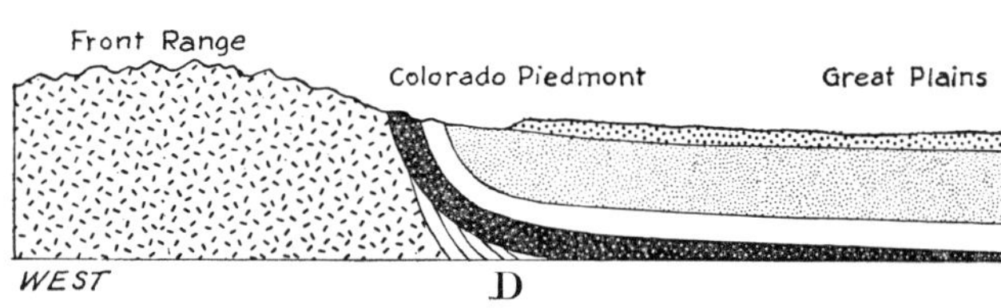

FIG. 58. Sketch sections illustrating structural evolution of the Front Range of Colorado. (A) In late Paleozoic time during formation of Colorado system. (B) In late Mesozoic time when the range was quiescent and buried. (C) In early Tertiary time immediately following Laramide orogeny. (D) Present relations following later Tertiary regional uplift and dissection.

by the plastic flow of the salt beds during later deformations; we will have more to say about them farther on in this chapter (Section 6).

By early Mesozoic time crustal activity had largely ceased in the Colorado system. Its ridges had been worn down to low relief and were finally buried by Triassic and Jurassic deposits. Nevertheless, the scar in the continental platform thus created remained a zone of weakness, ready to express itself in new forms when the region was again subjected to crustal forces. Although the area of the Eastern Ranges and Plateaus had returned to quiescence in latest Paleozoic and earliest Mesozoic time, this was merely an interlude between orogenies. Once Mesozoic time was under way, crustal forces again came into play and the great Cordilleran deformation had begun.

3. ROCKS AND EVENTS DURING THE GREAT DEFORMATION

MESOZOIC SEQUENCE NORTH OF THE GRAND CANYON. It will be appropriate to introduce the Mesozoic rocks of the Eastern Ranges and Plateaus by returning again to the Grand Canyon region (Fig. 59).

The Kaibab limestone of Permian and latest Paleozoic age forms the rim of the Grand Canyon. Away from the canyon it also forms the surface over wide areas, as though it were the last stratum to have been deposited in the region. Nevertheless, from Desert View at the eastern end of the canyon (Fig. 59) one may see the little knob of Cedar Mountain which is formed by a unit overlying the Kaibab—the redbeds of the Moenkopi formation. These redbeds, easily subject to erosion, have been removed except in scattered remnants such as this one.

More than canyon cutting is thus involved in the shaping of the Plateau country. There was also a *great denudation* by which weaker, higher formations were stripped from the resistant surface of the Kaibab limestone, not only along the canyon rim, but also over wide areas of the Kaibab and Coconino Plateaus.

Record of these higher formations can best be seen by passing northward from the canyon rim across the Kaibab Plateau toward the High Plateaus of Utah (Fig. 61). The strata dip gently in this direction but the land surface increases in altitude to heights of 10,000 feet or more, so that successively higher Mesozoic formations come in above the Kaibab, some poorly resistant, others strong and cliff-making; one thus ascends the sequence in a series of gigantic steps. This Mesozoic sequence is quite as wonderful in its way as the Paleozoic sequence which underlies it in the Grand Canyon, but instead of being exposed in one canyon wall it is spread out over a distance of twenty-five or thirty miles. The lines of cliffs are among the most striking scenic features of the plateau country and reappear in many places across its extent wherever the appropriate formations happen to be present.

In southern Utah these cliffs have been named according to their color (Fig. 60). Thus we find on ascending the section from south to north the low *Chocolate Cliffs*, the higher *Vermillion Cliffs*, and the very massive *White Cliffs* (which form the rock at Zion Canyon). Surmounting them are the less massive, better stratified *Gray Cliffs* which are capped in the higher plateaus by the *Pink Cliffs* (the rock at Bryce Canyon).

Each of these cliffs is held up by a formation of the succession of Mesozoic and later rocks. The Chocolate Cliffs are made by the Triassic Shinarump conglomerate which is underlain and overlain by less resistant Triassic redbeds, the Moenkopi and Chinle formations.

The Vermillion and White Cliffs are formed by the Wingate and Navajo sandstones, the lower one reddish, the upper without pigment. For the most part the Wingate and Navajo are great fossil sand dune de-

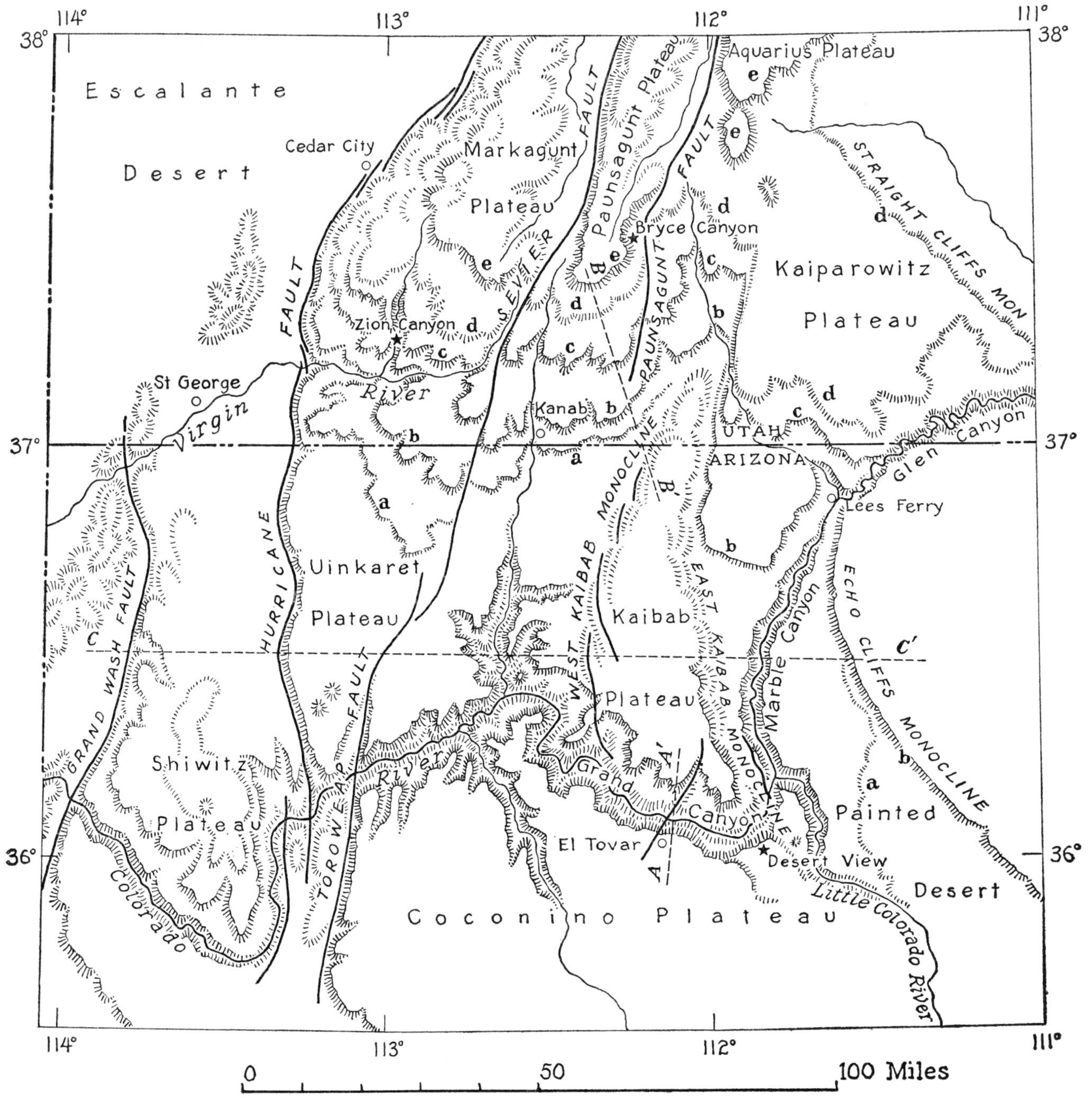

Fig. 59. Map of western part of Colorado Plateau in Arizona and Utah, showing cliff lines produced by erosion of formations, and fault lines. Letters indicate the following: (a) Chocolate Cliffs (Shinarump conglomerate, Triassic). (b) Vermillion Cliffs (Wingate sandstone, Jurassic). (c) White Cliffs (Navajo sandstone, Jurassic). (d) Gray Cliffs (Cretaceous rocks). (e) Pink Cliffs (Wasatch formation, Tertiary). Lines *A A′*, *B B′*, and *C C′* indicate locations of sections on Figs. 56, 60, and 74. Compiled from geologic maps of Arizona, 1924; of Utah, 1948; and other sources.

posits formed in a desert of Jurassic time, and by coincidence now exposed again in a desert region. In many parts of the plateau these sandstones have been sought out as dwellings and places of defense by the Indians, who have built their pueblos in great alcoves in the cliffs or on the tops of the sheer-sided mesas.

The succeeding Gray Cliffs are formed of interbedded sandstones and shales of the Cretaceous formations; the Pink Cliffs at the top are made by the continental beds of the early Tertiary Wasatch formation (Fig. 61).

Early Mesozoic rocks. Triassic and Jurassic rocks similar to those just described extend far across the Eastern Ranges and Plateaus. The lower part is nearly everywhere red colored, but changes to white above;

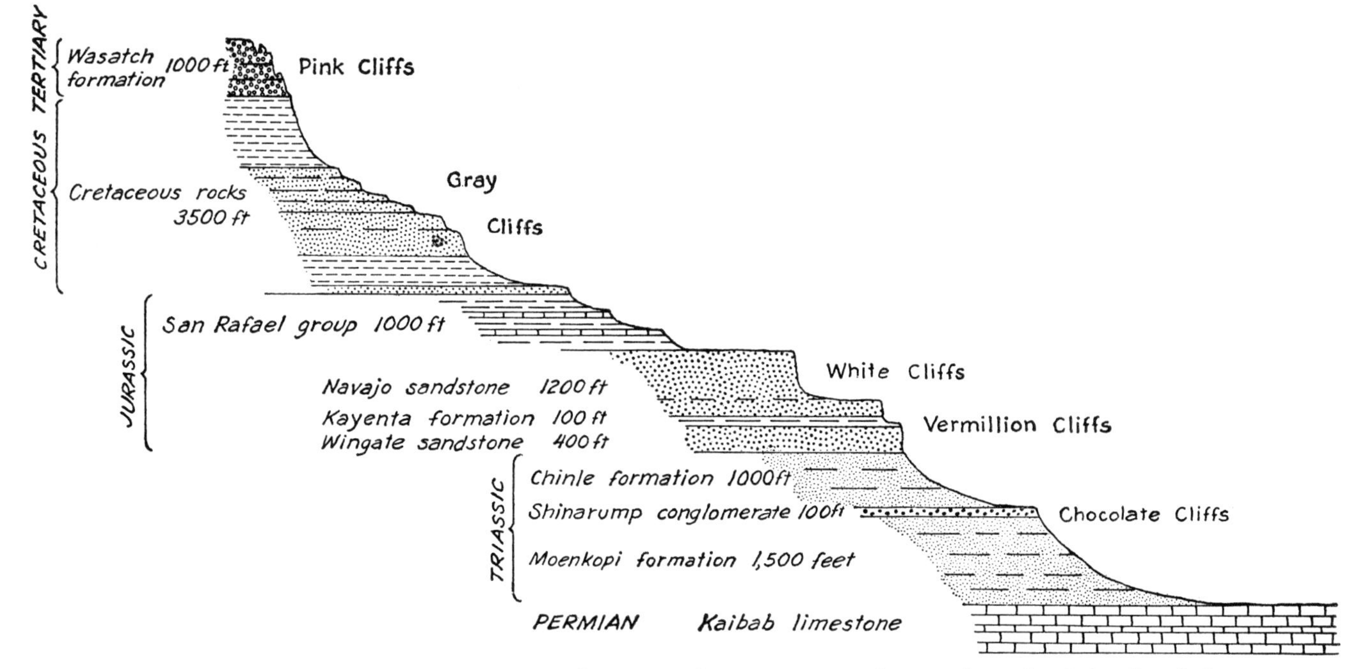

FIG. 60. Columnar section showing sequence of Mesozoic formations in southern Utah north of the Grand Canyon, and their topographic expression. Compiled from Gregory, 1950; and other sources.

both parts were formed largely in a continental and probably a desert environment. The deposits thin eastward across the Rocky Mountain region to a feather edge in the Great Plains, but thicken to 3,000 or 5,000 feet near the western edge of the Colorado Plateau toward the miogeosyncline, where some marine layers are interbedded. North of the plateau in the Central and Northern Rocky Mountains marine Jurassic deposits are widespread but attain no great thickness; they are known variously as the Sundance, Ellis, and Fernie formations.

These Triassic and Jurassic deposits somewhat resemble the clastic wedges which spread from the Appalachian geosyncline into its foreland. The comparison is not perfect but has some basis in the Jurassic, whose thick sandstones had their source in the west where the Nevadan orogeny was beginning. It is in the succeeding Cretaceous deposits that the clastic wedges of the Cordillera are best displayed, and to these we will now turn.

THE CRETACEOUS CLASTIC WEDGES. If the reader will examine the paleogeographic maps in any of the text books of historical geology, he will find that during Cretaceous time, and especially during its last half, a broad seaway extended northward from the Gulf of Mexico to the Arctic along the eastern side of the Cordilleran region, covering most of the area of the Eastern Ranges and Plateaus and overlapping the Interior Lowlands as far as Kansas, Iowa, and Minnesota. Westward, the seaway extended no farther than central Utah, or only a little beyond the Colorado Plateau.

Cretaceous deposits laid down in this seaway are, in effect, a series of clastic wedges, related to the growth of the Cordillera on the west in later Mesozoic time, and are worth examining for that reason. Our analysis can best begin at their eastern edge in the Interior Lowlands and proceed westward (Fig. 62).

The present eastern edge of the Cretaceous is probably not far from the original eastern limits of the deposit. In Iowa and Kansas the sequence is no more than a few hundred or thousand feet thick. It begins with a basal transgressive sandstone, the Dakota, above which are interbedded limestones and shales including the Greenhorn limestone and Niobrara chalk below and the Pierre shale above (Fig. 62).

Farther west near the front of the Rocky Mountains in Colorado the deposits have thickened to 8,000 or 10,000 feet, most of the increase being in the shales. The two limestone units, the Greenhorn and the Niobrara, persist but fade into shales a short distance to the west. At the top of the sequence is the marine Fox Hills sandstone, followed by the continental Laramie beds of latest Cretaceous age.

Beyond the front of the Rocky Mountains ledge-making beds of sandstone make their appearance in the shales, and are generally grouped at certain levels. One group in the upper part of the sequence is the widespread Mesaverde formation; a lower one, mainly developed in Wyoming, is the Frontier formation. The

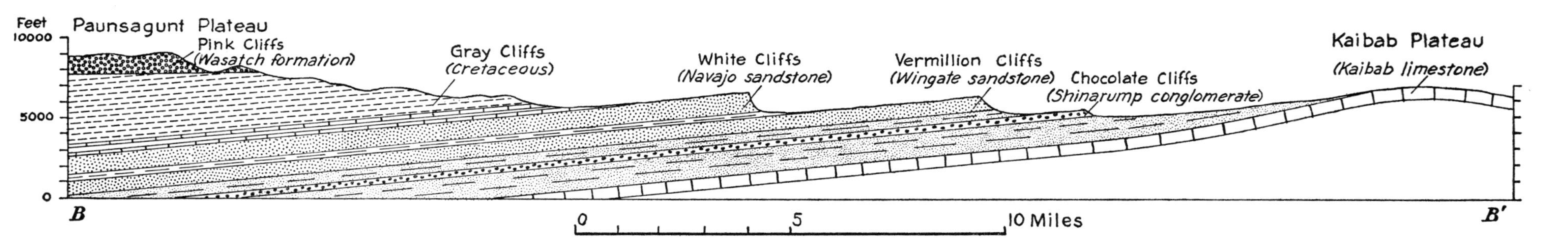

Fig. 61. Section showing sequence and structure of Mesozoic rocks north of Grand Canyon and the cliff lines which have been produced by their erosion. Section from Kaibab Plateau northward to Paunsagunt Plateau, one of the High Plateaus of Utah, along line *B B′* of Fig. 59.

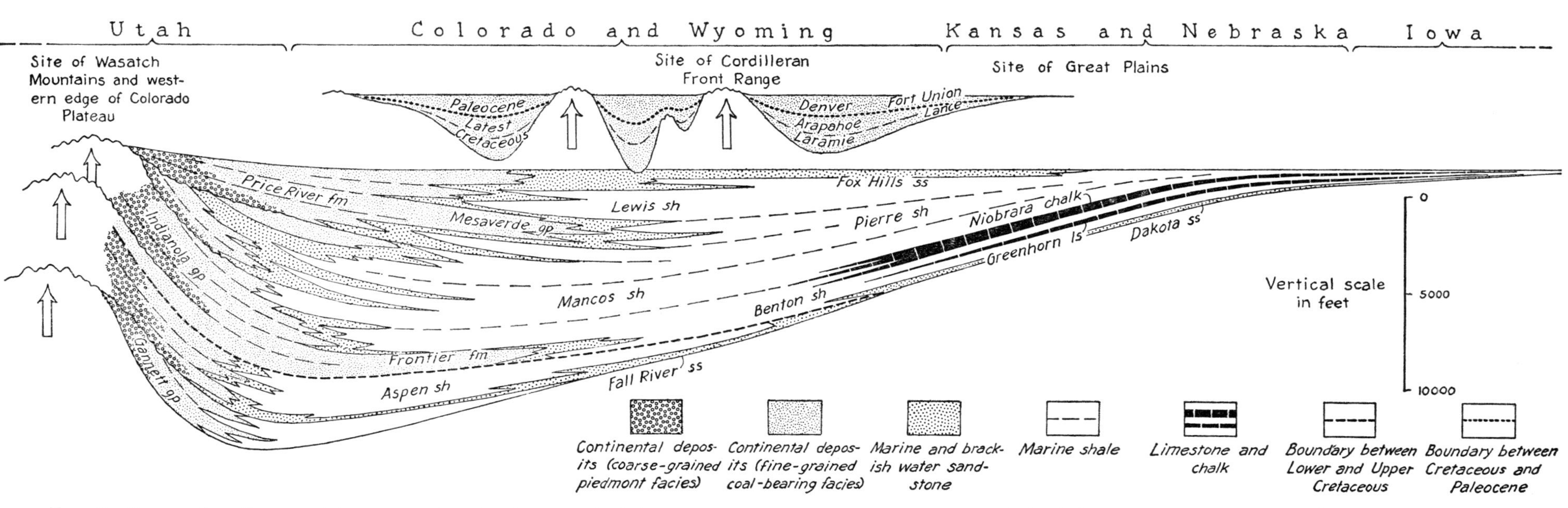

Fig. 62. Stratigraphic diagram of Cretaceous deposits that were laid down in the seaway along the eastern side of the Cordilleran region, in a section from Iowa to Utah. The latest Cretaceous and Paleocene deposits, which have a different habit from the remainder, are shown in a separate diagram above. In order to bring out general relations in the western half of the diagram, units are included which occur in a segment of the seaway several hundred miles broad. Compiled from Dorf, Cobban and Reeside, and other sources.

eastern edges of the sandstone layers are marine, but where they thicken westward as wedges they include continental beds with layers of coal.

At the west edge of the Colorado Plateau the Cretaceous deposits attain a thickness of 15,000 or 20,000 feet, and most of them are continental (Fig. 62). The marine shales have largely pinched out, and the sandstones and coal-bearing rocks have passed into coarse conglomerates made up of fragments of earlier Mesozoic and Paleozoic formations derived from the site of the Great Basin immediately to the west. Unconformities develop beneath some of the units, by which the conglomerates lie on the eroded and even upturned edges of the beds beneath. Along the edges of the Wasatch Mountains the unconformable beds overlap a rough topography of much deformed earlier rocks; such beds are, in fact, piedmont deposits laid down along the bases of newly upraised mountain areas.

In our traverse across the deposits of the Cretaceous seaway we have observed the thin transgressive marine deposits of its eastern edge with much limestone; the thick, dominantly shaly marine deposits at the present front of the Rocky Mountains, near the center of the seaway; and the even thicker, coarse continental beds at the western edge of the seaway, evidently laid down along the front of a mountain belt in process of growth.

Little information is available as to the extent of Cretaceous deposits farther west in the Great Basin. Here and there in its eastern half as far as central Nevada patches of coarse conglomerate and other continental beds lie on much deformed earlier rocks and in places contain fresh-water Cretaceous fossils. Very likely these deposits were not much more continuous than they now are, and formed in local depressions in the mountain belt rather than as widespread deposits; the main body of Cretaceous deposits with their clastic wedges probably thinned abruptly westward near the edge of the present Great Basin.

RELATION OF CRETACEOUS DEPOSITS TO OROGENY. The paleogeographic maps indicate a broad belt of land west of the Cretaceous seaway which is labeled the "Cordilleran geanticline" in some textbooks. The term "geanticline" is somewhat misleading as it implies merely a broad upwarp of the crust, whereas, as we have seen, the belt was a product of much stronger deformation—orogenic rather than epeirogenic. During much of Cretaceous time it was probably hilly or even mountainous and was undergoing vigorous erosion.

Moreover, the several unconformities in the Cretaceous at the western edge of the Colorado Plateau, the coarse piedmont deposits which overlie them, and the successive wedges of continental beds and sandstone which extend far eastward into the seaway indicate that deformation was accomplished in a series of pulses, probably with intervening periods of relative quiescence.

West of the Great Basin in the western part of the Cordillera, Nevadan orogeny had begun in Jurassic time. Orogenic pulses at the eastern edge of the Great Basin took place in Cretaceous time, and mainly in the last half. Deformation evidently progressed eastward from the area of Nevadan orogeny through the latter part of Mesozoic time, and near its close all the Cordillera up to the edge of the Colorado Plateau had been transformed into a mountain belt. Further progress of deformation eastward during latest Cretaceous and Paleocene time, we will now examine.

LARAMIE AND PALEOCENE DEPOSITS. Overlying the main body of the Cretaceous in the Eastern Ranges and Plateaus, and extending eastward beyond the front of the Rocky Mountains for some distance into the Great Plains, are continental and coal-bearing deposits of latest Cretaceous age, which are succeeded conformably by similar deposits of earliest Tertiary or Paleocene age.

During early geologic work in Wyoming and Colorado the latest Cretaceous deposits were termed the *Laramie formation* and this is still a useful general title for the whole. Farther north similar beds termed the *Lance* and *Fort Union formations* were also recognized, but their age was much disputed; the former is now generally placed in the latest Cretaceous like the Laramie, the latter in the Paleocene. For our purpose we need not burden ourselves with the multiplicity of other local names which have arisen.

These deposits have a different pattern from the widespread deposits of earlier Cretaceous time. They occur in great thickness in local basins—in part between the ranges, as in the small Hanna Basin and the larger Bighorn and Powder River Basins of Wyoming, in part in the Great Plains to the east, as in the Denver and Williston Basins (upper part of Fig. 62). As we shall see, the deposits, although conformable within the basins, were being deposited while the ranges about them were in process of growth, so that their present extent corresponds somewhat to their original areas of deposition. Deformation, whose eastward spread had reached the edge of the Colorado Plateau during earlier parts of Cretaceous time, was by now active in the plateau itself and in the ranges of the Rocky Mountains beyond it.

This is the so-called *Laramide orogeny* of latest Mesozoic and earliest Cenozoic time, once thought to have been distinct from the Nevadan orogeny of mid-Mesozoic time farther west. From the record presented here, however, we can see that the two orogenies, rather than being separate events in the shaping of the Cordillera, were parts of a continuing process.

LARAMIE QUESTION. These latest Cretaceous and earliest Tertiary deposits have given rise to one of the former great controversies of American geology, the "Laramie question." Actually, this involved two questions which are not entirely related.

(1) Where is the Cretaceous-Tertiary boundary? The deposits are conformable and much alike, but the lower part contains *Triceratops* and other representatives of the last of the dinosaurs, whereas the upper part contains mammals and plants of more modern or Tertiary aspect. At some time during this depositional period there was a change-over from the characteristic life of Mesozoic time to that of Cenozoic time, the former becoming extinct, apparently with some abruptness. This question need not concern us greatly; it is a matter for decision by paleontologists and stratigraphers and has now largely been resolved to their satisfaction.

(2) When did the Laramide orogeny take place, and when were the Rocky Mountains formed? The deposits are conformable, as stated, yet in their upper part they contain granitic debris from the cores of the ranges, indicating that the ranges had been greatly uplifted and deeply eroded (Fig. 63). Geologists have

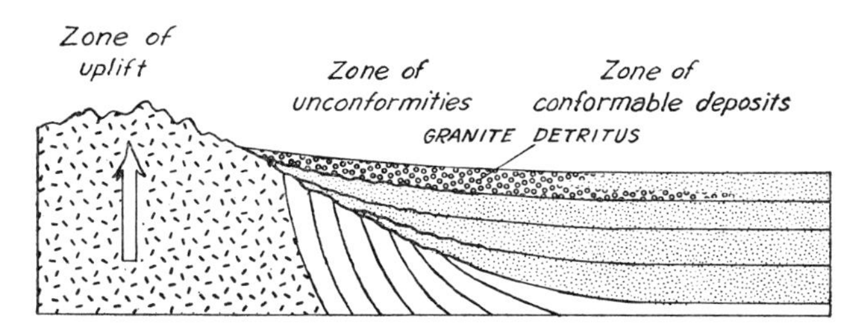

FIG. 63. Sketch section of the border of a typical mountain and basin in the Eastern Ranges during Laramie and Paleocene time, showing how conformable deposits could be laid down in the basin near the rising mountain area.

devoted much futile effort in a search for immense unconformities that were supposed to have been overlooked in the seemingly conformable basin deposits. It was thought that such unconformities should be a necessary consequence of the formation of the Rocky Mountains.

As we have seen, however, orogeny had been in progress for a long period of Mesozoic time, advancing progressively eastward across the Cordilleran region. In later Cretaceous time the ranges of the Rocky Mountains were gradually raised while deposition of the Laramie strata continued in the intervening basins. The ranges did not project in the manner they do now, but were eroded about as rapidly as they were uplifted, shedding detritus into the basins roundabout. Sediments in the basins and their contained fossils indicate a low-altitude, semi-tropical, moist climate; no great mountain barriers to the west shut off the flow of moisture-laden winds from the Pacific Ocean. The present lofty height of the Rocky Mountains did not develop until afterwards, and only by processes that were related to the waning stages of the formation of the Cordillera (see Section 7 of this chapter).

UNCONFORMITY AT BASE OF WASATCH FORMATION. In contrast to the conformable sequence of deposits from the Cretaceous through the Paleocene a considerable change took place from the Paleocene into the Eocene, or within the Tertiary. At the edges of the basins the coarse, bouldery Wasatch formation oversteps and overlaps all the earlier formations with marked unconformity, extending in places far into the ranges. This relation has sometimes been said to indicate the climax of the Laramide orogeny; more properly it was an anticlimax after the orogenic forces had largely spent themselves.

We have now carried our story of the Eastern Ranges and Plateaus through the great deformation, ending with the Laramide orogeny of late Cretaceous and Paleocene time. Events which followed the great deformation were many and varied, and are difficult to generalize because they are closely bound to local conditions. It is therefore appropriate that we now describe the structures of the several parts of the region that were formed mainly during the Laramide orogeny before resuming the history of the region from Laramide time to the present.

4. CENTRAL ROCKY MOUNTAINS

SUBDIVISIONS OF THE ROCKY MOUNTAINS. As noted earlier the "Rocky Mountains" are more of a geographical than a geological entity, the name being used for any of the frontal ranges of the Cordillera that face the Great Plains. To give the expression geological meaning it is necessary to break the Rocky Mountains into several parts. Geologists differ as to how this subdivision should be made and what the various parts should be called; it would be tedious here to review the many usages. For purposes of this book we will speak of a Northern, Central, and Southern Rocky Mountains, but the reader should be forewarned that these are units which differ in one or more particulars from those in other works.

The *Northern Rocky Mountains* extending from western Montana into Canada are ridges of sedimentary rocks that originated in the miogeosynclinal belt of the Cordillera; by subsequent deformation they have acquired a folded and faulted structure like that in the sedimentary Appalachians (see Chapter VIII, Section 3).

The *Southern Rocky Mountains*, by contrast, are Eastern Ranges which belong to that part of the Cordilleran system east of the front of the main geosynclinal belt. They are the high, closely packed ranges that center in Colorado, but whose prongs extend northward into Wyoming, westward into Utah, and southward into New Mexico. The Laramide and later structure of these ranges is complicated by its superposition on structures of the later Paleozoic Colorado system.

Between the Northern and Southern Rocky Mountains is a more heterogeneous set of ranges lying mainly in Wyoming but extending northward into Montana and westward into Idaho. This middle or central part of the Rocky Mountains includes on the west ranges of deformed miogeosynclinal rocks, and east of them a set of rather simple uplifts and basins which, like the Southern Rocky Mountains, are Eastern Ranges that formed from rocks outside the main geosynclinal belt. For the mountains on the west, formed of the miogeosynclinal rocks, we can use a synthetic term, the *Wyomide Ranges* (from *Wyom*ing and *Id*aho); those to the east we will term the *Central Rocky Mountains.* Unlike the Southern Rocky Mountains their rocks and structures show little evidence of Paleozoic disturbance, and their ranges originated largely during the Laramide orogeny. Exposition of the structure of the Eastern Ranges can best begin with this part, where the structures are less confused by earlier deformations.

General geology and geography. The Central Rocky Mountains are illustrated by the geological maps of Wyoming and Montana. On these one can see that the ranges are broad-backed uplifts, many of which expose wide areas of Precambrian basement rocks in their cores; also that they are of diverse trend and widely spaced, with broader basin areas between, most of whose surface rocks are of early Tertiary (Paleocene and Eocene) ages.

Some of the ranges of the Central Rocky Mountains rise to imposing heights; the Wind River Mountains, for example, include peaks more than 13,000 feet high. But the intervening basins form plains and plateaus much like the Great Plains farther east—so broad that in many places the mountains along their edges are only dimly visible. Although the state of Wyoming lies mainly in the Central Rocky Mountains, over half its area is plains country.

It was no accident, therefore, that most early western travel was through the present Wyoming country, avoiding the more massive ranges of the Rocky Mountains to the north and south. Emigrant wagons could roll across the plains with few obstacles and without passing through any mountains, until they came to the Wasatch Range in Utah, far to the west. So also, some years after the great western emigrations the first transcontinental railroad was laid out and built through the same area.

The ranges and their structure. In most of this account I have not burdened the reader with local geography, but to explain the Eastern Ranges it is necessary to name at least the major physical features.

On the east side of the Central Rocky Mountains are the *Black Hills* athwart the boundary between Wyoming and South Dakota—an isolated domical uplift well out in the plains, yet showing many features of the main mountain structure. Although the Black Hills rise neither as high as the ranges farther west nor show as much structural uplift, Precambrian rocks are laid bare in their core.

Farther west are the more lofty *Bighorn Mountains* with peaks of Precambrian rocks along their crests. The Bighorn Mountains extend into other ranges at their ends which together form a great arcuate uplift, convex toward the east.

Southwest of the Bighorn arc are the *Wind River Mountains*, again with high peaks of Precambrian rocks along their crests. The Wind River Mountains are merely a segment of a larger northwest-trending chain of ranges whose original Laramide structures were much modified later. Toward the southeast these are now low and half buried by Tertiary sediments, but to the northwest, south of Yellowstone Park, the chain rises in the *Teton Mountains.* These overlook the flanking basin of Jackson Hole to the east in towering precipices with some of the most spectacular alpine scenery in the United States. Here, as in the part of the chain southeast of the Wind River Mountains, the topography is greatly influenced by structures younger than the Laramide deformation; the Teton Mountains were raised by block faulting rather late in Tertiary time.

In northwestern Wyoming and extending into Montana are the *Beartooth Mountains* lying west of one end of the Bighorn arc and northeast of Yellowstone Park. Like the Bighorn and Wind River Ranges, they have a high-standing Precambrian core.

The Eastern Ranges in Montana, beyond, are generally lower than those in Wyoming, both topographically and structurally, and are differently aligned. Of these we need mention only the *Big Belt* and *Little Belt Mountains* on the west, whose cores are made up of sedimentary rocks of the later Precambrian Belt series.

The ranges are uplifts somewhat longer than wide, raised so high that in many places erosion along their crests has laid bare the basement of Precambrian rocks. Reconstruction of the sedimentary cover indicates that the uplifts are arch-like or block-like, with gentle dips over the crests and steep dips along the edges where the basement and immediately overlying strata descend to great depths in the adjoining basins (Fig. 64). The edges of the uplifts present several varieties of structure:

(A) The turned-up sedimentary cover may form a steeply-dipping border along the core of exposed Precambrian rocks, carved by erosion into lines of hogbacks (Fig. 65 A).

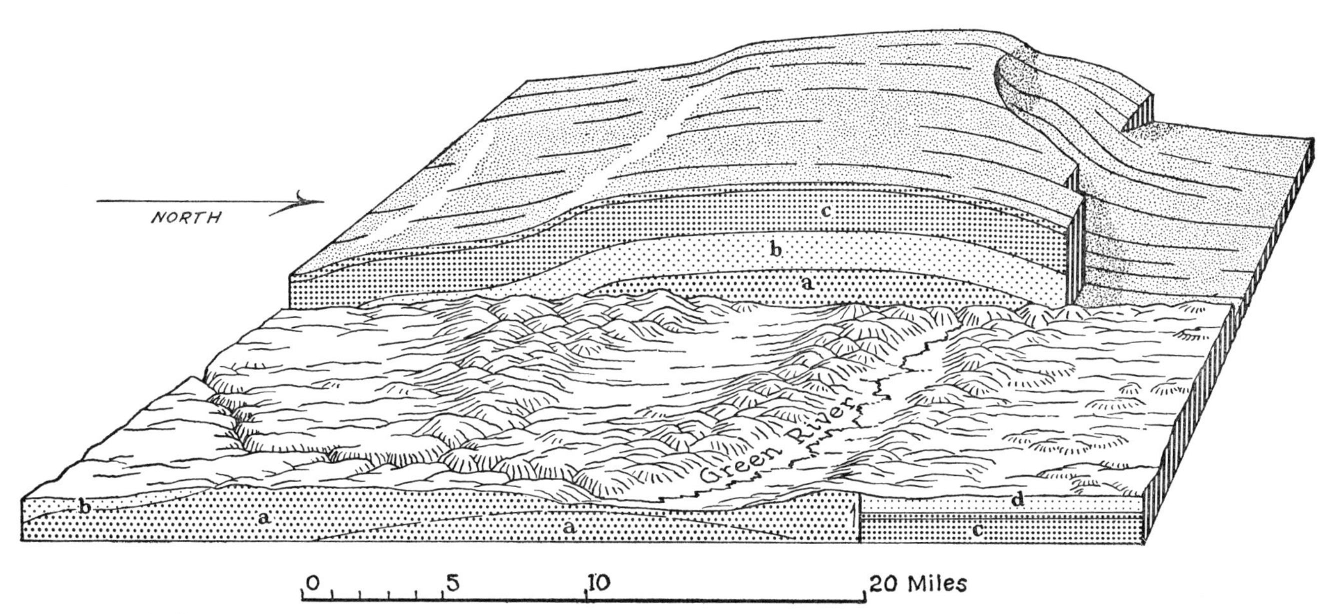

Fig. 64. Block diagram showing an uplift typical of those in the Central Rocky Mountains; eastern part of Uinta Mountains in northwestern Colorado and northeastern Utah. Rear half of block shows structural form of uplift without erosion; the front half, the present landscape. (a) Precambrian (Uinta Mountain group). (b) Paleozoic. (c) Mesozoic. (d) Tertiary. After Powell, 1876.

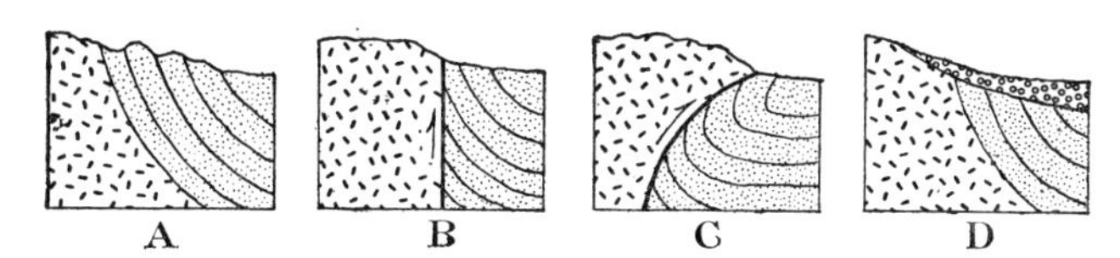

Fig. 65. Sketch sections illustrating four different sorts of structural features which occur along the edges of uplifts in the Central Rocky Mountains, as explained in text.

(B) In other places the sedimentary rocks are cut off by nearly vertical faults, along which Precambrian rocks of the core have been upthrust so that they adjoin directly the Cretaceous and early Tertiary rocks of the basins (Fig. 65 B).

(C) In a few places the faults dip under the mountains, and the Precambrian rocks of the core have been thrust over the rocks of the basins (Fig. 65 C). These thrusts differ from the lengthy low-angle thrusts in geosynclinal sedimentary rocks, as they are caused by very local bulging out of the uplifts as they arose.

(D) In still other places (notably along the southwest edge of the Wind River Mountains) the turned up rocks along the edges of the uplifts are masked by Eocene and later deposits younger than the uplifts; these overlap and overstep the older rocks from the basins toward the mountains (Fig. 65 D).

The basins and their structure. Between the ranges are the basins, which cover a greater area (Fig. 66). Between the Black Hills and Bighorn Mountains is the *Powder River Basin.* Enclosed within the arc of the Bighorn Mountains is the *Bighorn Basin.* Southwest of the arc between it and the Wind River Mountains is the *Wind River Basin.* Between the latter range and the northern prongs of the Southern Rocky Mountains is the very broad *Green River Basin.* Some smaller basins are enclosed between the ranges elsewhere, of which we need mention only the *Hanna Basin* on the southeast.

The basins have been more passive features than the ranges, and remained relatively stable while the latter were uplifted between them. Most of their surfaces are covered by Eocene formations which were laid down after the climax of the Laramide orogeny, hence have been little folded. But drilling indicates that in the central parts of the basins the older rocks—

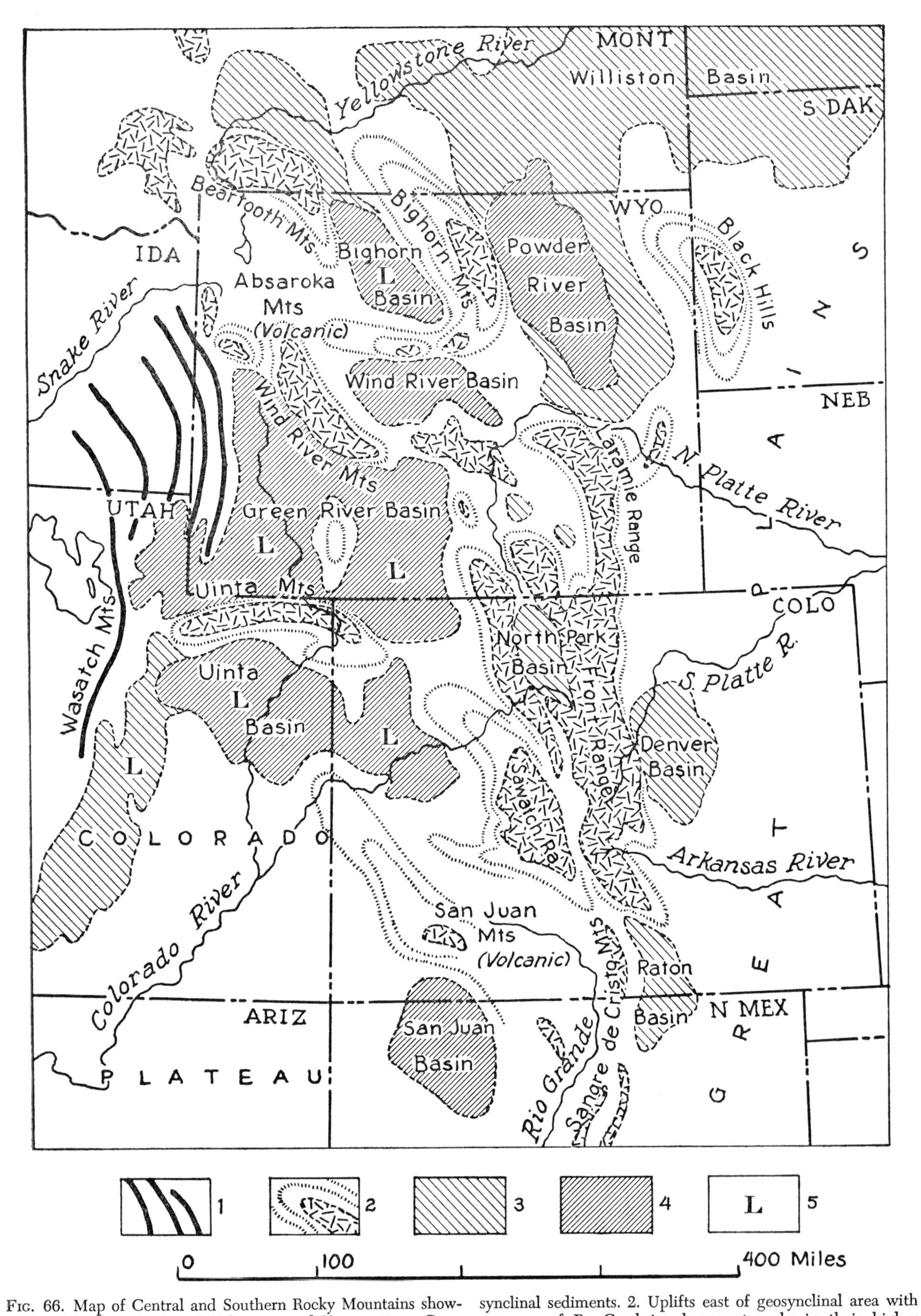

FIG. 66. Map of Central and Southern Rocky Mountains showing uplifts and basins of Paleocene and Eocene time. Compiled from Geologic Map of North America, 1946; and other sources; after King, 1958.

Explanation of symbols. 1. Folds and fault-blocks in geosynclinal sediments. 2. Uplifts east of geosynclinal area with outcrops of Pre-Cambrian basement rocks in their higher parts. 3. Basins which received Paleocene sediments. 4. Basins in which Eocene sediments were deposited over Paleocene sediments. 5. Areas of lake deposits, mainly of Eocene age.

Paleozoic, Mesozoic, and Paleocene—are less disturbed than in the uplifts.

Nevertheless, in many places along the edges of the basins close to the mountain uplifts the pre-Eocene rocks are folded into anticlines. These have served as traps for oil, and are the sites of many of the oil fields in this part of the Rocky Mountain region.

The Eocene deposits which form the surfaces of the basins are all of continental origin. The seas which had spread along the east side of the Cordillera in Cretaceous time had by then disappeared as a result of vicissitudes of the Laramide orogeny. This orogeny also for a time impeded the exterior drainage so that the deposits were laid down in enclosed basins. The Eocene sequence begins with the Wasatch formation, a reddish, sandy stream deposit which becomes coarse and bouldery where it overlaps the edges of the mountains. It is succeeded in many of the basins by deposits of a series of great lakes (Fig. 66)—well-bedded clays, silts, and sands with much oil shale—typified by the Green River formation of contiguous parts of Wyoming, Colorado, and Utah.

VOLCANIC AREAS. The Central Rocky Mountains are not a notable volcanic region. Wide areas lack any igneous rocks except those in the projecting Precambrian basement, but there are some widely separated areas of volcanic rocks and shallow intrusives which contrast with the dominant terrane of sedimentary and basement rocks. The volcanic areas are of several sorts, have had varied histories, and show different degrees of destruction by erosion.

An older center forms the *Crazy Mountains* of southwestern Montana (Fig. 67). These mountains are made by the Livingston formation, a mass of andesitic tuff derived from volcanic eruptions somewhere nearby, which was deposited during latest Cretaceous and Paleocene time and later downwarped gently into a basin. The relatively soft rocks of the Livingston formation project as bold mountains, because they are held together by several intrusive stocks and by great swarms of radiating dikes. Stocks and dikes were intruded during a second period of volcanism in early Tertiary time when the stocks were conduits that fed volcanoes and lava flows at a higher level, all traces of which have now been removed by erosion (section A A′, Fig. 67). Apparently no volcanic activity occurred in the Crazy Mountains area after early Tertiary time.

A more complex and persistent volcanic area occurs in the *Absaroka Mountains* and *Yellowstone National Park* between the Beartooth and Wind River uplifts in northwestern Wyoming, where flows, breccias, and tuffs were piled to a thickness of many thousands of feet and have not yet been removed by erosion; the area thus stands now as a high, deeply dissected lava plateau. Here volcanic eruptions began in late Eocene time toward the close of the sedimentary filling of the adjacent Bighorn and Wind River Basins. Eruptions continued at intervals through the Tertiary; the many hot springs and geysers in the National Park indicate that the volcanic fires have not yet cooled.

IMPINGEMENT OF GEOSYNCLINAL FOLDS ON WEST SIDE OF CENTRAL ROCKY MOUNTAINS. A final item completes our outline of the structure of the Central Rocky Mountains.

The western part of the area which is defined geographically as the Central Rocky Mountains is a set of ranges of very different character from those which we have considered; these are the *Wyomide Ranges* of westernmost Wyoming and southeastern Idaho. They consist of miogeosynclinal Paleozoic and Mesozoic sediments which have been thrown into closely packed folds and thrust slices without exposing any Precambrian basement. Their structures trend for long distances in the same direction, running northwestward into Idaho and Montana and southward into the Wasatch and other ranges of Utah.

The Wyomide Ranges contrast with those to the east with their erratic trends, broad backs, and Precambrian cores. In many places the latter strike westward toward the miogeosynclinal structures and seem to be overridden by them. In northwestern Wyoming the miogeosynclinal structures are thus deflected around the Gros Ventre and Teton Mountains at the northwest end of the Wind River trend, and in Utah the Wasatch Mountains meet the westward trend of the Uinta Mountains nearly at right angles.

Contrasts between the structures of the Central Rocky Mountains and the Wyomide Ranges on the west result from deformation of unlike rock masses. In the Central Rocky Mountains the sedimentary cover is rather thin with basement rocks fairly near the surface. The area was a foreland, more resistant to deformation than the miogeosynclinal area where the basement is thickly covered by sediments.

ECONOMIC PRODUCTS. General simplicity of the structure in the Central Rocky Mountains has influenced its economic products, inhibiting extensive metallic mineral deposits and enhancing the quantity of its mineral fuels.

Mineralized areas are few except in the uplifted bodies of Precambrian rocks. The Precambrian in some of the ranges of Wyoming contain beds of *iron ore* like those in the Huronian of the Lake Superior Region. In the northern Black Hills the Precambrian also contains the great ore body of the Homestake Mine, the *largest gold producer* in North America. General opinion is that this body was introduced into

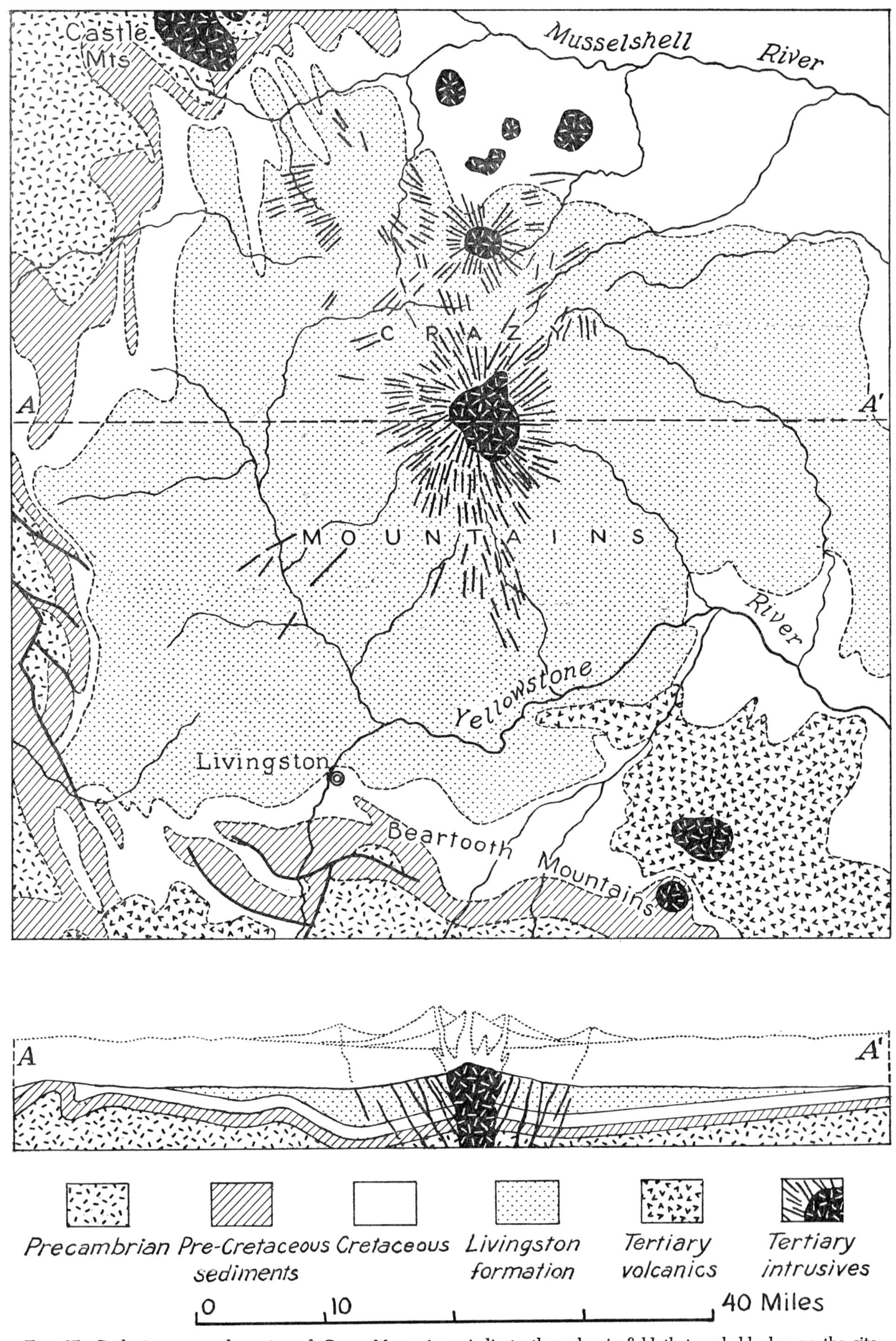

FIG. 67. Geologic map and section of Crazy Mountains, southwestern Montana showing Livingston formation of latest Cretaceous and Paleocene age, and the stocks and dikes which intrude it. Dotted lines above the section indicate the volcanic field that probably lay on the site of the Crazy Mountains in Tertiary time. After Weed, 1899; and Geologic Map of Montana, 1955.

the Precambrian rocks later on, perhaps during Tertiary time, when igneous rocks were intruded nearby as plugs and laccoliths.

By contrast, mineral fuels in the prevailing sedimentary rocks are widespread.

The Laramie and Paleocene continental deposits contain may beds of coal, mostly low-quality lignite but attaining bituminous rank where deformation was greatest. The Hanna Basin of southeastern Wyoming with thick coal-bearing deposits of these ages is fortunately placed along the line of the Union Pacific Railroad, and was a source of fuel for its locomotives before the days of the diesels.

Eocene lake beds such as the Green River formation, which overlie the Paleocene in the western intermontane basins, contain thick beds of *oil shale.* Technology for its extraction has been studied for many years but seems not to be economical now. Some day when our liquid petroleum supplies are further depleted and our automobiles are grinding to a halt, yet with uranium-powered family vehicles still a thing of the future, these great oil shale deposits may well come into their own as a source of petroleum.

The Central Rocky Mountains of Wyoming and Montana are also more prolific and widespread producers of *oil and gas* than the more complex and closely crowded ranges of the Northern and Southern Rocky Mountains. Many of the oil fields of the region occur in structural traps formed by the little anticlines around the edges of the major uplifts.

We will see presently that in the Southern Rocky Mountains the proportion between the values of metallic and non-metallic mineral deposits is reversed because of a somewhat different structure in the two regions.

5. SOUTHERN ROCKY MOUNTAINS

We have learned something of the characteristic style of Rocky Mountain structure from consideration of the relatively simple Central Rocky Mountains. Now let us turn to the Southern Rocky Mountains where many of the same kinds of structures occur, but with additional features which add to the complexity of the picture.

EXTENT OF SOUTHERN ROCKY MOUNTAINS. The main body of the Southern Rocky Mountains is in Colorado, where it forms a wide band across the central part of the state. Prongs extend northward into Wyoming and die out in a particularly extensive area of plains and basins. Another prong, the Uinta Mountains, extends westward into Utah and abuts against the Wasatch Mountains. Still others extend southward into New Mexico, one of which ends east of Santa Fe at Glorieta Pass, through which the Santa Fe Railroad has been built.

South of Santa Fe and Glorieta Pass shorter and more widely spaced ranges extend on each side of the Rio Grande to El Paso, Texas, but these are not truly part of the Rocky Mountains, as we will discover later in this chapter (Section 6).

COMPARISON OF CENTRAL AND SOUTHERN ROCKY MOUNTAINS. General features of the Southern Rocky Mountains may be seen on the geologic maps of Colorado, Wyoming, and New Mexico. Using the maps alone let us observe resemblances between the Central and Southern Rocky Mountains:

(1) Both sets of mountains are bordered on the east by the Great Plains or western edge of the Interior Lowlands.

(2) In both areas Precambrian rocks are extensively exposed along the cores of the mountain uplifts.

(3) Broad basins of Paleocene and Eocene rocks occur alongside both sets of ranges.

(4) Both regions contain dispersed areas of volcanic rocks and shallow intrusives. In the Southern Rocky Mountains these are best displayed in contiguous parts of Colorado and New Mexico where plugs and attendant dike swarms occur in the east, and dissected volcanic plateaus in the west (Fig. 71).

But let us note some obvious differences:

(A) The Southern Rocky Mountains are bordered on the west not by geosynclinal folds of the main Cordillera but by the Colorado Plateau.

(B) Ranges of the Southern Rocky Mountains are not widely dispersed; the whole mountain belt from Great Plains to Colorado Plateau is narrower than farther north.

(C) Individual ranges are more crowded, too, as shown by the close spacing of areas of Precambrian rocks which emerge along their cores. Crowding of the ranges is greatest at about mid-length in the latitude of Denver (Fig. 69).

(D) Basins between the ranges are correspondingly narrower than those farther north, and form high mountain valleys or "parks" rather than broad tracts of plains country.

(E) Large bodies of intrusive igneous rocks occur in the Southern Rocky Mountains, unlike any in the Central Rocky Mountains. The map legends call them "diorite porphyry," "quart monzonite porphyry," and "granite porphyry," suggesting a fairly deep-seated or plutonic habitat.

Other differences not evident from the maps alone will be apparent as the discussion proceeds.

RANGES ON THE NORTH. Let us observe the major physical features of the Southern Rocky Mountains in the same manner as we have those of the Central Rocky Mountains:

In the north half of Colorado and northward into Wyoming the Great Plains are faced on the west by the ramparts of the *Front Range* (Fig. 69), a broad-backed uplift along whose crest erosion has laid bare wide areas of Precambrian rocks; these project in such summits as Pikes Peak and Longs Peak, both named for leaders of early exploring expeditions. Southward near Canon City the massif of the Front Range plunges beneath the plains, although a small prong, the *Wet Mountains*, extends a little farther. Northward in Wyoming the main ridge forms the *Laramie Range*, but a western branch extends into the *Medicine Bow Mountains*. The Front Range owes its present form largely to uplift during the Laramide orogeny, but its position corresponds nearly, though not exactly, to the Front Range geanticline of the later Paleozoic Colorado system.

West of the Front Range in central Colorado is the shorter but somewhat higher *Sawatch Range*, also a broad uplift which lays bare wide areas of Precambrian rocks, one of whose peaks, Mount Elbert, is the highest summit in the Rocky Mountains. The Precambrian of this range is penetrated by extensive bodies of much later plutonic rocks such as the great stock of Mount Princeton at its south end.

Northward the Sawatch structure fans out (Fig. 69). A branch to the north forms the *Park Range*, which continues into southern Wyoming before plunging beneath younger rocks. A branch to the northwest forms the *White River Plateau*, also a broad-backed uplift but still largely sheeted over by Paleozoic rocks. It is trenched by the gorge of the Colorado (formerly "Grand") River above Glenwood Springs, traversed by the Denver and Rio Grande Railroad, where the arched structure of the uplift is wonderfully exposed.

This branch descends into a sag northwest of the White River Plateau, beyond which it rises again into the *Uinta Mountains*. These extend westward into Utah, approaching the Wasatch Mountains of the main Cordillera nearly at right angles. Like many other ranges of the Central and Southern Rocky Mountains, the Uinta Mountains are a broad-backed uplift in which Precambrian rocks are exposed along the crest (Fig. 64), but here the Precambrian is not crystalline basement and is the Uinta Mountain group, a thick mass of sedimentary rocks.

RANGES ON THE SOUTH. Near the latitude of Canon City the *Sangre de Cristo Mountains* develop behind the Front Range. Southward where the Front Range and Wet Mountains plunge beneath the Great Plains, the Sangre de Cristo Mountains form the frontal ridge and so continue into New Mexico. The Sangre de Cristo Mountains are not a broad-backed uplift like the Front and Sawatch Ranges (Fig. 70 B). Precambrian rocks emerge in places, but much of the range is formed of heavily folded late Paleozoic sediments that were derived from the Uncompahgre geanticline of the Colorado system, and were laid down in the trough east of it.

In front of the Sangre de Cristo Mountains stand the *Spanish Peaks*, whose structure resembles that of the Crazy Mountains of Montana (Fig. 71)—a basin of early Tertiary sediments intruded by central stocks, or volcanic necks, and by a remarkable swarm of radiating dikes.

West of the Sangre de Cristo Mountains are the *San Juan Mountains*, another volcanic center (Fig. 71). As in the Absaroka Mountains of northwestern Wyoming, volcanic activity was long-persistent here, beginning in the Paleocene, with a great climax in the Miocene, and with lesser volcanic activity in the Pliocene and Pleistocene. Its mass of eruptive rocks has been deeply dissected into a mountainous plateau.

The volcanics of the San Juan Mountains have been built on an eroded surface of disturbed earlier rocks, part of which belonged to the Uncompahgre geanticline of Paleozoic time. On the southwest side of the volcanic area Precambrian rocks project in the high peaks of the *Needle Mountains*. Northwest of the volcanic area the *Uncompahgre Plateau* extends into the Colorado Plateau past Grand Junction, Colorado. Like the White River Plateau, the Uncompahgre Plateau is largely sheeted over by sediments, here of Triassic age. Where these are breached by erosion the truncated earlier structure of the Uncompahgre geanticline is revealed with Precambrian beneath the Triassic along the crest and Paleozoic rocks intervening on the flanks (Fig. 68).

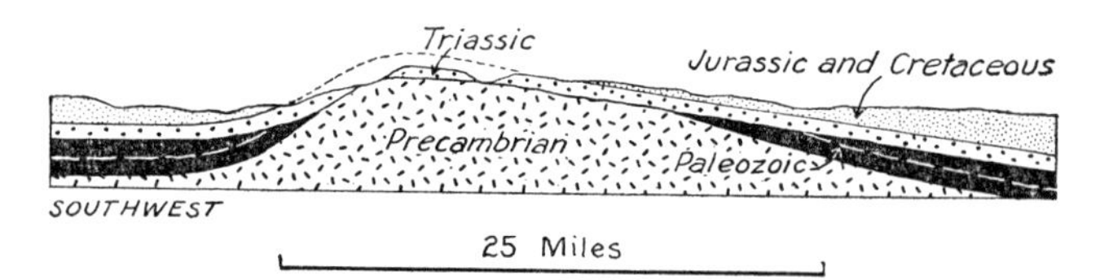

FIG. 68. Sketch section across Uncompahgre Plateau southwest of Grand Junction, Colorado showing truncation of Paleozoic rocks by Triassic along the site of the Uncompahgre geanticline of later Paleozoic time, and the arching of both Paleozoic and Mesozoic rocks by Laramide deformation.

BASINS OF SOUTHERN ROCKY MOUNTAINS. Basins of the Southern Rocky Mountains are most extensive east and west of the ranges.

East of the Front Range is the broad *Denver Basin*, thickly filled by Laramie and Paleocene deposits (Fig. 69). It became nearly inactive after Paleocene time,

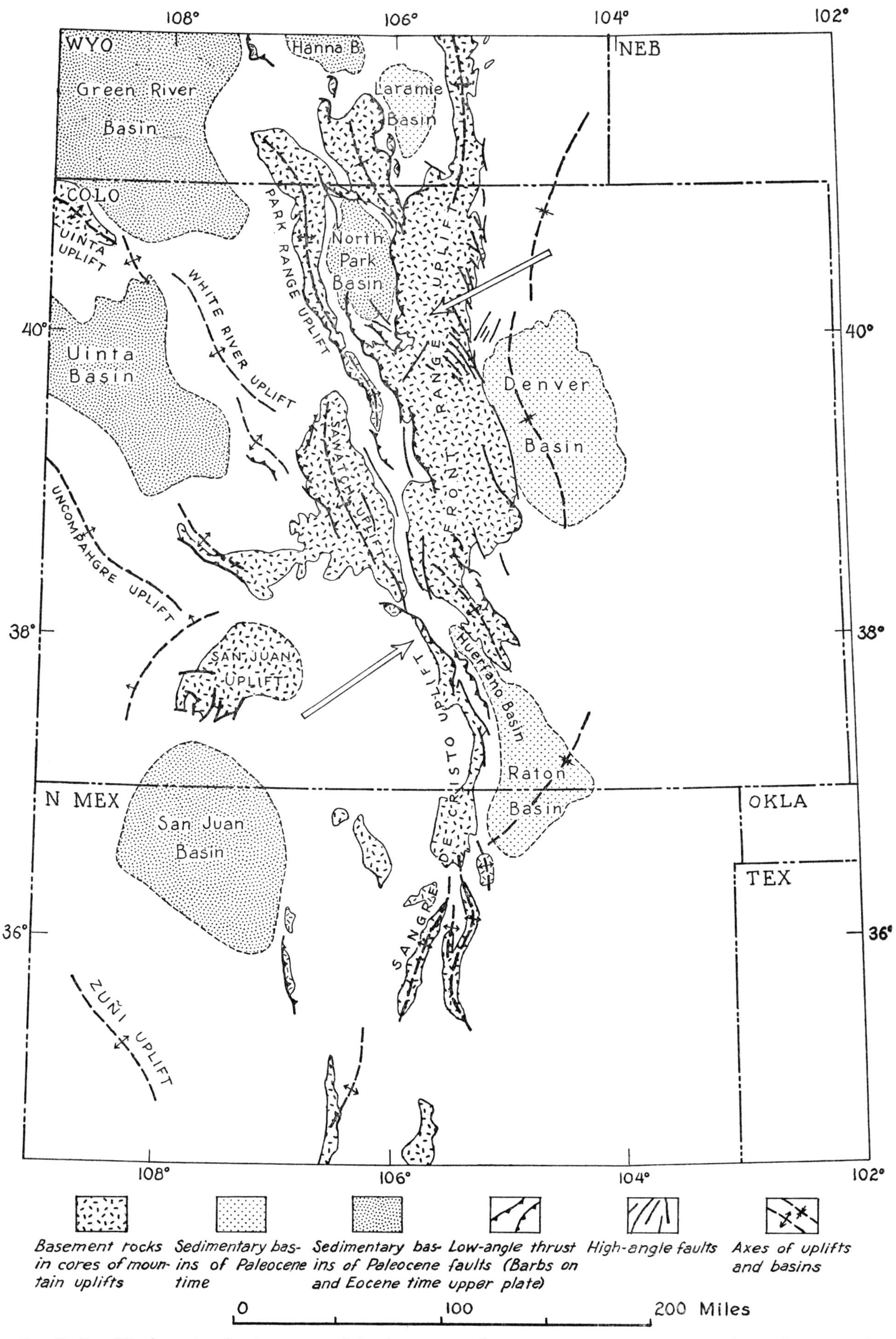

Fig. 69. Part III of a series showing structural development of Southern Rocky Mountains. Map showing structures formed by Laramide orogeny in late Cretaceous and early Tertiary time. Large arrows indicate directions of dominant thrusting in the north and south. After Burbank and Lovering, 1933; and other sources.

received no Eocene deposits, and is now partly masked by late Tertiary Great Plains deposits. Farther south in the angle between the Front Range and Sangre de Cristo Mountains is the similar but narrower *Huerfano Basin* which contains the stocks and dikes of the Spanish Peaks.

Along the edge of the Colorado Plateau west of the Southern Rocky Mountains are the broad *Uinta* and *San Juan Basins*, the first lying south of the Uinta Mountains, the second south of the San Juan Mountains. Their rocks and structures resemble those of the intermontane basins of the Central Rocky Mountains, as they contain Eocene as well as Laramie and Paleocene deposits; the Eocene of the Uinta Basin includes lake beds like those farther north.

Basins between the ranges form a chain of high mountain valleys or "parks" filled by early Tertiary sediments. *North* and *Middle Parks* west of the Front Range are parts of a single basin; *South Park* is a separate structure along the same trend. South of the "parks" is the *San Luis Valley* between the Sangre de Cristo and San Juan Mountains at the headwaters of the Rio Grande (Fig. 71). It differs from the basins so far described in that it is filled by Pliocene rather than earlier deposits, and has been shaped by block faulting which dropped the east side against the Sangre de Cristo Mountains and tilted the volcanic rocks of the San Juan Mountains beneath it. Similar block-faulted depressions of about the same age extend southward along the Rio Grande across New Mexico.

INFLUENCE OF EARLIER STRUCTURES ON PATTERN OF SOUTHERN ROCKY MOUNTAINS. So much for the "geological geography" of the Southern Rocky Mountains. Now let us analyse their structures more critically (Fig. 69).

In the Central Rocky Mountains the Laramide structures appear to have been newly born without deformational antecedents, but in the Southern Rocky Mountains they were much influenced by earlier features. Effects of the Precambrian structures cannot be evaluated, but it is worth recalling that their dominant trend is northeastward across the Laramide structures rather than parallel with them as farther north. Structures of the Colorado system of later Paleozoic time are more nearly comparable with those which developed later. Recall some of the details of this system: the Front Range geanticline which lies nearly on the site of the later Front Range, the Uncompahgre geanticline on the site of the later Uncompahgre Plateau and San Juan Mountains, and the deep sedimentary trough between, whose rocks were themselves raised into mountains later (Fig. 57).

LOW-ANGLE FAULTS. The rocks of the Southern Rocky Mountains are broken by low-angle faults as they are in places in the Central Rocky Mountains, but they are of more than local extent.

The eastern border of the Front Range is much like the borders of the uplifts in the Central Rocky Mountains. Paleozoic and Mesozoic strata are sharply upturned against the lofty Precambrian rocks of the range, and in places they are cut off by vertical upthrusts or high-angle thrusts so that the Precambrian lies directly against the strata of the Great Plains. But on the west side of the Front Range for much of its length Precambrian rocks are thrust westward on low-angle faults for as much as four or five miles over Cretaceous rocks. (Fig. 70 A). Similar low to high-angle faults occur in the Sawatch and Park Ranges beyond, again with the Precambrian thrust toward the west.

Farther south, by contrast, the rocks of the Sangre de Cristo and related ranges are thrust toward the east. In Huerfano Park on the east flank of the Sangre de Cristo Mountains complex thrusts in this direction have developed in the great mass of late Paleozoic clastic rocks, although these have been much confused by later upthrusts of the Precambrian basement (Fig. 70 B).

These varied directions of thrust during the Laramide orogeny appear to have been influenced by the Paleozoic structures of the Colorado system on which they were imposed. The westward thrust of the ranges of northern Colorado was from the site of the Front Range geanticline of Paleozoic time toward the trough that lay west of it; the eastward thrust of southern Colorado was from the site of the Uncompahgre geanticline of Paleozoic time toward the trough that lay east of it.

TRANSVERSE ZONE OF CENTRAL COLORADO. These structural peculiarities appear to be related in turn to another feature, a transverse zone which extends diagonally northeastward across the Southern Rocky Mountains (Fig. 71).

The transverse zone is most prominently expressed on the map by distribution of the larger bodies of intrusive igneous rocks. They begin on the southwest in the western part of the San Juan Mountains, extend through the Sawatch Range, and continue thence across the Front Range to the Great Plains north of Denver.

The intrusive rocks are diorite, quartz monzonite, and granite porphyries, and are of plutonic habit, unlike the shallow stocks and laccoliths that occur elsewhere in the region. Some have been dated radiometrically at about sixty million years, hence are vastly younger than the 1,100 million year old Pikes Peak granite of the Front Range—younger even than

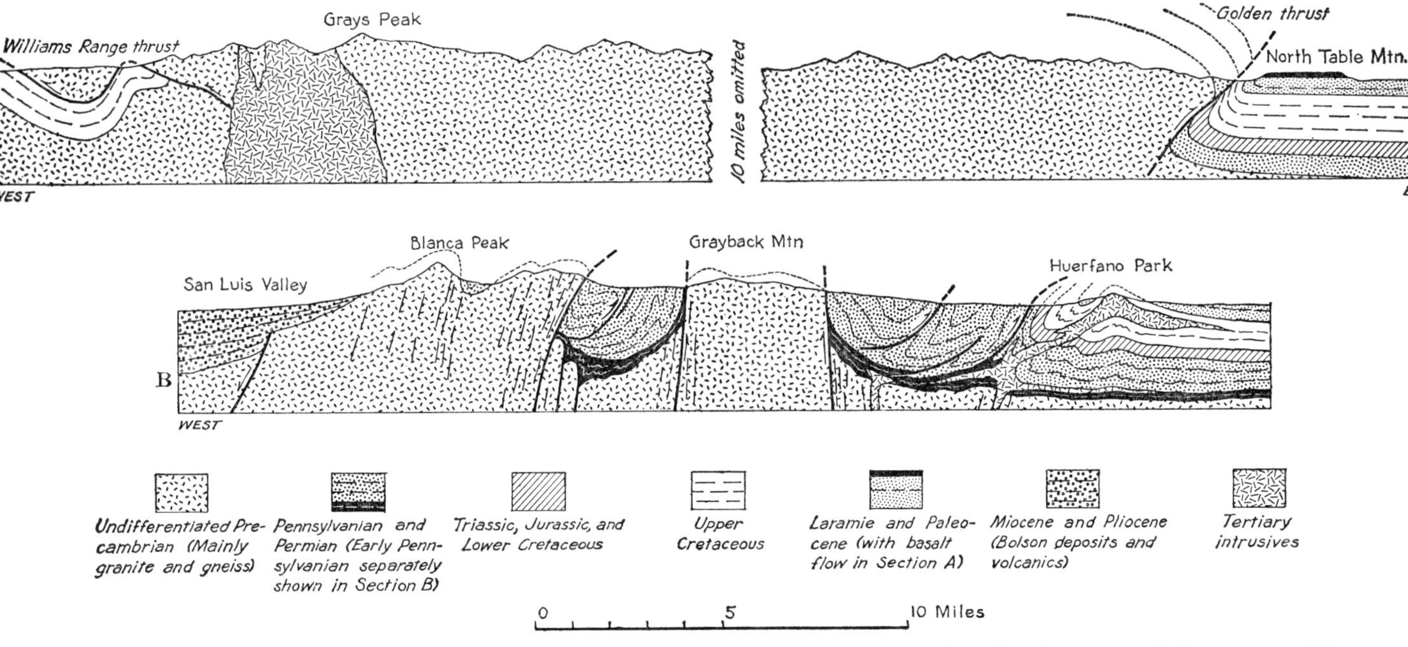

Fig. 70. Sections which compare the structure of the Front Range with that of the Sangre de Cristo Mountains in the north and south parts of the Rocky Mountains of Colorado. (A) Section of Front Range in latitude of Denver. (B) Section of Sangre de Cristo Mountains in latitude of Walsenburg. After Lovering, 1935; Van Tuyl and McLaren, 1933; and Burbank and Goddard, 1937.

the 400 to 185 million year old plutonic rocks of the Appalachians. They are probably of Paleocene or Eocene age, and were injected during the later stages of the Laramide orogeny.

Note that the two areas of thrusting in the Southern Rocky Mountains, one westward and the other eastward, are not quite opposite each other (Fig. 69). When compression and thrusting took place in Laramide time a rotational couple was set up, resulting in the part of the transverse zone in the Front Range in many minor faults with left-lateral strike-slip displacement. The lateral movements in the transverse zone, and the tensions thereby created, evidently produced conduits for the ascent of magmas from below.

Mineral deposits. The Southern Rocky Mountains are noted for their metallic mineral wealth which for many years made Colorado one of the great mining states of the west. Curiously, first discovery of gold (or of any other metal) in Colorado was in the alluvial deposits of Cherry Creek, a stream which heads in the Great Plains and enters the South Platte River near the present city of Denver. It was only after prospectors had flocked to Colorado on the strength of this find that the vastly richer deposits in the bedrock of the mountains to the west became known.

As discoveries followed each other through the years the pattern of the bedrock deposits became clearer. Almost all lie on or near the transverse zone just discussed (Fig. 71), as though its structures were such as to permit the rise of mineralizing solutions as well as of igneous magmas. Few important deposits lie north or south of the zone, the only exception being at Cripple Creek far to the south in the Front Range near Pikes Peak, where the throat of a former caldera has been mineralized.

That part of the transverse zone in the Front Range is known as the *mineral belt*, and includes such great mining camps as Central City and Georgetown. Farther southwest is Leadville between the Front and Sawatch Ranges; beyond are the camps of Ouray, Telluride, and Silverton in the western San Juan Mountains.

Unlike some other productive regions the Colorado mineral belt contains ores of a great variety of metals. The first attraction, of course, was the *gold* deposits, but in 1879 lead-carbonate silver ores were discovered at Leadville, and thereafter *silver* was extensively mined as well. It was only later that the value of the baser metals such as *lead, copper*, and *zinc* began to be appreciated, and it was found that much of the waste discarded in dumps and tailings was almost as valuable as the material that had been saved. Among later discoveries were the *molybdenum* deposit at Climax near Leadville, probably the richest occurrence of this metal in North America.

In its time, mining activity had a dominant influence in the shaping of the history of Colorado, but (sad to relate) it is now on the wane. Many of the large low-grade metallic deposits in which reserves

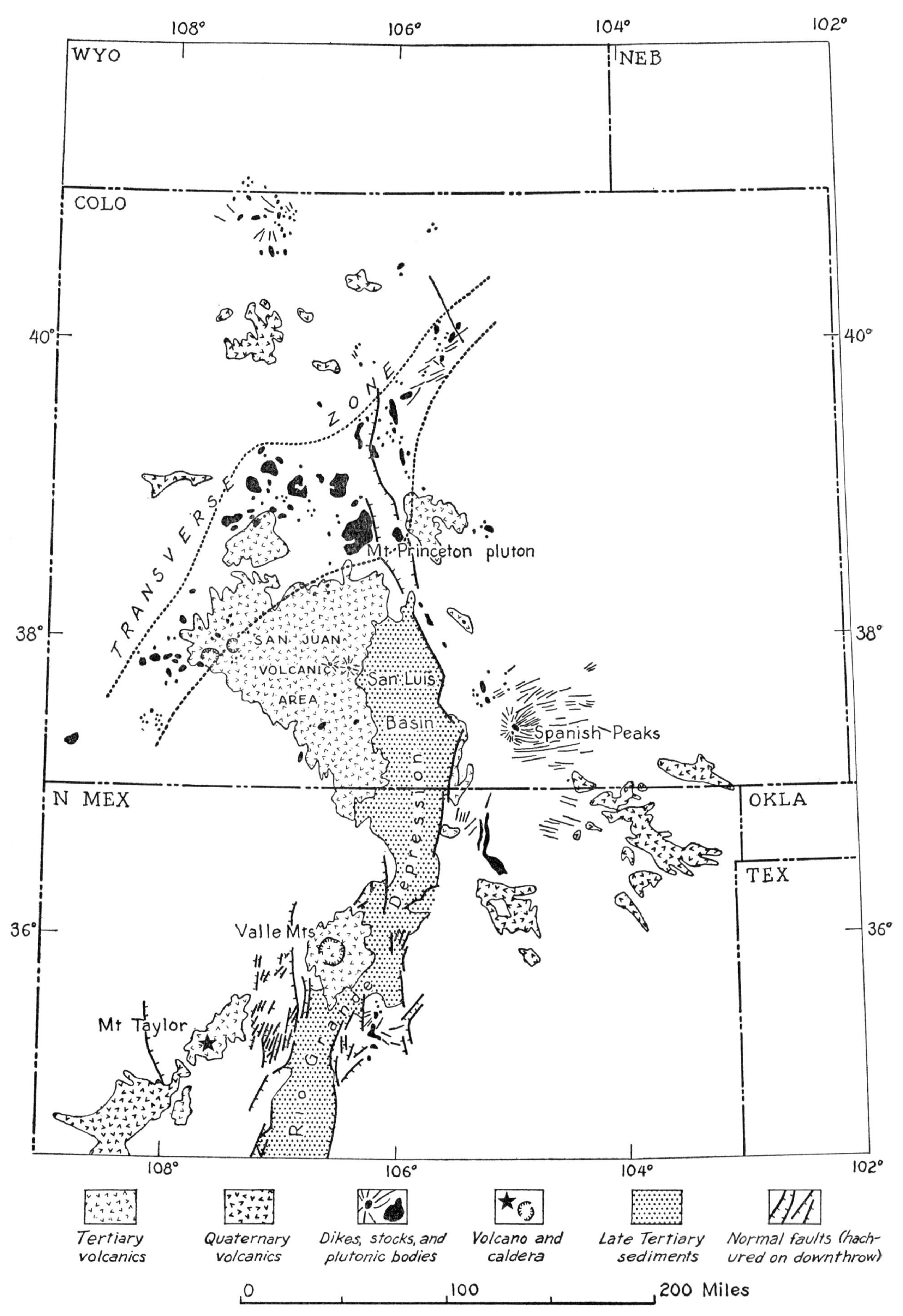

Fig. 71. Part IV of a series showing structural development of Southern Rocky Mountains. Tertiary and Quaternary structures, igneous rocks, and deposits. Some of the features shown are nearly contemporaneous with those of Fig. 69; others are much younger. After Burbank and Lovering, 1933; and other sources.

can be blocked out for years ahead are still being actively worked, but the days of the bonanza discovery and the small rich mine appear to be in the past. Mining camps picturesquely set in gorges far back in the mountains are now moribund; the narrow-gauge railroads with which they were once connected with the world have been dismantled. Colorado has learned to prosper on its less glamorous resources of agriculture, timber, and mineral fuels, and its superb mountain environment which each year draws increasing numbers of visitors.

6. LESS DEFORMED PARTS OF EASTERN RANGES AND PLATEAUS: NEW MEXICO RANGES AND COLORADO PLATEAU

We pass now to the western and southern parts of the Eastern Ranges and Plateaus which have been less deformed, on the whole, than the Central and Southern Rocky Mountains. Much of this region is table-land and constitutes the *Colorado Plateau*; on the southeast in New Mexico it is broken into a succession of block mountains which we can designate as the *New Mexico Ranges.*

New Mexico Ranges. Prongs of the Rocky Mountains extend into New Mexico but they die out southward; the Sangre de Cristo Mountains or easternmost prong ends east of Santa Fe at Glorieta Pass. Farther south shorter, lower, more dispersed ranges continue across New Mexico into western Texas where they merge with the Sierra Madre Oriental of Mexico. These ranges have little of the characteristic style of the Rocky Mountains and are not truly extensions of them.

Quite as significant as the ranges in New Mexico are the intervening basins, part of which are traversed by the Rio Grande in its course southward from the Rocky Mountains toward the Gulf of Mexico (Fig. 72 D). The river breaks out of the southernmost basins through a spectacular set of gorges in the Big Bend country of Texas.

We learned earlier that New Mexico was part of the continental backbone of Paleozoic time, and that over wide areas later Paleozoic strata directly overlie Precambrian basement. Compared with the sequence of Paleozoic and Mesozoic rocks in the Central and Southern Rocky Mountains, that in the New Mexico Ranges is relatively thin. Traces of Laramide orogeny with development of thrusts, uplifts, and basins may be observed in various parts of the area (Fig. 72 A, B, and C), but disturbances of this time affected it only lightly.

The modern topography of New Mexico, instead of being inherited from the Laramide orogeny, is largely a product of *block faulting* in late Tertiary time and afterward (Fig. 72)—one of the processes that modified the surface aspect of the Cordillera after the great deformations of Mesozoic time. The mountains were raised or tilted by movements along the faults which outline their bases, and the basins have been depressed and filled by great thicknesses of Pliocene and younger deposits derived from the wasting of the mountains.

We have already noted similar block faulting younger than the Laramide deformation in local areas farther north—in the Teton Mountains of Wyoming and the San Luis Valley of Colorado. We will encounter even more extensive block faulting of about the same age in the Great Basin of Utah, Nevada, and adjacent states. Features of this sort are termed *Basin and Range structure*; we will analyse them more fully when we discuss the Basin and Range province as a whole (Chapter IX, Section 2).

Geography of Colorado Plateau. The Colorado Plateau or "Plateau province" of Powell lies northwest of the New Mexico Ranges, and is a great, high-standing crustal block, nearly square and 500 miles broad on the sides, whose center is close to the common corner of the states of Colorado, Utah, Arizona, and New Mexico. Westward and southwestward the plateau breaks off in escarpments that overlook the more diversified and broken country of the Basin and Range province—the Great Basin of Utah and Nevada and the similar terrain in Arizona. The escarpments on these sides meet in northwestern Arizona south of the Grand Canyon, where they form a right-angled corner of the plateau. By contrast, the eastern and northeastern sides of the plateau are flanked by more elevated country, the ranges of the Southern Rocky Mountains.

The western and southwestern rims of the Colorado Plateau attain altitudes of 10,000 to 12,000 feet, and form the High Plateaus of Utah and the Mogollon Plateau of Arizona and New Mexico, large parts of which are heavily forested. Most of the remaining area is also lofty, but somewhat less so, and is a thinly inhabited, bare, desert country—the part of the region which ordinarily comes to mind when one hears the phrase "Colorado Plateau."

The Colorado Plateau is a region of plateaus, escarpments, and canyons, all laid out on a vast scale. Except where volcanic piles have been built on its surface or intrusives have disturbed its rocks, there are no mountains in the usual sense, but rather a series of *tables, benches*, and *steps.* Also, the streamways are not the familiar valleys of other regions, but steepsided *canyons*, narrow or wide, shallow or deep. The deep, wide Grand Canyon of the Colorado River

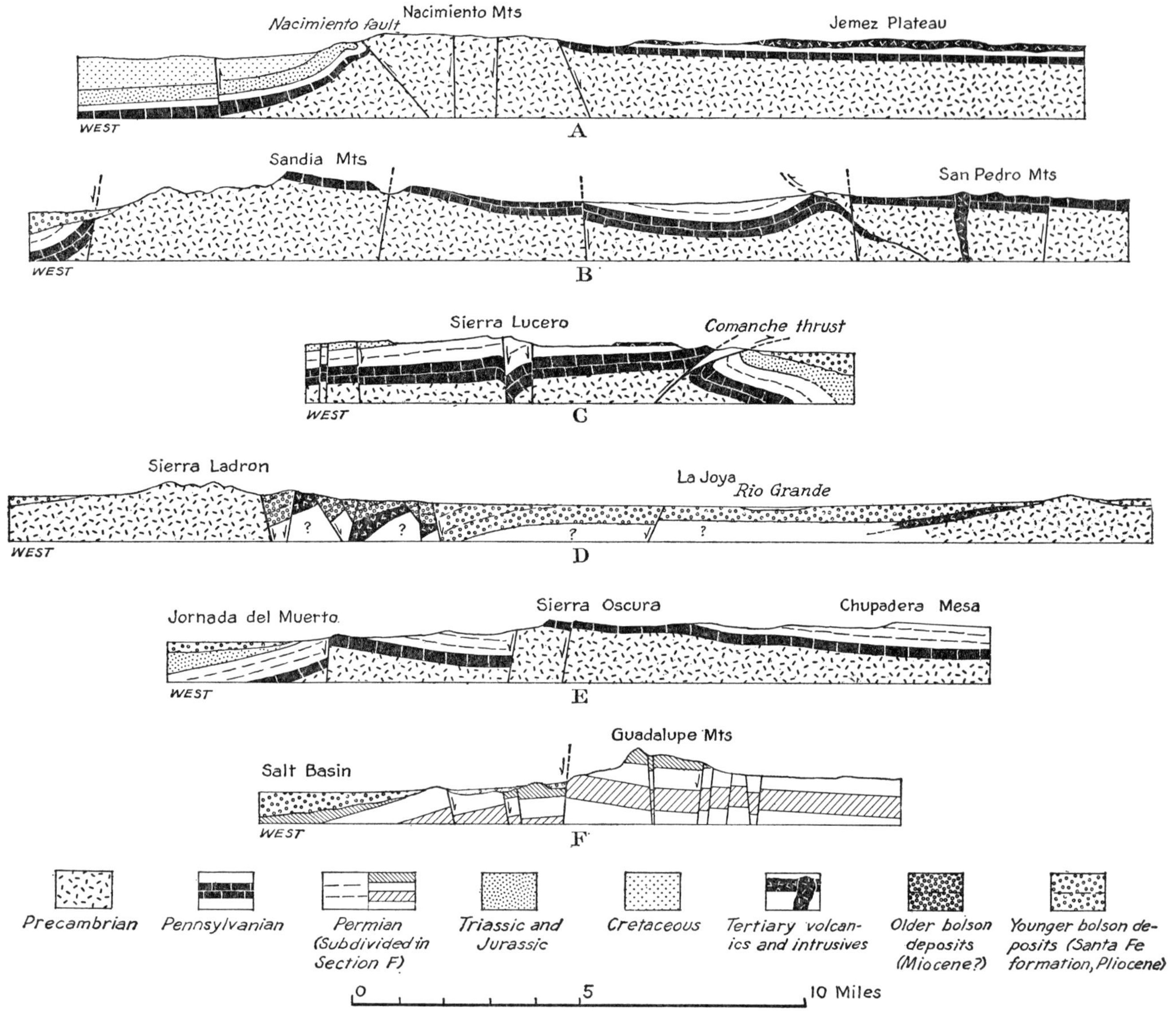

FIG. 72. Sections of the New Mexico Ranges to show their block-faulted structure produced by deformation late in Tertiary time. Section D shows faulting and tilting of later Tertiary sediments and volcanics. In sections A, B, and C are thrust faults older than the block faulting, probably formed during the Laramide orogeny. Note relatively small thickness of sedimentary cover in most of the sections, with basement rocks close to the surface. (A) Nacimiento Mountains, northern New Mexico. (B) Sandia Mountains northeast of Albuquerque, north-central New Mexico. (C) Lucero Mountains southwest of Albuquerque, central New Mexico. (D) Rio Grande depression east of Sierra Ladron, central New Mexico. (E) Sierra Oscura, south-central New Mexico. (F) Guadalupe Mountains, northern trans-Pecos Texas. Compiled from Wood and Northrop, 1946; Read and others, 1944; Kelly and Wood, 1946; Denny, 1940; Wilpolt and Wanek, 1951; and King, 1948.

is familiar to all, but the same river farther upstream in the plateau is canyoned likewise, as are all the tributaries which enter it, each canyon having a distinctive form conditioned by the country rocks and their structures.

Rocks of the plateau have been flexed or folded, but for the most part gently and on a grand scale. Wide areas of nearly flat-lying rocks are separated by abrupt bends in the strata along folds or monoclinal flexures (Fig. 74). Nearly all systems of the geologic column from Paleozoic to Tertiary occur at one place or another in the plateau but are not uniformally distributed. In the Grand Canyon, as we have seen, are many hiatuses or lost intervals in the record; other hiatuses occur elsewhere at the same or different levels.

HISTORIC SETTING. The distinctive features of the Colorado Plateau have been conditioned to a large extent by its prior history.

We have seen that the plateau in early Paleozoic time was as much a part of the stable continental platform as the Interior Lowlands, both of them sharing the overlap of Paleozoic strata from the Cordilleran geosyncline on the west (see Section 2 of this chapter). Later in Paleozoic time the plateau became separated from the rest of the platform by increasing mobility of the Colorado system on the site of the present Southern Rocky Mountains.

In Mesozoic time the plateau area was overspread, first by Triassic redbeds and Jurassic dune deposits, later by Cretaceous clastic wedges related to quickening orogeny in the main Cordillera farther west (see Section 3 of this chapter). Afterwards during the Laramide phase of Cordilleran orogeny, principal crustal activity shifted east of the plateau and deformed the Central and Southern Rocky Mountains. Nevertheless, much of the broad folding and flexing of the rocks in the plateau itself is also a product of Laramide movements, and growth of the Uinta and San Juan Basins on the border between the plateau and Rocky Mountains took place in the waning stages of that orogeny.

Basins and uplifts. The characteristic structure of the plateau is a series of broad uplifts and intervening basins.

To the northeast are the *Uinta* and *San Juan Basins*, thickly filled by Paleocene and Eocene sediments derived from the waste of the Southern Rocky Mountains (Fig. 69). Other shallower basins occur farther southwest, in which early Tertiary deposits, if they ever accumulated, have been removed by erosion. The *Black Mesa Basin* in the Navajo Country of northeastern Arizona is almost as large as the first two; the remainder are smaller.

Uplifts between the basins are not unlike those in the ranges of the Central and Southern Rocky Mountains, but they show less structural relief, so that they are still sheeted over by sedimentary formations. On the uplifts denudation has removed the poorly resistant higher strata down to some resistant stratum such as the Kaibab limestone, so that they are expressed topographically as broad swells or plateaus whose surfaces reflect the warping of the resistant rocks.

One of the most lofty of these is the *Kaibab uplift* which forms the Kaibab and Coconino Plateaus of the Grand Canyon district; it can serve as a type of the rest. It forms a great arch in the strata, one hundred miles long and twenty-five miles wide trending north-northwest, on whose crest the top of the Kaibab limestone rises to heights of 8,000 feet or more. Its original structural surface has been reduced by the great denudation of the plateau, which has removed the poorly resistant Mesozoic formations from its summit while they still remain in surrounding areas. Nevertheless, the surface of the uplift still projects several hundreds or thousands of feet above its surroundings. Other parts of the plateau project higher, it is true, and preserve younger sedimentary strata or volcanic rocks.

The Grand Canyon is "grand," in fact, because of the Kaibab uplift into whose south end it has been cut. Upstream and downstream the Colorado River also flows in canyons, but these are not as deep because the Kaibab limestone stands at a lower level and the formations beneath it are not as deeply penetrated by the river. Upstream from the Grand Canyon, for example, the river flows in the lower but narrower Marble Canyon whose walls are formed of Kaibab and Redwall limestones.

We are thus presented with a paradox. The Kaibab uplift is one of the highest in the Colorado Plateau, yet the main drainage of the region, the Colorado River, has cut its way across it. We will consider this problem in Section 7 of this chapter.

Monoclinal flexures. A striking feature of the uplifts of the Colorado Plateau is the manner in which the strata along their edges are flexed abruptly downward along monoclinal flexures. For the most part this is accomplished by bending of the strata, although locally they are faulted. Flexing and faulting seem to be closely related forms of yielding of the rocks; possibly the flexed sedimentary layers are draped over breaks in the stronger basement rocks beneath.

Along the east side of the Kaibab uplift the Kaibab limestone is bent down abruptly several thousand feet on the *East Kaibab monocline*, whose steeply tilted strata form an escarpment (Fig. 73 A); along the river the descent is from the rim of the high Grand Canyon to the rim of the lower Marble Canyon. On the downflexed side red Triassic formations of the Painted Desert are preserved above the Kaibab limestone.

Some miles farther east the strata are again bent down abruptly along the *Echo Cliffs monocline*, which carries Triassic formations beneath Jurassic sandstones at the edge of the Black Mesa Basin; here erosion has carved the flexed strata into an escarpment which faces the upraised side of the flexure rather than the depressed side (Fig. 73 B).

Elsewhere in the Colorado Plateau the monoclinal flexures also stand out well in the topography as a

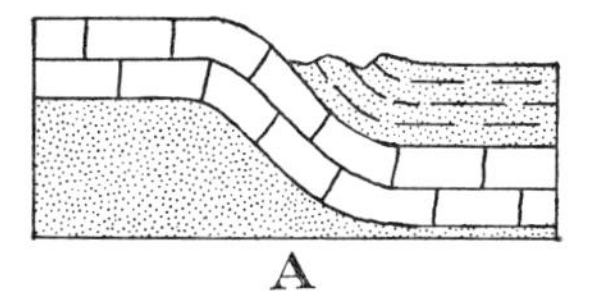

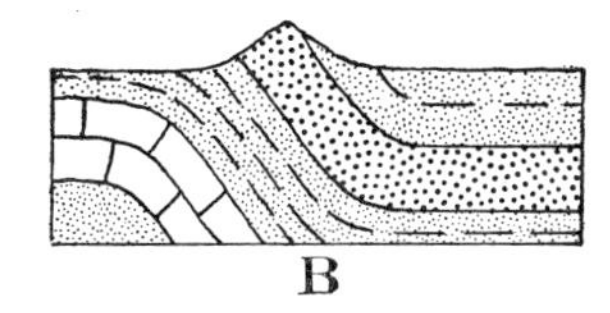

Fig. 73. Diagrammatic sections illustrating topographic expression of monoclinal flexures in Colorado Plateau. (A) With scarp facing downflexed side as in East Kaibab monocline. (B) With scarp facing upflexed side as in Echo Cliffs monocline.

result of etching by erosion of less resistant formations between the more resistant. Topographic expression is diverse, as in the two examples cited, and depends on the nature of the formations which happen to be exposed at any locality.

BLOCK FAULTING IN WESTERN PART OF PLATEAU. Westward from the Kaibab uplift toward the lower end of the Grand Canyon the strata descend also, but in a different manner; they form a set of giant steps, each dropped on the west side by a fault (Fig. 74). Proceeding in this direction from the Kaibab uplift the Colorado River thus crosses the *West Kaibab monocline* (in part faulted), and the *Toroweap, Hurricane*, and *Grand Wash faults*.

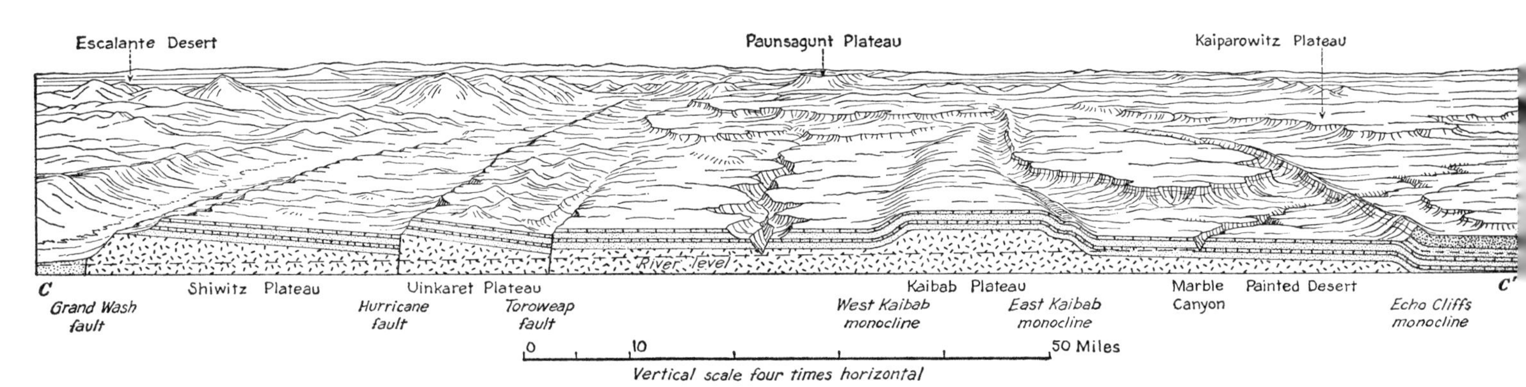

FIG. 74. Block diagram of western part of Colorado Plateau north of Grand Canyon showing monoclinal flexures to east and fault blocks to west. Section on front of block is a little north of Grand Canyon along line C C' of FIG. 59; landscape behind extends across northern Arizona to High Plateaus of Utah. After Powell, 1876.

The faults are followed by escarpments on their upthrown sides that extend far north and south of the river. Northward in Utah the escarpments on the faults intersect and offset the variously colored cliff lines (the Chocolate, Vermillion, White, Gray, and Pink Cliffs of Section 3 of this chapter), which are the products of erosion of tilted strata. Northward in cental Utah where the highest units of the succession are preserved, the faults outline various parts of the High Plateaus.

The fault blocks of the western part of the Colorado Plateau resemble those of the Basin and Range province to the west, which we will consider later (Chapter IX, Section 2). The structure of this western part of the plateau appears, in fact, to be transitional between that of the typical plateau on the east and the Great Basin part of the Basin and Range province on the west.

LACCOLITHS AND SALT STRUCTURES. Two other items complete the roster of significant structures of the Colorado Plateau:

Several clusters of mountain peaks in its central part are carved from intrusive bodies or the sedimentary rocks which were arched above the intrusives. One of the clusters near the Colorado River was sighted by Powell during his exploration of the canyons and named the *Henry Mountains* after his sponsor Joseph Henry of the Smithsonian Institution. Somewhat later Powell's colleague, G. K. Gilbert, determined that the intrusives were "pools of rock" of a hitherto unknown structural type—*laccoliths* that were formed by magmas injected at shallow depth along the bedding planes of the sedimentary rocks, doming the strata above and leaving a horizontal floor beneath. Later work has shown that many of the larger intrusives of these mountains are actually stocks without floors, which were conduits that fed the adjacent laccoliths.

Similar stocks and laccolithic clusters form the La Sal, Abajo, and other groups of mountains in the vicinity. They were intruded rather late in the history of the plateau, all probably at about the same time and in much the same manner. They formed after the gross features had been developed, and they bear slight relation to them.

The *salt structures* were not observed during the early explorations and were only discovered later. The Paradox formation, an evaporite deposit of Pennsylvanian age, underlies part of the east-central plateau southwest of the Uncompahgre uplift. During Laramide deformation the salt and other incompetent layers of the evaporite sequence were wrinkled into anticlines, and intruded or flowed into the overlying rocks along the crests; subsequent collapse and solution greatly increased the complexity of the structures. Two or three of these long narrow anticlines fringe the Uncompahgre Plateau along its southwestern side.

ECONOMIC PRODUCTS. The central part of the Colorado Plateau is bare desert country, probably one of the most empty and thinly inhabited regions of the United States. I recall vividly a plane flight over the region some years ago when, hour after hour, as far as the eye could see, not much was in view of the works of man—at most a sad little road to some out-of-the-way house or field.

Now, to some extent, there is no longer such desolation; a *uranium* boom has changed even this remote corner of the country. Roads have been built into far places, and prospectors have fanned out over the whole region. Indian tribes of the plateau have acquired the same heady feeling as their brothers in Oklahoma of a quarter of a century ago when oil was discovered there; tribal lands are being leased for extraction of uranium which occurs in many of the sedimentary formations. Some years ago the Navajo Indians were reported to be in a sad plight—in possession of a barren land which could not support their increasing population. Now the Navajos and other tribes have acquired sizeable funds from uranium leases, and are investing these in an improvement of their domains.

Despite other discoveries elsewhere the rocks of the central Colorado Plateau remain the chief source of uranium in the United States. The metal occurs in the form of the yellow mineral *carnotite*, a hydrous vanadate of potassium and uranium, which impregnated various sandy sedimentary layers mainly of Mesozoic age; it is mined by following out the strata. Probably the carnotite was not laid down at the time when the sediments were deposited but was introduced later by circulating and mineralizing solutions which sought out and followed the more porous strata.

7. EVENTS AFTER THE GREAT DEFORMATION

Let us now return to the history of the Eastern Ranges and Plateaus which we left at the close of the great deformation of the Cordillera that ended with the Laramide orogeny of late Cretaceous and Paleocene time.

Laramide orogeny rather fully shaped the structures of the region—the folds and faults, the uplifts and basins—although these were accentuated later and other structures were superposed upon them. Compared with the modern landscape of the Eastern Ranges and Plateaus, however, that produced by the climactic orogeny was still primitive. The modern landscape evolved during later phases of Cordilleran history in Tertiary and Quaternary time. Again it is worth emphasizing the distinction between geological and topographic "mountains."

EARLY TERTIARY ENVIRONMENTS. As we indicated in our discussion of the rocks formed during the great deformation (Section 3 of this chapter), the ranges produced by Laramide orogeny did not project in the manner they do today, but were eroded about as rapidly as they were uplifted. Regional altitudes remained low; floors of the basins stood no more than a thousand feet above sea level, and the intervening mountains projected only a few thousand feet higher. The climate was uniformly moist and semi-tropical; no great mountain barriers to the west cut off the flow of moisture-laden winds from the Pacific Ocean.

This record may be deduced from the Paleocene and Eocene sediments of the Central and Southern Rocky Mountains and from their contained plant and mammalian fossils. The somber-colored, coal-bearing Paleocene deposits of the basins formed in a region of forests and swamps, and their mammals were a forest-dwelling community. The red-banded deposits of the succeeding Eocene contain a larger number of mammals that lived in open grasslands, indicating a spreading of that environment, but apparently without a significant change in altitude or climate. Eocene deposits of the different basins contain the same species of mammals, indicating that the mountain uplifts between them did not project high enough to create barriers to migration.

Laramide orogeny expelled the seas from the region and marine deposits appear no more except for a brief incursion of the sea during later Paleocene time in the northern Great Plains (represented by the Cannonball formation of North and South Dakota). All succeeding deposits were laid down in a continental environment.

The extensive basin deposits of Eocene time, partly laid down on flood plains of streams, partly in a series of great lakes, might suggest that Laramide orogeny also made the region one of interior drainage. A little reflection suggests that this is implausible. The region was not made lofty by the orogeny, yet its summits must have projected to some height above the Interior Lowlands on the east and formed a drainage divide; the probable rainfall of Paleocene and Eocene time was such that it could not have been trapped entirely by basins without outlets. Even the deposits of the lakes, which were formed when basin subsidence was so great as to pond the drainage, are such as to indicate that these lakes possessed outlets through much of their history. *Laramide orogeny initiated the modern drainage pattern of the Eastern Ranges and Plateaus.*

Compare, now, these Paleocene and Eocene landscapes of the Eastern Ranges and Plateaus with the landscape of present times. Today the region is one of rugged mountain ranges whose summits project to heights of 13,000 and 14,000 feet. Not only are the mountains high, but so also is the remainder of the region; the edge of the Great Plains at Denver is a mile above sea level, and many of the plains in Wyoming and Montana are even higher. The modern rivers cut through the ranges of the Rocky Mountains in rugged gorges, and those which traverse the Colorado Plateau are canyoned through most of their length. The climate of the Eastern Ranges and Plateaus is marked by harsh contrasts; plains, basins, and lower plateaus are arid or semi-arid, and only the mountains and higher plateaus receive heavy rainfall.

How was this change brought about? Clearly the whole region—from Great Plains across the Rocky Mountains into the Colorado Plateau—was raised as a unit 5,000 feet or more since Eocene time. Geologists are not fully agreed as to the manner in which this was accomplished, or by what stages, as their conclusions depend on interpretation of the elusive evidence furnished by the Tertiary and Quaternary land forms and deposits. The pages which follow outline what appears to be the most plausible story.

Middle and late Tertiary deposits. During middle and late Tertiary time deposits were laid down as widely in the Eastern Ranges and Plateaus, or more so, than during early Tertiary time. The pattern is different, however; they are preserved as erosion remnants of original broad sheets of sediments rather than as downfolds in original depositional and structural basins. Total thickness at any locality is generally a thousand feet or less rather than more than a thousand feet as with the earlier deposits.

Our discussion can best begin with the middle and later Tertiary deposits of the Great Plains, which extend from South Dakota to Texas and four hundred miles eastward from the mountain front.

During Laramide orogeny the Great Plains were a foreland area much like the Allegheny synclinorium and other foreland areas on the continentward side of the Appalachians. During Laramie and Paleocene time a series of foreland basins such as the Williston and Denver Basins developed east of the mountain front; these became inactive during the Eocene and received few or no deposits during that period.

The Laramie and Paleocene deposits have little to do with the present surface configuration, which is a long slope on the later Tertiary deposits away from the mountains and toward the Missouri River. Some years ago when I traveled westward with my family, the slope of the Great Plains was a matter of puzzlement to them. How could we leave the Missouri River at an altitude of less than a thousand feet, travel over a featureless, level country for six hundred miles, yet arrive in Denver at the west edge of the plains at an altitude of more than a mile above sea level? Without apparently climbing we had been climbing all the way—up the great slope of the plains.

Much of the surface of the plains is covered by the Ogallala formation of late Miocene and Pliocene age. Despite its extent it is a relatively thin sheet of deposits, in part fine-grained, but containing lenses of coarser gravels carried eastward from the Rocky Mountains in the channels of withering streams. In many places its layers are peculiarly cemented by *caliche* as a result of soil-forming processes in an arid climate, so that they are resistant to erosion and create the *cap rock* of the plains.

Deposits beneath the Ogallala emerge in the northern part of the plains, the most extensive of which form the White River group of Oligocene age, eroded into the extensive *bad lands* of South Dakota. The White River group also extends westward as smaller remnants between and into the Central Rocky Mountains of Wyoming.

During the first geological investigations of the Great Plains it was assumed that the later Tertiary deposits there—the Ogallala, the White River, and the rest—were laid down in a series of vast shallow lakes, hence that their depositional surfaces were nearly level and were later tilted regionally eastward. Later investigations proved that the deposits were dominantly of fluviatile origin, and also that the texture of their gravels was such, under semi-arid conditions, that they must have moved down a gradient nearly as great as the present slope of the country. The plains were tilted eastward as a part of the regional arching of the Eastern Ranges and Plateaus, but *this tilting preceded the laying down of the later Tertiary deposits.*

During later Tertiary time deposits were spread, not only over the Great Plains east of the Rocky Mountains, but over the Colorado Plateau southwest of it. These deposits, the Bidahochi formation, are preserved as remnants in the Navajo and Hopi country of northeastern Arizona, and contain gravels derived from at least as far away as the San Juan Mountains. Part of the Bidahochi is of lacustrine origin, indicating that on this side of the mountains the drainage was at times obstructed.

The fossil plants and mammals contained in the middle and late Tertiary deposits record an increasing regional altitude and aridity. By late Miocene time many of the early browsing herbivorous mammals had disappeared; those which survived such as the horses were adapted to feeding on the harsh grasses of semi-arid plains. The transition to more arid conditions in later Tertiary time resulted partly from a world-wide secular change toward cooler, more arid, more contrasting climates which foreshadowed the onset of the Pleistocene ice ages. In part it was a result of the raising of higher mountains on the west which produced a rain shadow over the plains country. By

Pliocene time the climate of the Great Plains was apparently about as arid as that of the present.

Erosion surfaces in the ranges. So much for middle and late Tertiary conditions in the plains and lowlands around the mountain ranges. What were conditions in the ranges which were the sources of the rivers and hence of the detritus deposited roundabout?

Most of the ranges in the Central and Southern Rocky Mountains exhibit a widespread *subsummit surface*—a series of accordant crests which extend across the deformed bedrock structures, above which chains and clusters of higher peaks project on the divides, and below which the streams have cut valleys and gorges to depths of thousands of feet. It apparently marks a time of widespread planation of the ranges, and it can be proved that this time was the same as that during which the later Tertiary formations were deposited in the Great Plains. Where the original profile of the subsummit surface can be reconstructed, it is steeper and more concave than would have been formed during an erosional regime in a humid climate; like the Great Plains deposits it was formed in a semi-arid or arid climate.

In later Tertiary time during the climax of deposition in the Great Plains and of planation in the mountains, a landscape had developed in the Eastern Ranges and Plateaus which differed not only from that of earlier Tertiary time, but from that of the present. The region had by then attained an altitude only slightly lower than today's, yet the local relief was probably as subdued, or more so, than that of earlier Tertiary time. Extensive areas not now covered by later Tertiary deposits were then buried, including the earlier Tertiary intermontane basins and the lower mountain ends and spurs. Most of the emergent areas were planed off to form the subsummit surface; the only mountainous areas remaining were the unreduced peaks along the axes of the ranges.

Quaternary denudation and dissection. At some stage late in Tertiary or early in Quaternary time this subdued landscape began to be destroyed by more vigorous stream erosion. Formerly this erosion was believed to have been activated by a regional uplift greater than any which preceded it, but the steep gradients, which we now know must have existed on the depositional surface in the plains and the erosional surfaces in the mountains, suggest that such regional uplift was less than that which occurred earlier in the Tertiary. Quickening of stream erosion at this time was actuated not only by uplift, but by the more rainy glacial periods of Pleistocene time.

Regardless of causes, the drastic effects of the change in stream habit on the landscape are evident to all. Later Tertiary deposits were stripped off of large areas where they had been laid, and the streams excavated the rocks beneath by varying amounts according to their resistance to erosion. Weak Cretaceous and early Tertiary rocks of the basin areas were etched out; strong Precambrian crystalline rocks and Paleozoic stratified rocks of the ranges were left projecting above them.

It was at this time, apparently, that the rivers of the region acquired their anomalous courses. We have noted that Powell was puzzled by the crossing of the Green River through the great arch of the Uinta Mountains, which the river could have avoided by making a slightly longer traverse around the end of the uplift (Chapter VI, Section 3). Powell concluded that the Green and similar rivers were *antecedent*—that they were in existence before the orogeny and were able to continue cutting down as the uplifts were raised across them. We now believe, however, that most such streams in the region were *superimposed*—that they had wandered at will down the slopes of the subdued erosional and depositional surfaces of later Tertiary time, but when they again cut downward they excavated wide valleys where they discovered weak rocks beneath them, and were confined to narrow gorges where they encountered belts of stronger rock.

History of river systems in Eastern Ranges and Plateaus. All this is very well, the reader may say, but what was the specific history of the rivers in the Eastern Ranges and Plateaus? The more specific we become the more doubts and confusions multiply.

In general, modern drainage of the region was initiated by Laramide orogeny, but there is little assurance that any particular stream held its course through the vicissitudes which followed that orogeny. Some rivers can be proved to have shifted greatly through time; for example, rivers on the eastern slope of the Northern and Central Rocky Mountains once flowed northeastward into Hudson Bay, but were deflected southward by the Pleistocene continental glaciers and coalesced to form the Missouri River which now drains to the Gulf of Mexico. Could there not have been equally great shifts of other streams which have left less obvious records?

Especially puzzling is the history of the two great rivers which flow south and southwest from the Rocky Mountains—the Rio Grande and the Colorado—which pass through long reaches of mountains, plateaus, and deserts on their way to the sea.

Waters no doubt flowed southward and southwestward from the mountain area for much of the time since the Laramide orogeny; the Miocene and Pliocene basin deposits of northern New Mexico contain gravels of such south-flowing rivers. But the basin deposits of this age farther south in New Mexico and Texas and in Nevada and Arizona below the end of the Grand Canyon are made up of locally derived detritus, and contain no gravels of the sort that would have been brought in by any large, through-flowing river.

The greatest problem pertains to that part of the Colorado River in the southwestern Colorado Plateau, where it has carved the Grand Canyon across the lofty Kaibab uplift. The river could not have been antecedent to this uplift; the uplift was produced by Laramide orogeny and passes unconformably beneath early Tertiary deposits at its north end. Moreover, Miocene and Pliocene deposits downstream from the Kaibab uplift are all locally derived and show no evidence of the presence of a large, through-flowing river.

It has been suggested that the river coursed across such uplifts as that of the Kaibab Plateau when they were in an early state of growth, that renewed uplift ponded the drainage on their upstream sides, until the river overflowed through its original valley and cut this to its present depth. Such a sequence of events is possible, but field relations suggest otherwise; so far as known the Kaibab uplift was folded entirely during Laramide orogeny. It has also been suggested that the Colorado River formerly flowed southward through the area of the Bidahochi formation, and so to the sea, but was later diverted westward by uplift of the southern rim of the plateau. It is believed that filling of the area upstream from the Kaibab Plateau to a depth of about 600 feet would be sufficient to allow the river to drain westward, utilizing the smaller consequent and subsequent stream valleys that had already been established between the Kaibab Plateau and the Grand Wash Cliffs.

The problem of the course of the Colorado River across the southwestern part of the Colorado Plateau is as yet unsolved, and will no doubt be debated for years to come.

EFFECT OF MODERN ENVIRONMENT ON WATER ECONOMY OF THE WEST. Bearing in mind the low relief, low altitude, and well-distributed rainfall of early Tertiary times in the Eastern Ranges and Plateaus, let us consider how the changes of later Tertiary and Quaternary time affected the economy of the region—of how people must live and adjust themselves to this environment and especially to its unevenly distributed water resources.

The heights of the Rocky Mountains are areas of abundant rainfall and give rise to the great rivers which flow east and west from the continental divide. The plains roundabout are semi-arid, yet they are the places where people must farm and build their cities. Water in the mountains must be conserved and carried by flumes and ditches to the plains country for irrigation and water supply.

This is not merely the dilemma of a few mountain states. Even in faraway Los Angeles an increasing amount of water must come from the Colorado River which originates in the Central and Southern Rocky Mountains. If Denver, to meet its growing needs as a city, plans to tap the streams flowing westward from the continental divide, it comes into immediate conflict with Los Angeles, because this would use water that otherwise would be available there. The peculiar rainfall and water supply conditions of the Rocky Mountains affect the economy of the whole western country.

REFERENCES

2. *Rocks and events before the great deformation*

Burbank, W. S., and Lovering, T. S., 1933, Relation of stratigraphy, structure, and igneous activity to ore deposition of Colorado and southern Wyoming, *in* Ore deposits of the western states (Lindgren volume): *Am. Inst. Min. Met. Eng.*, p. 272-316.

McKee, E. D. 1931, *Ancient Landscapes of the Grand Canyon*, 51 p.

Noble, L. F., 1914, The Shinumo quadrangle, Grand Canyon district, Arizona: *U.S. Geol. Survey Bull. 549*, 100 p.

3. *Rocks and events during the great deformation*

Baker, A. A., Dane, C. H., and Reeside, J. B., Jr., 1936, Correlation of the Jurassic formations of parts of Utah, Arizona, New Mexico, and Colorado: *U.S. Geol. Survey Prof. Paper 183*, 66 p.

Gregory, H. E., 1950, Geology and geography of the Zion Park region, Utah and Arizona: *U.S. Geol. Survey Prof. Paper 220*, 200 p.

Spieker, E. M., 1946, Late Mesozoic and early Cenozoic history of central Utah: *U.S. Geol. Survey Prof. Paper 205*, p. 117-161.

4. *Central Rocky Mountains*

Bradley, W. H., 1948, Limnology and the Eocene lakes of the Rocky Mountain region: *Geol. Soc. America Bull.*, v. 59, p. 635-648.

Chamberlin, R. T., 1940, Diastrophic behavior around the Bighorn Basin; *Jour. Geol.*, v. 48, p. 673-716.

Darton, N. H., and Paige, Sidney, 1925, Description of the central Black Hills: *U.S. Geol. Survey Geol. Atlas*, folio no. 219, 34 p.

Horberg, Leland, Nelson, Vincent, and Church, Victor, 1949, Structural trends in central western Wyoming: *Geol. Soc. America Bull.*, v. 60, p. 183-216.

Rouse, J. T., 1937, Genesis and structural relationships of the Absaroka volcanic rocks; Wyoming: *Geol. Soc. America Bull.*, v, 48, p. 1257-1296.

Thom, W. T., Jr., 1923, The relation of deep-seated faults to the surface structural features of central Montana: *Am. Assoc. Petrol. Geol. Bull.*, v. 7, p. 1-13.

5. *Southern Rocky Mountains*

Boos, C. M., and Boos, M. F., 1957, Tectonics of eastern

flank and foothills of Front Range, Colorado: *Am. Assoc. Petrol. Geol. Bull.*, v. 41, p. 2603-2676.

Behre, C. H., Jr. 1953, Geology and ore deposits of the west slope of the Mosquito Range, Colorado: *U.S. Geol. Survey Prof. Paper 253,* 176 p.

Burbank, W. S., and Goddard, E. N., 1937, Thrusting in Huerfano Park, Colorado, and related problems of orogeny in the Sangre de Cristo Mountains: *Geol. Soc. America Bull.*, v. 48, p. 931-976.

Knopf, Adolph, 1956, Igneous geology of the Spanish Peaks region, Colorado: *Geol. Soc. America Bull.*, v. 49, p. 1727-1784.

Larsen, E. S., Jr., and Cross, Whitman, 1956, Geology and petrology of the San Juan region, southwestern Colorado: *U.S. Geol. Survey Prof. Paper 258,* 303 p.

Lovering, T. S., and Goddard, E. N., 1950, Geology and ore deposits of the Front Range, Colorado: *U.S. Geol. Survey Prof. Paper 223,* 319 p.

Powell, J. W., 1876, Report on the geology of the eastern portion of the Uinta Mountains, and a region of country adjacent thereto: *U.S. Geol. Geograph. Survey Terr.*, 2nd Div., 218 p., atlas.

6. *Less deformed parts of Eastern Ranges and Plateaus*

Hunt, C. B., and others, 1953, Geology and geography of the Henry Mountains region, Utah: *U.S. Geol. Survey Prof. Paper 228,* 234 p.

———, 1956, Cenozoic geology of the Colorado Plateau: *U.S. Geol. Survey Prof. Paper 279,* 99 p.

Kelley, V. C., 1952, Tectonics of the Rio Grande depression of central New Mexico: *New Mexico Geol. Soc. Guidebook, 3rd Ann. Field Conf.* (Rio Grande country, central New Mexico), p. 93-105.

———, and Silver, Caswell, 1952, Geology of the Caballo Mountains: *New Mexico Univ. Publ. Geol. no. 4,* 286 p.

———, 1955, Regional tectonics of the Colorado Plateau and relationship to the origin and distribution of uranium: *New Mexico Univ. Publ. Geol. no. 5,* 120 p.

Powell, J. W., 1873, Geologic structure of a district of country lying to the north of the Grand Canyon of the Colorado: *Am. Jour. Sci.*, 3rd ser., v. 5, p. 456-465.

Strahler, A. N., 1948, Geomorphology and structure of the West Kaibab fault zone and Kaibab Plateau, Arizona: *Geol. Soc. America Bull.*, v. 59, p. 513-540.

7. *Events after the great deformation*

Atwood, W. W., and Atwood, W. W., Jr., 1938, Working hypothesis for the physiographic history of the Rocky Mountain region: *Geol. Soc. America Bull.*, v. 49, p. 957-980.

Johnson, W. D., 1901, The High Plains and their utilization: *U.S. Geol. Survey 21st Ann. Rept. (1900),* pt. 4, p. 601-741.

Mackin, J. H., 1937, Erosional history of the Bighorn Basin, Wyoming: *Geol. Soc. America Bull.*, v. 48, p. 813-894.

Van Houten, F. B., 1948, Origin of red-banded early Cenozoic deposits in Rocky Mountain region: *Am. Assoc. Petrol. Geol. Bull.*, v. 32, p. 2083-2126.

CHAPTER VIII

THE MAIN PART OF THE CORDILLERA: ITS GEOSYNCLINE AND THE MOUNTAIN BELT THAT FORMED FROM IT

1. COMPARISONS AND COMPLICATIONS

In Chapter VII we explored an extensive part of the western Cordillera, yet this part is a set of Eastern Ranges and Plateaus which did not originate from a true geosynclinal area—as a well-behaved mountain belt should—and whose features are a novel structural element, not exactly comparable to any in the Paleozoic mountain belt on the southeastern side of the continent.

We turn now to that part of the Cordillera farther west which has had a more conventional history, originating as a geosyncline along the border of the continent and developing through time into a mountain belt. Some of the features will prove to resemble remarkably those of the Appalachian mountain belt, and we can account for many of the differences as the local peculiarities one would expect on passing from one far-separated mountain belt to another. We will find that greatest differences are in the happenings after the climactic orogeny, which have been more profound in the western region than the eastern. For this reason these post-orogenic features are reserved for discussion in Chapter IX.

Even the fundamental features of the geosyncline and the orogenic belt possess such a wealth of detail that our discussion had best be restricted to instructive samples of the whole region selected from various parts of the United States and southern Canada.

2. CORDILLERAN MIOGEOSYNCLINE

GENERAL RELATIONS. During Paleozoic time the Cordilleran miogeosyncline, like the Appalachian miogeosyncline, was the scene of a long-continued, little interrupted sedimentation, largely in shallow water. Sedimentation in the miogeosynclinal area thus differed little from that on the continental platform, but there was more subsidence of its floor so that deposits accumulated to greater thickness and formed a more complete sequence.

In the Cordilleran miogeosyncline, as in the Appalachian miogeosyncline, the side toward the continent (here to the east rather than the west) is marked by an abrupt thinning of the sedimentary section, partly by overlap and wedging out of the Cambrian deposits at the base, partly by thinning and disappearance of the systems above.

Within the miogeosynclinal area itself, as in the Appalachians, the lower part of the Cambrian consists of clastic deposits and the upper part of limestones and dolomites. Other and younger limestones and dolomites follow in great thickness and the sequence is topped by clastic wedges which spread across the miogeosynclinal area into the foreland; these were derived from quickening orogeny in the eugeosynclinal part of the system.

Nevertheless, the parallelism between these two miogeosynclinal areas on opposite sides of the continent is not complete:

(1) In the Appalachian geosyncline some sediments and volcanics of late Precambrian age underlie the Cambrian. In the Cordillera late Precambrian sediments attain much greater thickness and extent, especially in the north.

(2) In the Cordilleran miogeosyncline climax of orogeny was considerably later than in the Appalachian miogeosyncline; the characteristic miogeosynclinal sequence—from basal clastics into carbonates into clastic wedges—is offset upward in the section. Crustal disturbances in the latter part of Paleozoic

time were minor; in places they gave rise to clastic formations, but in others the carbonate part of the sequence continued high into the Paleozoic. Mississippian limestones spread over most of the Cordilleran miogeosyncline, and in places limestones extend through the Pennsylvanian and even into the Permian. True clastic wedges make their appearance only above the Paleozoic in the Triassic and Jurassic and especially in the Cretaceous.

BELT SERIES. Let us consider first the segment of the miogeosyncline in the Northern Rocky Mountains—from west-central Montana northward across the International Boundary into the Canadian Rockies of western Alberta and eastern British Columbia.

A striking feature of this segment is the wide surface extent of the *Belt series.* The Belt is named for the Big Belt and Little Belt Mountains of west-central Montana which are part of the Eastern Ranges of the Cordillera, but these mountains lie near the edge of its sedimentary basin; the Belt series wedges out between basement and Paleozoic rocks a short distance farther east. Westward in the Cordillera the Belt series has a much greater thickness and extent and forms most of the ranges of the Northern Rocky Mountains in northwestern Montana, northern Idaho, southwestern Alberta, and southeastern British Columbia. Here it forms the surface over an area of at least 35,000 square miles, and thickens from about 13,000 feet on the east to more than 45,000 feet in the Purcell Mountains on the west (Fig. 75).

Toward the east the Belt series includes thick limestone formations, but farther west the limestones thin out and various beds of sandstone make their appearance. Nevertheless, the greater bulk of the series is *argillite* or slightly altered shaly rocks, mostly gray or dull green but red-colored in certain parts. The Belt contains no indications of life except for some curious limestone reefs which are believed to have been built by lime-secreting algae. It is mainly a sedimentary and non-volcanic series, but a flow, the Purcell lava, occurs high in the sequence in many of the ranges.

The sedimentary rocks of the Belt series much resemble those of later ages—so much so that the early geological explorers supposed the series to be of Paleozoic or even younger age, and many geologists today doubt that it can be much older than the Cambrian. So far as its relation to overlying rocks is concerned it might be Early Cambrian, as the first strata which lie on it through most of its extent are Middle Cambrian or younger. Nevertheless, several lines of evidence indicate that it is probably considerably older:

(A) A hundred miles north of the main area of the Belt series along the strike of the mountains is a thick sequence of fossiliferous Cambrian rocks, the Lower Cambrian part of which consists of the usual clastics below and carbonates above. Nothing in this sequence is comparable to the Belt series.

(B) About the same distance west of the front of the Rocky Mountains, in the Purcell Mountains far back in the geosynclinal area, the Belt is overlain unconformably by another thick mass of clastic rocks, the *Windermere series* (Fig. 75). Fossils of various Paleozoic ages have been found at different levels above the middle of the Windermere as it was originally described, but its lower part is unfossiliferous and is probably older than the earliest fossiliferous Cambrian. The Belt series is therefore not the youngest Precambrian in the Northern Rocky Mountains.

(C) At several places the ages of veins which cut the Belt series have been determined radiometrically and have proved to be surprisingly ancient (1,000 to 1,200 million years). While these results require further evaluation, they suggest that Belt sedimentation took place much earlier than anyone had anticipated.

These considerations indicate that the geosyncline in which the Belt series was deposited was distinct from and older than the true Cordilleran geosyncline, with the deposits of the latter lying unconformably upon it. Along the eastern border of the Cordilleran miogeosyncline the first deposits above the Belt are of Middle Cambrian age. Farther west nearer the eugeosyncline, the first deposits above the Belt are the Windermere series which are much older—probably older than the fossiliferous Cambrian. In the Windermere area sedimentation in the true Cordilleran geosyncline must have begun before Cambrian and Paleozoic time in a sort of "Eocambrian" epoch.

YOUNGER ROCKS OF THE NORTHERN SEGMENT. Above the Belt series lie the Paleozoic miogeosynclinal deposits. In northwestern Montana, except along the mountain front, they are preserved only as occasional outliers or infolds. It is thus difficult to determine their former extent, although it may be presumed that they once covered the area of the present exposures of the Belt series.

Across the International Boundary in Canada exposures of the Belt series become narrower, and the series plunges northward beneath younger strata. Paleozoic miogeosynclinal rocks form most of the surface of the Northern Rocky Mountains about 150 miles north of the boundary along the main line of the Canadian Pacific Railroad west of Calgary, Alberta.

We can best begin our discussion of these miogeosynclinal rocks at Banff, the beautiful mountain resort

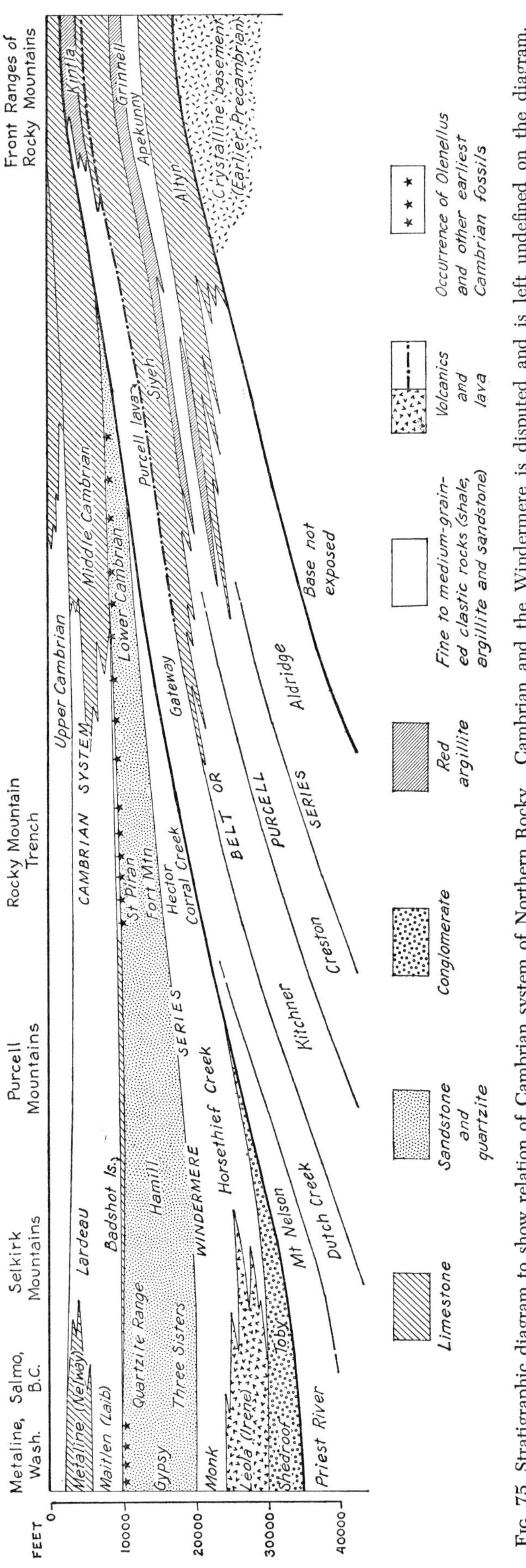

FIG. 75. Stratigraphic diagram to show relation of Cambrian system of Northern Rocky Mountains to Windermere and Belt series. Diagram covers a zone about 200 miles in length immediately north of International Boundary. The boundary between the Cambrian and the Windermere is disputed and is left undefined on the diagram. Compiled from many sources including Okulitch, 1949, 1956.

on the railroad, lying in a valley between the high eastern ranges of the mountains. The ranges are formed of the upper part of the Paleozoic succession, about 8,000 feet thick and largely limestone (as one might expect in a miogeosynclinal area). The most prominent formation is the massive Rundle limestone, whose 2,500-foot bulk crowns all the mountain ridges in the vicinity (shown by the black pattern on Fig. 76). It is of middle Mississippian age, or about the equivalent of the Madison limestone of the northwestern states and the Redwall limestone of the Grand Canyon. Below are more than 5,000 feet of shales, thin-bedded limestones, and dolomites of earlier Mississippian and Devonian ages which rest on Middle Cambrian strata; the latter come to the surface only in narrow exposures at the bases of the thrust blocks. Above the Rundle is the 700-foot Rocky Mountain quartzite, a sandy deposit of Pennsylvanian age which forms the top of the Paleozoic section. It is overlain by Triassic and Jurassic shaly rocks which have been eroded to form the valleys between the ridges; we will say more about them shortly.

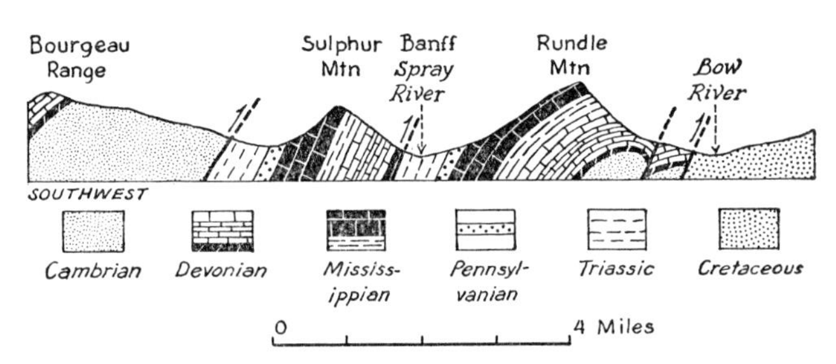

FIG. 76. Section across ridges and valleys of the Rocky Mountains in the vicinity of Banff, Alberta to show miogeosynclinal sequence. After Warren, 1927.

Eastward the Paleozoic formations thin and wedge out. Even at Banff, note the large breaks in the record between the Devonian and Middle Cambrian, between the middle Mississippian and the Pennsylvanian, and between the Pennsylvanian and the Triassic. In the Alberta basin, the foreland east of the miogeosyncline, many of the units of the Banff section are thin or missing. Only the Devonian is surprisingly persistent and contains some of the great productive horizons in the oil fields of the basin.

West of Banff the lower part of the Paleozoic is raised to the surface and forms all the ranges of the inner part of the Rocky Mountains (Fig. 78). Devonian, Silurian, and Ordovician rocks occur some distance to the west, but the most extensive rocks are of Cambrian age and attain thicknesses of 10,000 feet or more. The Cambrian begins with a few thousand feet of sandstone, but the remainder is a great mass of limestone and dolomite broken only in places by shaly beds.

Paleozoic miogeosynclinal rocks in this segment of

the Cordillera thus total about 20,000 feet, of which the Cambrian accounts for nearly half. Whether all this mass accumulated at any one place may be questioned, as the later Paleozoic is thick in the frontal ranges where the Cambrian has thinned, and the later Paleozoic is now missing where the Cambrian is thick.

Be this as it may, the Paleozoic rocks possess many features like those in the Appalachian miogeosyncline —the Early Cambrian sandstones at the base, the great mass of carbonates above them, and the thinning and wedging of all the units toward the foreland. The sequence differs in that the carbonates continue well through the Paleozoic instead of ending in the middle of the Ordovician; the Pennsylvanian is sandy, yet we look in vain for Paleozoic clastic wedges of the sort characteristic of the upper part of the miogeosynclinal section in the Appalachians.

The change in sedimentation from carbonates to clastic deposits, derived from erosion of the interior parts of the Cordilleran geosyncline, began after Paleozoic time. Valleys between the frontal ridges of the Northern Rocky Mountains are underlain by shales of Triassic and Jurassic age—the marine Spray River and Fernie formations. Drilling in the plains east of the mountains indicates that these do not extend far into the foreland east of the miogeosyncline.

Some of the deeper downfolds in the frontal ranges of the Rocky Mountains also contain remnants of lowest Cretaceous strata—the Kootenay continental deposits. These and higher Cretaceous beds are more completely developed in the foothills along the mountain front. The Cretaceous has a true clastic wedge structure—a replica of that which we have described earlier in the Eastern Ranges and Plateaus—with tongues of continental and marine sandstones projecting eastward into marine shales.

Finally, in the foreland along the mountain front in the deeper parts of the Alberta basin are the latest Cretaceous and Paleocene continental deposits (Edmonton and higher formations), which are comparable to the Laramie and overlying formations farther south, and like them were laid down while the Laramide orogeny was still in progress.

Miogeosynclinal rocks of southern Great Basin. To continue our story, yet keep it within bounds, let us transfer our attention to another area about a thousand miles south of the International Boundary—the region west of the Grand Canyon and Colorado Plateau extending across the Great Basin into the Inyo Mountains of California. (Fig. 77).

In the Grand Canyon the whole Paleozoic sequence is encompassed within the walls of the gorge and is less than 4,000 feet thick. As we have seen (Chapter VII, Section 2), strata of Cambrian, Mississippian,

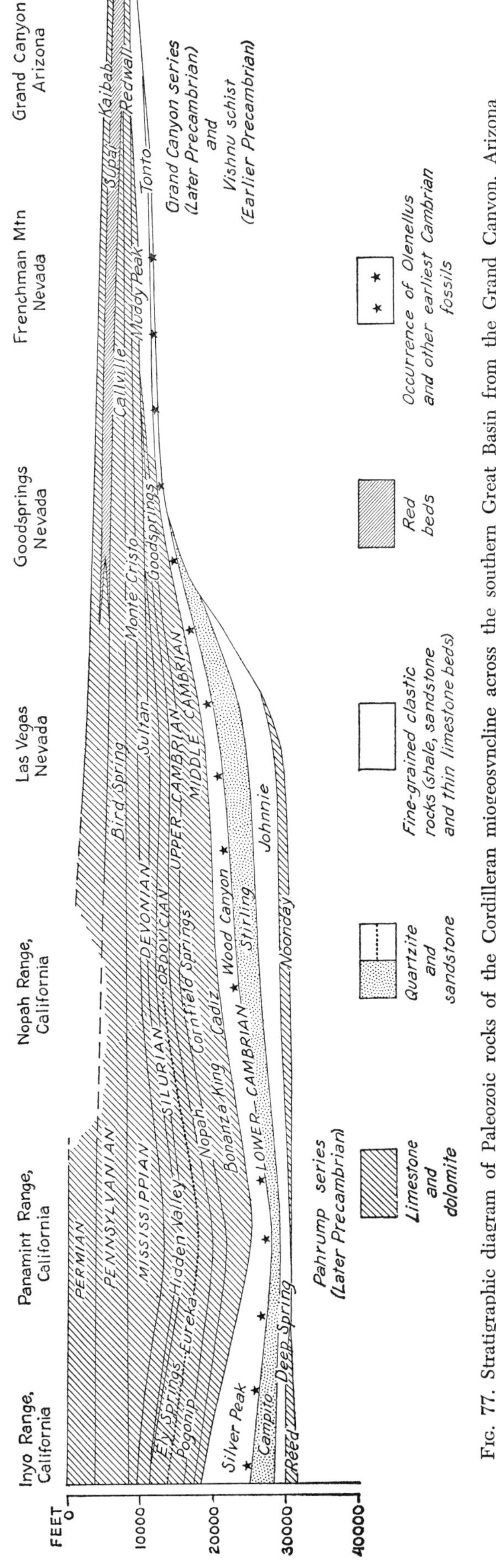

Fig. 77. Stratigraphic diagram of Paleozoic rocks of the Cordilleran miogeosyncline across the southern Great Basin from the Grand Canyon, Arizona to the Inyo Mountains, California. Based on Longwell, 1950; and other sources.

and Permian ages are represented. While there are minor occurrences of other systems in the region, most of the Ordovician, Silurian, Devonian, and Pennsylvanian are missing, their places being taken by disconformities below and above the Redwall limestone.

The familiar Grand Canyon formations extend with little change westward down the Grand Canyon past the Grand Wash Cliffs and thence across the eastern Great Basin to Las Vegas, Nevada, but the section thickens by wedging in of formations at the positions of the disconformities in the Grand Canyon. Between the Supai formation and the Redwall limestone the thick body of Bird Spring limestone of Pennsylvanian and early Permian age appears. Between the Redwall limestone and the Tonto group a Devonian formation occurs. Finally, on the unconformity at the base of the Paleozoic, Lower Cambrian rocks wedge in beneath the Middle Cambrian part of the Tonto group (Fig. 77).

Still greater thickening of the Paleozoic rocks into the miogeosyncline takes place west of Las Vegas where they attain more than 25,000 feet. Typical sections are exposed in the Nopah Range east of Death Valley and in the Inyo Range east of Owens Valley, both in California.

Nearly half the great thickness of Paleozoic rocks in the Nopah and Inyo Ranges is Cambrian and most of this is Lower Cambrian (Fig. 77). In the Nopah Range the latter consists of the Noonday dolomite at the base (lying unconformably on earlier rocks), the Johnnie formation (shales and thin sandstones), the Stirling quartzite, and the Wood Canyon formation (shales and thin limestones); in the Inyo Range equivalent beds are termed Reed dolomite, Deep Spring formation, Campito quartzite, and Silver Peak group. The units crop out widely in the ranges of this part of the desert region and are recognizable even from a distance by their characteristic forms and colors. *Olenellus* and other Early Cambrian fossils occur in the Wood Canyon and Silver Peak strata, but the beds beneath are unfossiliferous; the lower unfossiliferous beds are thus not certainly provable as part of the Cambrian. Like the lower part of the Windermere series north of the International Boundary they mark the beginning of development of the Cordilleran geosyncline during a sort of "Eocambrian" epoch.

Unconformably beneath the Paleozoic strata of this region there are, further, about 7,000 feet of late Precambrian beds of the Pahrump series which closely resemble the Grand Canyon series. Like the Belt series farther north, the Pahrump series evidently formed in a trough whose development preceded that of the main Cordilleran geosyncline.

Paleozoic rocks elsewhere in eastern Great Basin. Farther north in the Great Basin the earliest deposits of the Cordilleran geosyncline merge into a single body of sandstone, the Prospect Mountain quartzite, which again contains Early Cambrian fossils only near its top.

The remainder of the Paleozoic sequence in the eastern Great Basin is largely limestone and dolomite, which includes representatives of the Middle and Upper Cambrian and the higher geologic systems into the Permian. In east-central Nevada the Devonian is represented by the Nevada dolomite and Devils Gate limestone, 4,000 to 5,000 feet thick. Occasionally the succession of limestones and dolomites is broken by other deposits: The Eureka quartzite of Middle Ordovician age occurs nearly everywhere in the eastern Great Basin, and the Upper Devonian and Mississippian includes two shale units, the Pilot and Chainman, which merge westward into a single body, the White Pine shale.

In parts of the region peculiarities develop in the upper part of the Paleozoic section. Part of the Mississippian passes westward into the Diamond Peak quartzite which contains chert pebbles eroded from eugeosynclinal Paleozoic rocks farther west. Above are other pebbly and conglomeratic beds, and many of the units are separated by unconformities. These features are related to belts of disturbance in central Nevada along the boundary between the miogeosyncline and eugeosyncline which we will discuss later in Section 4 of this chapter.

Farther east, also, in the vicinity of the Oquirrh Range south of Great Salt Lake the Pennsylvanian and upper Mississippian strata thicken to 20,000 feet, most of which are included in the Oquirrh formation, a body of interbedded sandstones and limestones. This great mass of deposits seems to have accumulated in a local area of exceptional subsidence within the miogeosyncline and attests its instability in later Paleozoic time. Sediments of the Oquirrh formation, like the later Paleozoic clastics farther west, may be related to disturbances along the boundary between the miogeosyncline and eugeosyncline.

Mesozoic rocks of eastern Great Basin. Above the Paleozoic miogeosynclinal sequence in the eastern part of the Great Basin the record becomes more fragmentary. In places the Paleozoic is succeeded by marine Triassic, and farther east by Jurassic. There is, however, a wide gap between these Triassic and Jurassic rocks and those of the western part of the Great Basin; strata of the two regions are so different that they were probably laid down in separate seaways.

Here and there in the eastern part of the Great Basin patches of continental and fresh-water sandstone, shale, and conglomerate lie on the eroded surfaces of the deformed earlier rocks. Some of these are proved by fossils to be of Cretaceous age; others are probably Paleocene or Eocene. Unlike the Paleozoic and earlier Mesozoic deposits, they were formed in discontinuous basins during or after the main orogeny of the region and thus were not a part of the deposits of the Cordilleran miogeosyncline.

3. STRUCTURE OF EASTERN PART OF MAIN CORDILLERA

During the Laramide orogeny and the disturbances which preceded it—from Paleocene until well back into Cretaceous time—the stratified rocks which were laid down in the Cordilleran miogeosynclinal area were strongly deformed. Before taking up details of the structures thereby produced it will be instructive to compare them with those of other regions which we have already considered. In this comparison we will use one segment as representative of the whole: the Northern Rocky Mountains.

COMPARISON WITH SOUTHERN ROCKY MOUNTAINS. The eastern edges of the Northern and Southern Rocky Mountains resemble each other in several respects:

(1) In both areas mountainous topography ends toward the east along an abrupt front which faces the Great Plains or edge of the Interior Lowlands.

(2) Structurally this front corresponds to a boundary between much deformed and little deformed rocks.

(3) The Great Plains were the foreland of both mountain areas, received Cretaceous and Paleocene deposits, and were warped down into basins.

(4) In both mountain areas rocks lower stratigraphically than those of the plains were raised so high that they have been uncovered by erosion. These lower rocks are of earlier Mesozoic, Paleozoic, and Precambrian age.

Here the resemblance ends and we must note some significant differences:

(A) In front of the ranges of the Southern Rocky Mountains the rocks of the Great Plains are little disturbed almost to the mountain base. Between the front of the Northern Rocky Mountains and the Great Plains the rocks have been intensely crumpled and sliced in a foothills belt.

(B) At the edges of the uplifts of the Southern Rocky Mountains the sedimentary section has been turned up steeply or cut off by high-angle faults associated with the upthrust of the range. In the Northern Rocky Mountains the older sedimentary rocks have been carried eastward over the younger rocks for many miles along great low-angle thrust faults.

(C) In the Southern Rocky Mountains the Paleozoic sedimentary rocks are a relatively thin sequence that was laid down on the continental platform; those of the Northern Rocky Mountains are a thick sequence that formed in the Cordilleran miogeosyncline.

(D) In the Southern Rocky Mountains the Precambrian rocks are a metamorphic and plutonic basement; those of the Northern Rocky Mountains are geosynclinal sediments.

These differences between rocks and structures along the edges of the Northern and Southern Rocky Mountains are matched by contrasts farther back in the two systems. The ranges of the Southern Rocky Mountains are mainly broadbacked uplifts which raise Precambrian basement rocks to view, separated by narrow to wide basins containing much younger rocks. The ranges of the Northern Rocky Mountains are formed from long, narrow, closely spaced folds and thrust blocks made up of geosynclinal sedimentary rocks alone, not of basement rocks. These contrasts reflect the manner in which the unlike rocks of the two areas have responded to deformation.

COMPARISON WITH SEDIMENTARY APPALACHIANS. The Northern Rocky Mountains are a region of lofty peaks and ranges unlike the subdued landscape of the Appalachians whose Valley and Ridge province consists of low, rounded ridges and broad limestone valleys. Moreover, the Appalachian ridges are bordered toward the continental interior by plateaus as high or higher, rather than by plains.

We should not be deceived by these superficial contrasts, as the geological resemblances are much greater. The Northern Rocky Mountains and the Valley and Ridge province are analogous parts of mountain systems on opposite sides of the Central Stable Region; both are made up of sedimentary rocks which were laid down in a miogeosyncline and were afterwards folded, faulted, and thrust toward the interior of the continent. Both regions contain great low-angle thrusts of the same sort, and in neither do basement rocks emerge, suggesting that deformation was largely confined to the sedimentary rocks above it.

The differences in topography are partly climatic, as the great limestone masses have been worn down more readily in the genial southeastern climate of the Appalachians than in the harsh northwestern climate of the Rocky Mountains. Partly, too, the Appalachians were deformed several periods earlier than the Cordillera, so that they have been exposed to the weather for a longer period and have attained greater topography maturity. Finally, the crustal movements which followed the deformation of the rocks in the sedimentary Appalachians were of less magnitude, for the most part, than those in the eastern part of the Cordillera.

SECTION IN NORTHERN ROCKY MOUNTAINS OF SOUTHERN CANADA. Let us now take up details of the structure of the Northern Rocky Mountains, or that part of the eastern Cordillera which we have used for comparison with structures of other regions.

We will deal first with a cross-section about 150 miles north of the International Boundary along the main line of the Canadian Pacific Railroad west of Calgary, Alberta—the segment whose miogeosynclinal Paleozoic and Mesozoic rocks we have already discussed in Section 2 of this chapter. Proceeding west from Calgary we encounter in succession the following structures (Fig. 78 B):

(1) The *Great Plains* east of the mountains, whose rocks are warped down into the *Alberta basin.* The surface rocks are flat-lying continental deposits of late Cretaceous and Paleocene age, beneath which (shown by drilling) is a foreland section of Mesozoic and Paleozoic rocks.

(2) A *foothill belt* as much as twenty miles wide, whose surface rocks are mainly Upper Cretaceous shales and sandstones which, in contrast to the rocks of the Great Plains, have been steeply tilted and sliced into a multitude of thrust blocks, each displaced by small amounts toward the east (Fig. 78 B and Fig. 79).

(3) West of the foothill belt the *topographic eastern front* of the Rocky Mountains, a bold escarpment of Paleozoic limestone. Near the railroad the contact between the Paleozoic of the mountains and the Cretaceous of the foothills is the *McConnell thrust,* a major fault dipping at a low angle to the west, along which the Paleozoic has been carried eastward over the Cretaceous at least five miles and probably much farther.

(4) Behind the mountain front a belt of *Front Ranges* about thirty miles broad (Fig. 78 A and B). These are high, parallel mountain ridges separated by longitudinal valleys, in one of which Banff is located (Fig. 76). Each ridge is supported by later Paleozoic limestone and especially the massive Rundle limestone; each valley is cut on Mesozoic sandstones and shales. The whole has been thrown into a succession of long narrow folds and thrust slices that repeat the section many times, the thrusts dipping westward and probably joining the sole fault (or McConnell thrust) at depth.

(5) Behind the Front Ranges the higher, more massive *Main Ranges* about forty miles broad (Fig. 78 A and B). They are made up of earlier Paleozoic miogeosynclinal rocks, mainly Cambrian, which have been less faulted and more openly folded than those of the Front Ranges, so that many of the mountains have the form of lofty plateaus. The powerful forces to which the rocks of the Main Ranges have been subjected are attested, however, by their much greater uplift than those in the Front Ranges, and by presence of slaty cleavage in the more argillaceous layers.

(6) At the west edge of the Rocky Mountains in this latitude, and separating them from the Purcell and other interior ranges beyond, a peculiar, long, narrow valley, the *Rocky Mountain Trench.* This is no mere erosional feature, for it is not drained by a single river; instead, most of the great rivers of the northern Cordillera flow in it at one place or another—the Kootenai, the Columbia, the Fraser, the Peace, and many more. The trench extends at least a thousand miles northward from the International Boundary, nearly parallel to the trends of the ranges (Plate I).

Along the line of the railroad the Rocky Mountain Trench seems to be a synclinorium in which Ordovician and younger strata are downfolded between the Cambrian of the Rocky Mountains on the east and the Precambrian (Belt and Windermere) of the Purcell Mountains on the west (Fig. 78 A). But the true nature and origin of the trench remain elusive, as the visible structures along it differ from place to place. Some geologists have suggested that it marks a fault zone between mountain structures of different ages, the interior ranges having been deformed earlier during Mesozoic time, and the Rocky Mountains later during Laramide orogeny. Neither this explanation of the trench nor any other can be proved in the present state of our knowledge.

SECTION AT INTERNATIONAL BOUNDARY. Farther south near the International Boundary is another significant section of the Northern Rocky Mountains; its eastern part is splendidly exposed in Glacier National Park and adjacent parts of the Lewis, Livingston, and Clark Ranges (Fig. 78 C and D).

Here the mountains are composed of Precambrian sedimentary rocks of the Belt series which have been raised so high that nearly all the former cover of Paleozoic miogeosynclinal strata has been removed by erosion. The Belt series has been broadly folded at most; characteristic features of the landscape are the lines of gently dipping ledges on the mountain sides. The mountains face eastward in bold escarpments upon a foothill belt where, as farther north, intensely faulted Cretaceous rocks lie at the surface.

The Belt series of the mountains overlies the Cretaceous rocks of the foothills on the *Lewis thrust,* one of the most prominently exposed of the great low-angle faults of North America. The low dip of the fault and the superposition of Belt on Cretaceous can be proved by following the fault out onto the points of the ridges or up the intervening valleys. Besides, Chief Mountain,

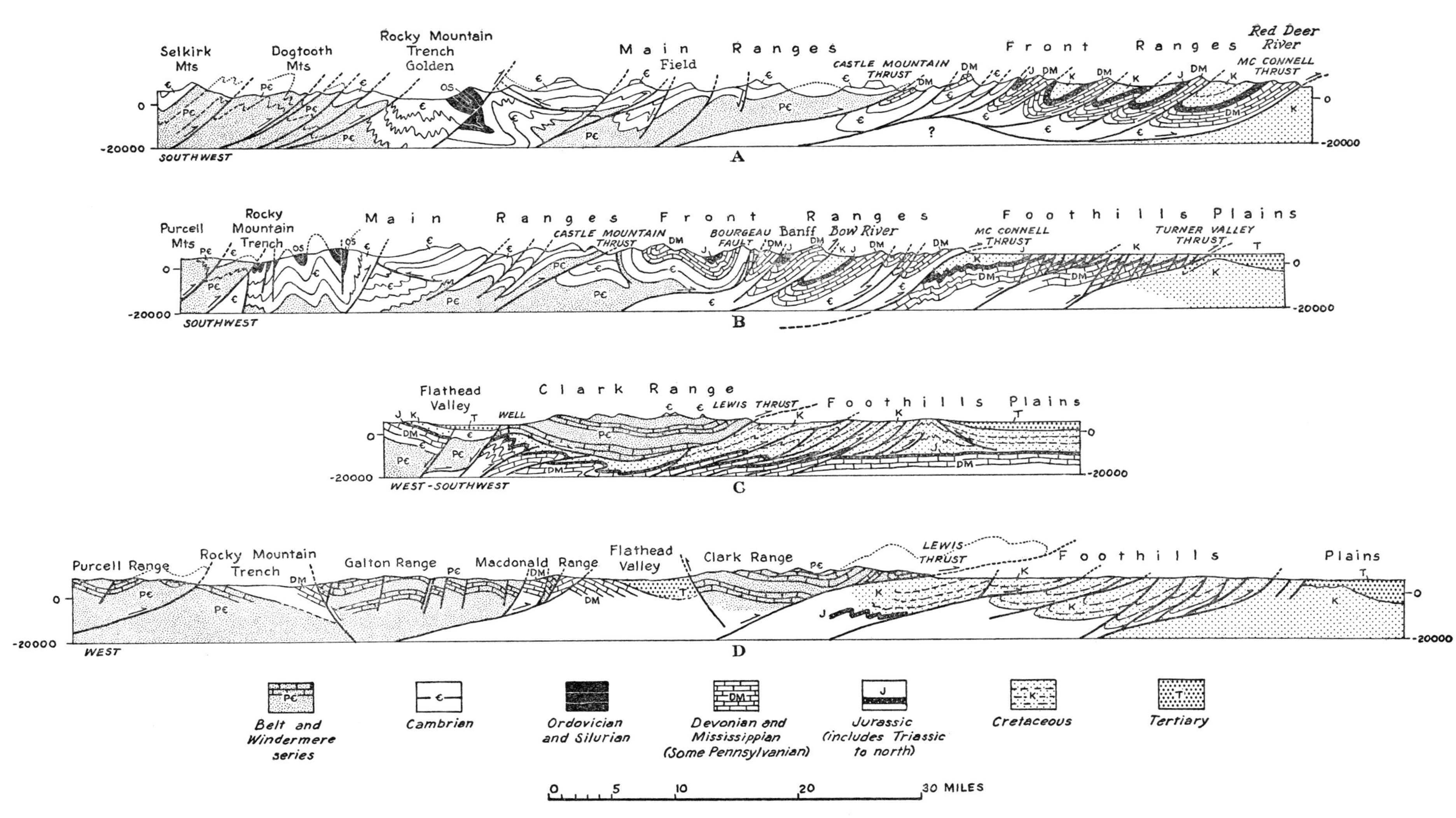

Fig. 78. Sections of Northern Rocky Mountains, Alberta and British Columbia. (A) Northeast from Golden about 175 miles north of International Boundary. (B) Through Banff about 150 miles north of International Boundary. (C) In northern part of Clark Range 5 to 30 miles north of International Boundary. (D) Along International Boundary. The unlike structures shown in Flathead Valley in sections C and D represent different interpretations of subsurface relations. Compare with sections in the comparable Valley and Ridge province of the Appalachians in Fig. 25. After North and Henderson, 1954; Clark, 1954; Daly, 1912 (with interpretation by Link, 1935).

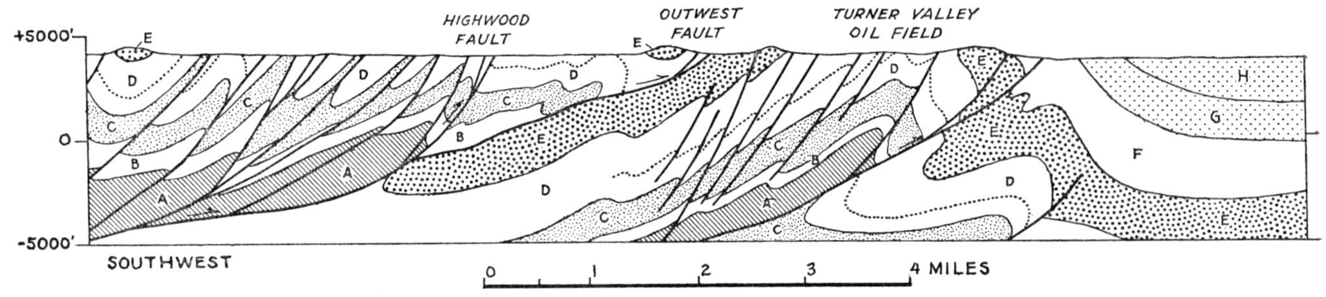

Fig. 79. Section across Turner Valley area 25 miles southwest of Calgary, Alberta showing typical foothill structures of Northern Rocky Mountains where subsurface structure has been extensively determined by drilling. After Gallup, 1951; and Hume, 1957.

Explanation of symbols. (A) Banff and Rundle limestones (Mississippian). (B) Fernie formation (Jurassic). (C) Blairmore formation, (D) Alberta shale, (E) Belly River formation, (F) Bearpaw shale, (G) Edmonton formation (Cretaceous). (H) Paskapoo formation (Paleocene).

a peak projecting from the foothills some miles in front of the mountains, is a *klippe* or erosional outlier of the thrust sheet (Fig. 80). Like the main mountain front it is composed of Belt strata which lie on the Cretaceous. Sinuosities in the trace of the thrust—from its exposures farthest up the valleys to those on the points of the ridges and in the klippe—indicate that the rocks above the thrust have ridden eastward over those beneath them for a distance of at least seven miles.

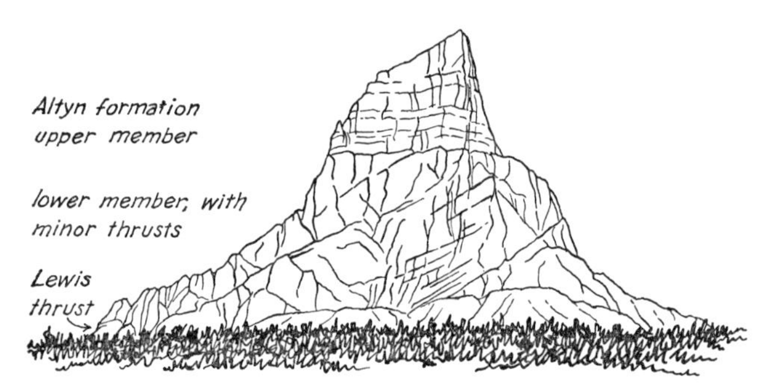

Fig. 80. Chief Mountain at the eastern edge of the Rocky Mountains in northwestern Montana. View is from the south, the mountains being to the left and the plains to the right. This is probably the most striking example in the United States of a klippe or thrust outlier. The Altyn formation of the Belt series (Precambrian) lies on Cretaceous strata on the sole of the Lewis thrust. After Willis, 1902.

Actual movement on the Lewis thrust is certainly much greater. North of the International Boundary well back in the mountains are several *windows* in the thrust sheet where deep valleys have cut through the Belt strata into those of the Cretaceous beneath the thrust. In many parts of the mountains oil seeps issue from the Belt rocks and probably originated in the Mesozoic and Paleozoic rocks beneath. On the strength of these seeps several holes have been drilled for oil in the mountains. One of these, the Pacific-Atlantic, Flathead no. 1 well passed from Belt series at 4,500 feet into highly disturbed Paleozoic and Mesozoic strata (Fig. 78 C); it lies nineteen miles southwest of the nearest emergence of the Lewis thrust along the mountain front.

Windows, oil seeps, and drill holes indicate that the Lewis thrust underlies the whole of the Lewis, Livingston, and Clark Ranges, and that the rocks above it have been transported more than nineteen miles eastward. They have, further, moved without rupture, in contrast to the rocks in the segment farther north where those in the Front Ranges above the McConnell thrust have been split into many slices.

One might infer a comparison and a correlation between the Lewis thrust and the McConnell thrust, as both are master low-angle faults along the mountain front. The comparison is apt but the correlation is not. To the north the Lewis thrust changes into a high-angle fault which passes behind the front of the mountains; the McConnell thrust develops out of folds and fault blocks east of it. Thrusting of rocks of the mountains over those of the foothills is a continuous feature, but at one place movement is concentrated on one low-angle fault, elsewhere on another.

Lewis and Clark transverse zone. Between Helena and Missoula, Montana on the east, and Spokane, Washington on the west, the continuity of the Cordillera is interrupted by a series of west-northwest-trending transverse valleys. This topographic interruption has made its mark on human affairs, as it furnished the route for the historic journey of *Meriwether Lewis* and *William Clark* in 1805 to the Pacific Coast, by which the Oregon Country was gained for the United States; today it is followed by U.S. Highway 10. The topographic interruption expresses a fundamental discontinuity in Cordilleran structure which appropriately has been termed the *Lewis and Clark line* or transverse zone.

The transverse valleys are followed by discontinuous high-angle faults of the same trend, of which the best known is the *Osburn fault* of northern Idaho.

Along the same trend in the Eastern Ranges zones of abrupt flexing and belts of en echelon faults seem to express a draping of the sedimentary cover over faults in the basement rocks. On the Osburn fault and others near it there is a proved *right-lateral strike-slip displacement*. Curiously, however, the en echelon fault belts farther east indicate a movement in the opposite sense, suggesting that there were many complex adjustments in different directions in the transverse zone from one place to another and from one time to another.

Interesting though these faults may be, they are merely one feature of a much larger discontinuity. North of the transverse zone the ranges and valleys of the Cordillera are like those which we have described along the International Boundary and northward. They extend north-northwest and consist of geosynclinal sediments of Belt and later ages which have been thrown into folds and thrust blocks, or were transported eastward on low-angle faults like the Lewis and McConnell thrusts.

South of the transverse zone the ranges are more disordered and some trend north-northeast—an odd direction in the Cordillera. Folded and faulted miogeosynclinal sediments occur in patches, but many of the mountains consist of Precambrian basement rocks older than the Belt series, of acidic plutonic rocks and of Tertiary volcanics. A part of the character of the area south of the transverse zone was determined early in geologic time, as the Belt series appears to have been deposited over it only thinly, if at all, and near it overlaps abruptly toward a shore line with coarse marginal deposits.

PLUTONIC ROCKS SOUTH OF LEWIS AND CLARK ZONE. The most striking of the new features which appear south of the Lewis and Clark zone in the eastern Cordillera are great masses of acidic plutonic rocks.

To the west is the vast *Idaho batholith* which forms the rough mountain country of most of central Idaho. It had its inception during the Nevadan orogeny, but plutonic activity in the mass was prolonged; parts have yielded radiometric ages of about 100 million years (about the same as those of the plutonic rocks in the Sierra Nevada) but others have yielded later ages. The batholith may have been sufficiently consolidated by the time of the Laramide compression to disorder the structures of the eastern part of the Cordillera which were forming in front of it.

During Laramide orogeny the region east of the Idaho batholith became a plutonic realm also; numerous masses of granitic rocks, large and small, were emplaced in western Montana south of the transverse zone.

Largest of these is the *Boulder batholith* southwest of Helena, a through-breaking mass which has ascended high in the crust from its place of origin. At the surface it has an oval outline sixty miles long and twenty wide, but it probably originated at its southwestern end as a broad stock and spread thence northeastward as a floored intrusive in a downwarp of the latest Cretaceous deposits. The Boulder batholith is notable because it can be dated more closely by both stratigraphic and radiometric means than most of the plutonic bodies of the Cordillera:

(a) It invades lavas and tuffs which contain fossil plants of late Cretaceous age, and is overlain unconformably by other tuffs whose fossil plants are of early Oligocene age.

(b) Radiometric determinations on the plutonic rock itself indicate an age of sixty to seventy million years.

METALLIC MINERAL DEPOSITS OF THE LEWIS AND CLARK ZONE AND TO THE SOUTH. Western Montana and northern Idaho—with their transverse zone, the discontinuity which it expresses, and the plutonic realm south of it—are also the locus of the largest metallic mineral deposits of this part of the Cordillera. Exploitation of these deposits has contributed much to the wealth of the region.

The great copper deposits of Butte, Montana occur in veins in the fractured western part of the Boulder batholith; they have been worked for many years and are still actively producing. Farther west are the mining camps of Philipsburg, Montana and those in the Coeur d'Alene Mountains of northern Idaho.

It is worth recalling that the principal mineral deposits of Colorado are in another zone of transverse structures and plutonic rocks which crosses the Southern Rocky Mountains. Relation between the mineral deposits and the other features of these transverse zones is probably more than coincidence; the weakness in the crust which permitted an emplacement of magmas must also have allowed an ascent of mineral-bearing solutions.

WYOMIDE RANGES OF SOUTHEASTERN IDAHO AND WESTERN WYOMING. South of the region just discussed, in south-central Idaho, structures of the Mesozoic and older rocks are obscured across the *Snake River Plain*. This is another transverse zone that extends across the Cordillera, but it is much younger than the Lewis and Clark zone in west-central Montana—a broad downwarp which has been filled by Pliocene and Pleistocene lava flows.

South of the Snake River Plain the structures of the eastern part of the main Cordillera emerge again in the Wyomide Ranges—the western part of the Central Rocky Mountains as defined physiographically. Here the topography and structure are again like those in the Northern Rocky Mountains—a series of parallel ridges made up of folds and thrust blocks of Paleozoic Mesozoic miogeosynclinal sediments. The Wyomide Ranges, however, do not make the front of the Cordilleran system but pass behind the uplifts and basins of the Central Rocky Mountains; we have already mentioned their impingement on the latter (Chapter VII, Section 4).

Various west-dipping, low-angle thrust faults occur in the Wyomide Ranges, the easternmost of which face the Green River Basin in southwestern Wyoming. But the most famous is the *Bannock thrust* farther west in the system, which has been credited by its probably overenthusiastic discoverers with an eastward displacement of more than thirty miles. At its type area near the southeastern corner of Idaho the rocks above the Bannock have clearly been moved for many miles, but it probably changes character and loses displacement northward and southward in the same manner as the Lewis thrust. The significant feature here as in the Northern Rocky Mountains is not a single dominant low-angle thrust, but the whole complex of thrusts along which eastward movement has been distributed.

EASTERN GREAT BASIN. Southward from Idaho and Wyoming in northern Utah structures of the eastern part of the main Cordillera run out into the Great Basin, through which they extend into southern Nevada and southeastern California.

This is, however, a region of Basin and Range topography and structure, of disconnected ranges separated by broad alluvial basins produced by block faulting late in geologic time—one of the modifications of Cordilleran structure of which we have spoken and of which we will have more to say in Chapter IX, Section 2. Here the pattern of fundamental earlier structures has been obscured and disordered like bits and pieces of a jigsaw puzzle, but with essential parts missing where older rocks are covered by the younger in basins between the mountain ranges.

In our discussion of the miogeosynclinal rocks of this region in Section 2 of this chapter we selected the segment from the lower end of the Grand Canyon across southern Nevada as representative of the whole. In this segment the fundamental structures are also better displayed than farther north because of greater dissection of the ranges and basins by the Colorado River and its tributaries.

Here we have already recorded the westward thickening of the Paleozoic rocks from the Colorado Plateau or continental platform into the miogeosyncline on the west (Fig. 77). With this change there is an increasing complexity of the structure as well, from the nearly flat-lying strata on the Grand Wash Cliffs at the lower end of the Grand Canyon into strong folds and faults in the thicker sediments to the west. From the Muddy Mountains between the Grand Wash Cliffs and Las Vegas, Nevada to the Spring Mountains southwest of Las Vegas are at least seven west-dipping, low-angle, major thrusts above each of which the rocks have been transported for miles to the east.

In the Appalachians and Northern Rocky Mountains nearly all the thrust sheets have been eroded so far back that only their subsurface features are preserved; materials which were at the surface during the thrusting have been removed. Along the thrusts of southern Nevada such materials are preserved in many places, permitting inferences as to what the surface was like at the time of deformation.

Some of the thrusts in the northern Muddy Mountains are bordered on the east by coarse fanglomerates (Overton formation) of probable late Cretaceous or early Tertiary age, derived from erosion of the advancing thrust sheets. The thrust sheets override fossiliferous Cretaceous rocks (Willow Tank and Baseline formations) which themselves contain fragments of earlier formations, indicating that crustal activity began well before the end of Cretaceous time. Another thrust in the Spring Mountains farther west brings Cambrian beds over Jurassic sandstones. Between the two are water-worn cobbles and boulders, now sheared, which must have been laid on an erosion surface; evidently the thrust sheet later overrode this surface. The fault at the base of the thrust sheet was thus an *erosion thrust* of a sort which has been anticipated theoretically but seldom substantiated elsewhere.

To pursue details of the structures in other parts of the eastern Great Basin would probably only bewilder the reader without profit. Many folds and thrusts are known, but the larger pattern is for the most part undetermined. Not only have the fundamental structures been obscured over wide areas by Basin and Range structure, but many of the ranges have been little explored geologically.

This lack of knowledge may seem odd, as the region has been prospected and mined for generations, and has been examined by many mining geologists. Ordinarily, however, the local mineral deposit or ore body has been the chief claimant of interest, and little exploration has been done in the less mineralized intervening areas.

A new approach to understanding of the region came during the last decade, when it was conceived that this was a fruitful field for petroleum exploration because of its thick masses of miogeosynclinal rocks. Exploration of such a potential oil province must be made on a regional basis, with investigation of the structure and stratigraphy of wide areas.

Whether the eastern part of the Great Basin will become a major oil province still remains to be determined. Hopes that it may be were raised in 1954 when the Shell Oil Company, Eagle Springs Unit no. 1 well, put down in Railroad Valley in east-central Nevada to a depth of 10,356 feet, obtained 375 barrels of oil per day at a depth of 6,737 feet. Curiously, this first oil strike in the province was derived from rhyolite tuff of Tertiary age; the oil probably migrated there from Paleozoic or Mesozoic marine sediments beneath. With the impetus of possible oil production in the eastern Great Basin, we will probably learn much about its fundamental structures before many years have passed.

CONCLUSION. Characteristic structures of the miogeosynclinal area extend southward into California and Arizona, but seemingly with changes in style that are as yet poorly understood; it would unduly prolong the story to attempt an interpretation of these southern areas. Characteristic structures also extend westward about halfway across the Great Basin, where they meet the rocks and structures of the eugeosynclinal area; it will be most fruitful now to pass on to these features.

4. CORDILLERAN EUGEOSYNCLINE AND STRUCTURES THAT FORMED FROM IT

COMPARISON WITH CRYSTALLINE PART OF APPALACHIANS. In our earlier summary of the Cordillera (Chapter VI) we pointed out that the oceanward part of the system away from the continental interior possesses many of the same features as the oceanward part of the Appalachians—it was a eugeosynclinal area during the geosynclinal stage, and its rocks were afterwards consolidated by metamorphism and plutonism into a terrane like that in the crystalline Appalachians.

The differences are merely the expectable ones between two widely separated mountain systems. Here the oceanward side is on the west rather than the southeast, and the ocean is the Pacific rather than the Atlantic. In the Appalachians eugeosynclinal conditions largely ceased by the end of Devonian time; here they continued into the Jurassic, so that the Cordilleran eugeosyncline contains large volumes of late Paleozoic and early Mesozoic deposits.

The crystalline area of the Appalachians is concealed in many places by later surficial deposits; that of the Cordillera is equally so. Up to its climactic orogeny this western belt of the Cordillera must have been continuous along the Pacific border. Now it is widely covered by sediments and lavas, mainly continental in origin and Tertiary in age, so that it is exposed only in fragments. To visualize its original character and extent requires much reconstruction.

RELATIONS ON THE MAP. Some idea of the original extent of the western belt of the Cordillera can be gained from geological maps, which indicate the disconnected exposures of its altered geosynclinal rocks and the extensive bodies of acidic plutonic rocks that invade them (Plate I).

To the north is the mighty *Coast batholith*, an assemblage of plutonic rocks in the high mountains along the Pacific coast, which extends continuously for 1,100 miles from Yukon Territory through the panhandle of Alaska to southern British Columbia. A little to the east in British Columbia are other plutonic bodies, large of themselves but dwarfed by their mighty neighbor on the west. They extend southward across the International Boundary into the state of Washington where they pass beneath the Tertiary lavas of the Columbia Plateau and Cascade Range so that their extensions are lost.

East of the Columbia Plateau and farther east than the other plutonic bodies of the belt is the great *Idaho batholith* of central Idaho, referred to earlier (Section 3 of this chapter). Southwest of it, with the connection partly masked by lavas, eugeosynclinal rocks and small plutonic bodies emerge in the *Blue Mountains*, *Wallowa Mountains*, and other highlands of northeastern Oregon. Beyond is a wide gap covered by Tertiary lavas, but the southwestward strike of the rocks of northeastern Oregon carries them toward the *Klamath Mountains* of southwestern Oregon.

Taken together, the eugeosynclinal rocks and associated plutonic bodies of northern Washington, central Idaho, and the Blue and Klamath Mountains of Oregon seem to define an *east-bulging arc* of the eugeosynclinal rocks and structures 400 miles across, now more than half concealed by the Tertiary lavas of the Columbia Plateaus (Fig. 89 and Chapter IX, Section 3).

The eugeosynclinal and plutonic rocks of the Klamath Mountains extend from Oregon into northern California, where their strike turns southeastward toward the *Sierra Nevada* from which they are separated by only a short lava-covered gap. In the Sierra Nevada the eugeosynclinal rocks, with great masses of embedded acidic plutonic rocks, are exposed along 400 miles of strike to the *Tehachapi Mountains*; outlying plutonic masses also occur well to the east in the western part of the Great Basin.

South of the Tehachapi Mountains the eugeosynclinal and plutonic rocks are much broken up and confused by the complex structures of southern California—those trending east and west in the *Transverse Ranges*, and those trending southeast along the San

Andreas fault. South of this confused area, however, they reappear as a more continuous body in the *Peninsular Ranges* of southernmost California, which extend southward into Baja California.

THE SIERRA NEVADA. To characterize the eugeosynclinal rocks and the structures that formed from them we will discuss more fully a typical segment, that in the Sierra Nevada of California.

We must do so, however, with a word of caution. The present Sierra Nevada is a great tilted block (Fig. 82), shaped accidentally in Tertiary and Quaternary time out of a more extensive earlier eugeosynclinal and crystalline terrane that extended far northward and southward as well as eastward and westward. The present range is merely a fragment of the original structure.

Let us recall how gold was discovered in the western foothills of the Sierra Nevada in 1848—in the tailrace of Sutter's Mill at Coloma—and how it was exploited by the Forty-niners. Most of the Forty-niners obtained their gold from *placers*; that is, from the alluvial deposits of streams flowing out of the Sierra. But as the placers became depleted the more experienced prospectors and miners began following the alluvial gold up the rivers to its source in the older rocks.

They found a great concentration of gold in quartz veins along a narrow zone in the foothills where the bedrock was intensely sheared and faulted, and so gave it the name of Veta Madre or *Mother Lode.* Here they began the more difficult task of vein or lode mining, which their successors profitably carried on for generations. More considered judgment in later years has indicated that this zone was only one source of the placer gold; much of it came from other sources in the bedrock or from older placer deposits of Tertiary age. The name Mother Lode has persisted, nevertheless, in popular legend.

Geologic investigation of the gold belt in the Sierra Nevada lagged far behind the feverish searchings of the prospectors. The belt was first studied between 1860 and 1874 by the Geological Survey of California under *J. D. Whitney*, and later by parties of the U.S. Geological Survey under *G. F. Becker*; the U.S. Survey produced a dozen geologic folios covering the western slope of the Sierra Nevada which are still classic.

The geologists found that the rocks of the gold belt fell into two great subdivisions which were called in the folios the *Superjacent series* and the *Bedrock or auriferous series.*

SUPERJACENT SERIES. The Superjacent series comprises those deposits which lie with little disturbance on the eroded edges of the Bedrock series. Along most of the western slope of the Sierra Nevada these are of Tertiary age, but at the north end of the Sacramento Valley somewhat older strata, the Chico formation of Late Cretaceous age, have a similar relation to the rocks beneath, hence may be included. The Chico establishes the climax of the deformation of the Bedrock series as having occurred before the end of the Mesozoic time and as having been produced by an orogeny earlier than the one which deformed the Rocky Mountains farther east.

The Tertiary deposits of the Superjacent series fringe the western base of the Sierra Nevada, and also extend far up its slopes in old swales in the topography or on ridge tops between the present streams (Fig. 82A, B). They include gravels which were laid down along the courses of vanished rivers which flowed westward from sources somewhere east of the modern crest line, across the site of the Sierra Nevada before it was raised to its present heights. The gold prospectors found that these old river gravels contained placer gold like the modern ones, and they worked them extensively by hydraulic mining.

Volcanic eruptions occurred from time to time in the region, and lavas flowed down the river valleys; these formed a capping of the gravels which resisted erosion; when further downcutting took place the former valleys were thus left standing in the higher parts of the country (Fig. 81).

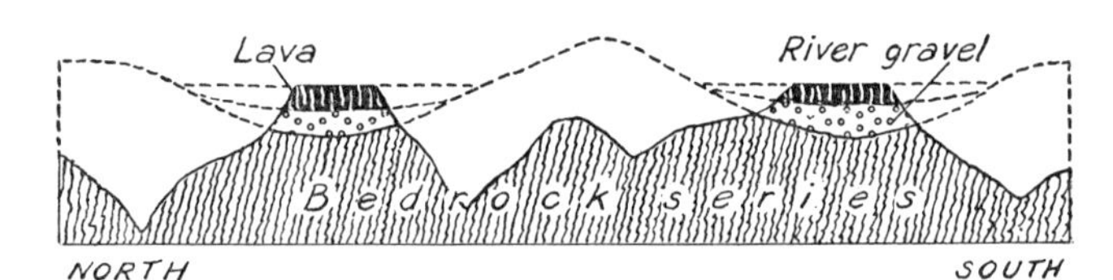

FIG. 81. Sketch section in the western part of the Sierra Nevada showing relation of Tertiary gravels and lavas (Superjacent series) to the bedrock and to present topography.

Detailed study of the Superjacent gravels and lavas indicates that they had a more complex history than was first supposed, and were formed at intervals from Eocene to Pliocene time. But that is another story; our interest here is in the Bedrock series.

BEDROCK SERIES. The Bedrock series comprises the eugeosynclinal rocks of the Sierra Nevada—now deformed, upended, and altered—and the plutonic rocks which invade them. The stratified eugeosynclinal rocks were early recognized as of both Paleozoic and Mesozoic age, the former being termed the *Calaveras formation* and the latter, the *Mariposa slate.* Besides these, the folios indicate numerous bands of "diabase porphyrite," "augite porphyrite," "quartz porphyrite," and "amphibolite," which one would suppose were igneous intrusives; they are, instead, surface lavas and pyroclastics that are interbedded in the Paleozoic and

Mesozoic stratified succession. Taken together, the sediments and lavas are a typical eugeosynclinal assemblage most of which was laid down beneath the sea.

THE PALEOZOIC ROCKS. The rocks mapped as *Calaveras* on the west slope of the Sierra Nevada comprise a great thickness of slate and phyllite with interbedded layers of graywacke and conglomerate, lenses of chert and limestone, and in places much lava and tuff. We may suppose that the Calaveras accumulated in deeply subsiding troughs separated by tectonic and volcanic ridges, probably to the accompaniment of much crustal unrest.

The age of the Calaveras is indicated by occasional fossils only. During the folio work it was supposed to be of early Carboniferous (Mississippian) age, but later discovery of fossils as young as Permian indicates that it embraces a considerable age span and probably includes a number of subdivisions as yet undifferentiated.

The Calaveras itself may contain no rocks older than the Carboniferous, but earlier Paleozoic beds have been identified in nearby areas—fossiliferous Devonian and Silurian in the Klamath Mountains, fossiliferous Silurian at the north end of the Sierra Nevada, and fossiliferous Ordovician on the east flank of the Sierra Nevada near the head of Owens Valley. In the Klamath Mountains, moreover, the Devonian and Silurian rest on still earlier sediments and volcanics from which no fossils are known, but which may be Ordovician or Cambrian. Eugeosynclinal conditions began in this western part of the Cordillera at a very early period.

The basement on which the eugeosynclinal deposits were laid is undetermined. Certain belts of schist and gneiss in the Klamath Mountains were called Precambrian in early reports but may well be equivalent to other rocks in the succession altered in zones of unusual disturbance. There is, in fact, no proved Precambrian anywhere near the Pacific coast except the radiometrically dated anorthosite of the San Gabriel Mountains north of Los Angeles; this lies in a special structural situation not closely related to that of the rocks of which we are speaking. Although the evidence is inconclusive, the eugeosynclinal deposits may not have formed on sialic continental crust but on simatic oceanic crust, so that the eugeosynclinal belt of the Cordillera was built out from the continent at the expense of the Pacific Ocean basin.

Occasional layers of conglomerate in the eugeosynclinal Paleozoic rocks give testimony of crustal instability in the area during their formation. Besides, the Calaveras is more complexly metamorphosed than the Mesozoic rocks; it is therefore thought by some geologists to have been deformed and metamorphosed before the latter were deposited over it. This difference may, however, reflect only the greater depths to which it was depressed and buried, and there is no convincing evidence of major late Paleozoic orogeny in the Sierra Nevada.

THE MESOZOIC ROCKS. The Mesozoic rocks of the gold belt include not only the *Mariposa slate*, but the even thicker *Amador group* beneath, also partly sedimentary but with extensive flows of basic lava (the "porphyrites" of the folios). Amador and Mariposa are of Middle and Late Jurassic age and lie with hiatus on the Paleozoic rocks. This hiatus is largely filled in nearby areas where rocks of Early Jurassic and Triassic age are present; like the Mariposa and Amador these are a succession of slates, graywackes, cherts, and volcanics of eugeosynclinal character.

The Mariposa is assigned to the Kimmeridgian or next-to-the-last stage of the Jurassic on the basis of ammonoids, pelecypods, and other marine invertebrate fossils. It has been claimed, in fact, that no rocks younger than Kimmeridgian are present in the Bedrock series, but the paleontological basis for this is dubious. Fossils of still younger ages have been reported occasionally; for example, the Monte de Oro formation at the north end of the gold belt is said to contain fossils of Portlandian age (Fig. 83 B).

DEFORMATION OF THE EUGEOSYNCLINAL ROCKS. A striking feature of the bedrock in the gold belt is the steep dip of its Paleozoic and Mesozoic strata. In other regions steep or vertical beds are common enough on the flanks of folds, but here they are well-nigh universal. The rocks are, moreover, seldom folded, and the sequence is almost invariably upward toward the east, toward the heart of the Sierra Nevada (left-hand ends of sections, Fig. 82). The only duplication is created by faults of the Mother Lode zone and others west of it, which raise the older rocks on their eastern sides by complex strike-slip and dip-slip movements.

On the opposite or eastern side of the Sierra Nevada south of Mono Lake the structure is similar but in the reverse sense. Paleozoic and Mesozoic strata dip steeply, with the sequence upward toward the west, again toward the heart of the Sierra (right-hand end of section C, Fig. 82). The two steeply dipping and opposing belts appear to be the flanks of a much greater feature—a crustal downfold whose trough is now obliterated by the masses of acidic plutonic rocks that form the core of the range.

The stratified rocks were metamorphosed, probably at the time of the deformation, so that their constituents were recrystallized into chlorite and other low-grade metamorphic minerals, and the more argil-

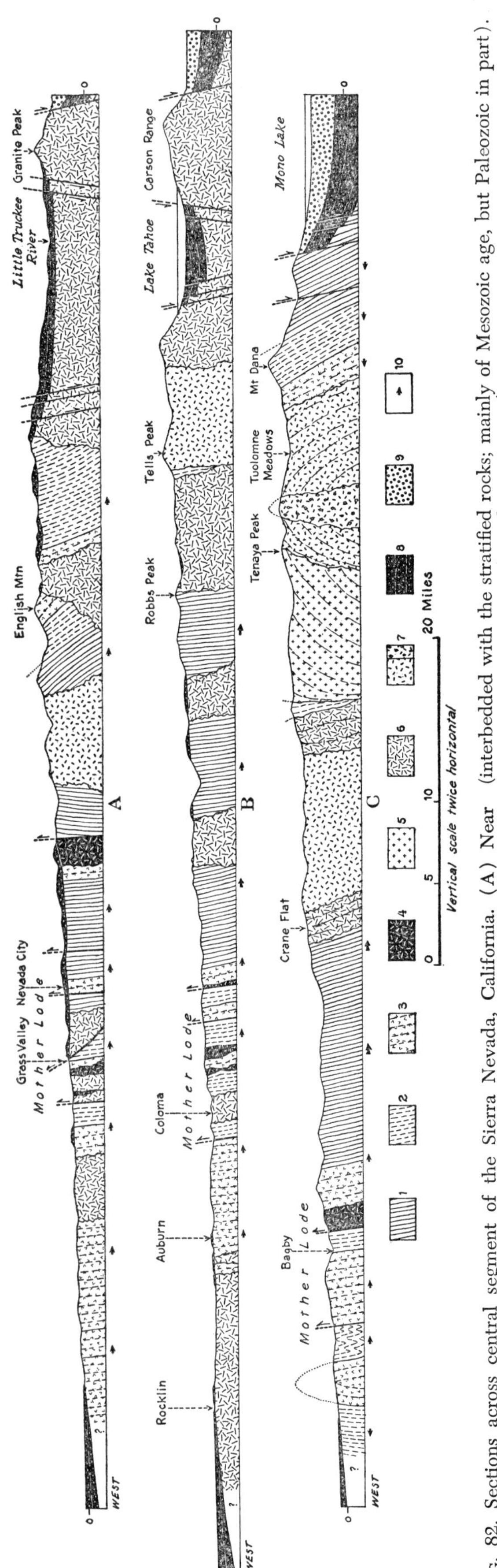

Fig. 82. Sections across central segment of the Sierra Nevada, California. (A) Near Yuba River. (B) Near American River. (C) Near Merced River a little north of Yosemite Valley. Compiled from sections in the geologic folios and various later sources.

Explanation of symbols. 1. Paleozoic strata (Calaveras formation on west, unclassified units on east in section C). 2. Mesozoic strata (mainly Amador group and Mariposa formation; includes Milton formation on east in section B). 3. Volcanic rocks (interbedded with the stratified rocks; mainly of Mesozoic age, but Paleozoic in part). 4. Ultrabasic intrusives (mainly serpentine).
Rocks of Sierra Nevada batholith: 5. quartz monzonite, 6. granodiorite, 7. granite and porphyritic granite.
8. Tertiary sediments and lavas ("Superjacent series"). 9. Quaternary sediments (mainly in depressed areas east of Sierra Nevada). 10. Direction of tops of sedimentary sequences (indicated by arrows).

laceous rocks were converted into slates and phyllites. During the plutonic invasions which followed, a higher-grade thermal metamorphism was imposed which locally converted the rocks into hornfels or even into gneiss.

The plutonic rocks. Earliest plutonic rocks of the Sierra Nevada were ultrabasic (ultramafic) and are largely represented now by *serpentines* which are embedded in the eugeosynclinal rocks as pods or masses a few miles in diameter; they remind us of the early ultrabasics that invade comparable rocks in the crystalline area of the Appalachians.

Of much greater extent and volume are the succeeding acidic plutonic rocks. Broadly, these may be called "granitic," although true *granites* in the technical sense are rather minor, most of them being the somewhat more basic *quartz monzonites, granodiorites,* and *quartz diorites.* Unlike many acidic plutonic rocks of other regions, those of the Sierra Nevada are of prevailing light gray or white color from lack of reddish coloration in their feldspars. Their shining peaks and bare rock faces inspired John Muir's poetic designation of the region as the "range of light."

The acidic plutonic rocks are embedded in the earlier rocks as small to large masses throughout the length of the Sierra Nevada (Plate I). They have been referred to as the *Sierra Nevada batholith,* but this is an oversimplification. One set of plutonic masses extends along the western side west of the Mother Lode belt, at least as far south as the latitude of Sacramento (Fig. 82 A and B); it forms an extension of the plutons of the Klamath Mountains and appears to be the oldest of the series (see below). Farther east acidic plutonic rocks form a nearly solid body along the crest of the range (near the axis of the crustal downfold above referred to). One should not infer that even this was a single great lake of solidified magma. Close study indicates that it consists of a whole succession of intrusions of varying shapes and sizes, each of different texture and composition, the younger cross-cutting the older (Fig. 82 C). We may presume that emplacement of the whole must have occupied a long period.

Nearly all the acidic plutonic rocks of the Sierra Nevada are massive and cross-cutting rather than foliated and concordant, and were largely emplaced after the deformation of the eugeosynclinal rocks had ceased. Apparently they rose as true magmas from the depths to their present levels in the crust, breaking through the strata or forcing them aside. Elsewhere in the western part of the Cordillera structural relations of the plutonics are more varied; they include not only massive and cross-cutting bodies such as those in the Sierra, but also foliated and concordant bodies and magmatic permeations in a prevailing, highly altered country rock.

Viewing the western belt of the Cordillera as a whole from Alaska to Mexico the acidic plutonic rocks occupy a much greater surface area than their counterparts in the Appalachians and pose, even more compellingly, some fundamental geologic questions:

(A) What was the source of the enormous volumes of granitic material?

(B) Why are granitic rocks characteristically emplaced only in the former eugeosynclinal areas?

(C) What became of the eugeosynclinal rocks whose place they have taken?

These questions are much debated, and final answers are still elusive. To this writer the ultimate source of the granitic material would seem to be in the eugeosynclinal rocks themselves, their transformation taking place when they were carried down to great depths in the crust. The plutonic rocks of different structural habits in the various parts of the western Cordillera seem to be members of a *granite series* —progressing from static, partly transformed material at depth and during early phases of the orogeny to mobile magmas which invaded higher levels of the crust during closing phases of the orogeny. Clearly, the acidic plutonic rocks of the Sierra Nevada fall into the latter part of the series; perhaps they had their origin earlier and at greater depths in the trough of the great downfold of eugeosynclinal rocks which followed the axis of the range.

Age of Nevadan orogeny. These events—deformation of the eugeosynclinal rocks and their subsequent invasion by plutonic rocks—represent the *Nevadan orogeny* in its type area. When did this orogeny take place?

Within the Sierra Nevada itself the time of the orogeny is bounded only by wide limiting dates; youngest strata involved are of Late Jurassic age (Mariposa formation and its associates), and oldest strata not involved are of Late Cretaceous age (Chico formation); Nevadan orogeny occurred during the intervening hiatus (columns for Sierra Nevada, Fig. 83). Farther west beyond the Sacramento Valley the Late Cretaceous lies conformably on earlier Cretaceous strata (Horsetown and Paskenta formations), and these on Late Jurassic strata (Knoxville formation), none of which are metamorphosed or invaded by acidic plutonic rocks (columns for west side of Sacramento Valley, Fig. 83). The Knoxville has been thought, as well, to intergrade with and overlie the great mass of sediments of the Franciscan group which form much of the Coast Ranges farther west (Fig. 83 A). The Knoxville is of Portlandian age; it and the Franciscan group have been supposed to be altogether younger than the deformed rocks of the Sierra Nevada, the latest of which were believed to be Kimmeridgian.

These relations are the basis for a proposal that deformation of the eugeosynclinal rocks of the Sierra Nevada occurred during a relatively brief interval toward the end of, but not at the close of Jurassic time (between the Kimmeridgian and Portlandian stages), with invasion of the plutonic rocks following shortly after. It is worth commenting that this particularistic view was not shared by the men who did the basic field work; judgment of the U.S. Geological Survey in summarizing the folio work and investigations which succeeded it was that the plutonic invasions and related orogeny were of "Late Jurassic or Early Cretaceous" age.

The picture of orogenic processes which we have already constructed in this book would lead us to suspect that the latter view was the more likely. It would appear incredible that the strata in the Sierra Nevada should all be Kimmeridgian and older, and those west of it, all Portlandian and younger—especially as these stages are adjacent in the Jurassic sequence and their differentiation rests on subtle differences in the plications or sutures of invertebrate fossils, frequently of the same genera. Later paleontological work indicates, in fact, that the two sequences do overlap; absence of beds later than Kimmeridgian in the Sierra Nevada has not been proved and there are indications that the Monte de Oro formation of that area is Portlandian. Moreover, the Franciscan group, which had been thought to underlie the Knoxville and to be Portlandian also, has proved to include fossiliferous beds not only of Jurassic age, but of Early and early Late Cretaceous age (Fig. 83 B).

The Late Jurassic and Early Cretaceous sedimentary rocks west of the Sierra Nevada which were thought to have been formed in a new sedimentary trough after the close of the Nevadan orogeny thus appear in new light. They might, instead, have been deposited before and during the orogeny in the Sierra Nevada, with intensity of deformation diminishing westward. Moreover, as we shall see presently, deformation and plutonism began earliest in the western part of the eugeosynclinal belt and advanced eastward with time.

Radiometric determinations now indicate that the invasions of plutonic rocks which followed shortly after the deformation occupied a sizeable fraction of later Mesozoic time. Plutonic rocks on the west are earliest; those west of the Mother Lode belt in the Sierra Nevada have yielded ages of 131 to 143 million years and are related to those of the Klamath Mountains which have yielded ages of about 134 million years; these are probably of Middle Jurassic age. On the other hand, a suite from the Sierra Nevada batholith proper east of the Mother Lode near Yosemite Valley have yielded ages of 95 to 83 million years, with the

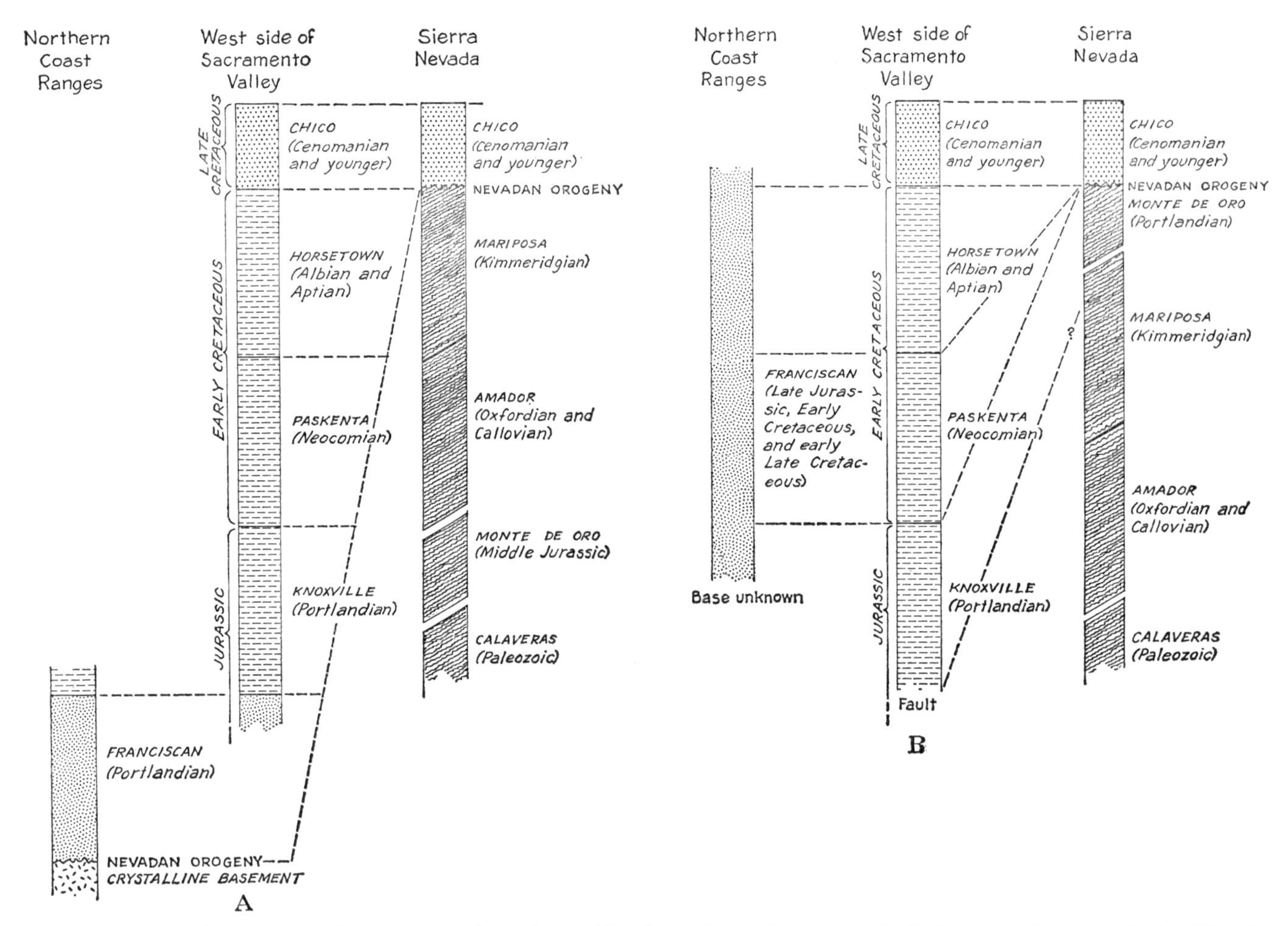

Fig. 83. Stratigraphic sections of Mesozoic rocks in Sierra Nevada and to the west, which bear on the age of the Nevadan orogeny. (A) According to correlations previously proposed (based on Anderson, 1938; and Taliaferro 1942). (B) According to recently revised correlations (based in McKee and others, 1956; and Irwin, 1957).

oldest on the west and youngest on the east; they are probably of Early to Middle Cretaceous age.

One can therefore infer that intrusion of plutonic rocks was in progress beneath the surface along the site of the Sierra Nevada at the same time as Late Jurassic and Cretaceous sedimentary rocks were being deposited conformably west of it. During this epoch the site of the Sierra was probably being uplifted and eroded, but it was not until much later that the plutonic masses were unroofed; sedimentary products of their deeply weathered surfaces first appear in the high-grade clays of the Eocene Ione formation along the west flank of the Sierra Nevada.

Elsewhere in the western belt of the Cordillera where deformed and plutonized eugeosynclinal rocks occur it has commonly been assumed on the basis of relations in California, that they were likewise deformed in late Jurassic time. In many places decisive evidence is lacking, the limiting dates being as great or greater than in California. But where evidence is obtainable the record turns out to be complex and varied. In the Hawthorne-Tonopah area of southwestern Nevada major thrust faulting was in progress during deposition of adjacent Early Jurassic sediments; in northern Baja California Early and Middle Cretaceous rocks are involved in the orogeny and are overlain unconformably by Late Cretaceous rocks; in the northern Cascade Range of Washington there was major deformation and regional metamorphism before the Middle Jurassic, followed by folding and thrusting in Middle Cretaceous time. Nevadan orogeny in the western belt of the Cordillera was thus a prolonged event which began in mid-Mesozoic time and approached or overlapped the time of Laramide orogeny in the eastern belt of the Cordillera.

Conclusion. This review of the features of the Sierra Nevada and its environs is perhaps a sufficient sampling of the rocks and structures of the western belt of the Cordillera to illustrate how it was transformed from a eugeosyncline into a crystalline and plutonic region. Descriptions of other areas in the belt would add further details to the local history and would underscore the broader principles here set forth, but the essential picture would not be altered.

Relation of euogeosynclinal and miogeosynclinal rocks in Great Basin. Before leaving the subject of the eugeosynclinal belt of the Cordillera let us drop back to the east and discuss its relation to the miogeosynclinal belt with which we dealt in Section 2 of this chapter.

It is worth reminding ourselves that the topographic Sierra Nevada originated in Tertiary time, and that the realm of the Cordilleran eugeosyncline and the structures which formed from it extended over a much wider area, not only along but across the strike. Somewhere east of the Sierra Nevada in the Great Basin this realm must impinge on the realm of the miogeosyncline.

One of the sequences of miogeosynclinal rocks which we discussed was that in the Inyo Mountains of the southern Great Basin, a range which is separated from the Sierra Nevada on the west merely by the trough of Owens Valley. Here the change from miogeosynclinal rocks with their dominant carbonates to the eugeosynclinal rocks with their dominant clastics and volcanics takes place across a single desert valley in a zone immediately east of the present topographic front of the Sierra Nevada. North of the Inyo Mountains, however, the boundary between miogeosynclinal and eugeosynclinal rocks diverges northward or northeastward from the front of the Sierra and passes into the middle of the Great Basin.

A contrast between the pre-Tertiary rocks of the eastern and western Great Basin has been known since the days of Clarence King's Fortieth Parallel Survey—rocks of the eastern half are largely Paleozoic carbonates, those of the western half are largely Triassic and Jurassic sediments with considerable volumes of interbedded volcanics and intrusions of acidic plutonics like those in the Sierra Nevada. These Triassic and Jurassic rocks are of the sort that one might expect along the margin of a eugeosyncline.

With further investigation it was found that the Paleozoic changed in character toward the west as well. At various places in the central Great Basin the place of the Ordovician limestones was taken by graptolite shales; other shales also developed westward in the Devonian and Mississippian. At some localities

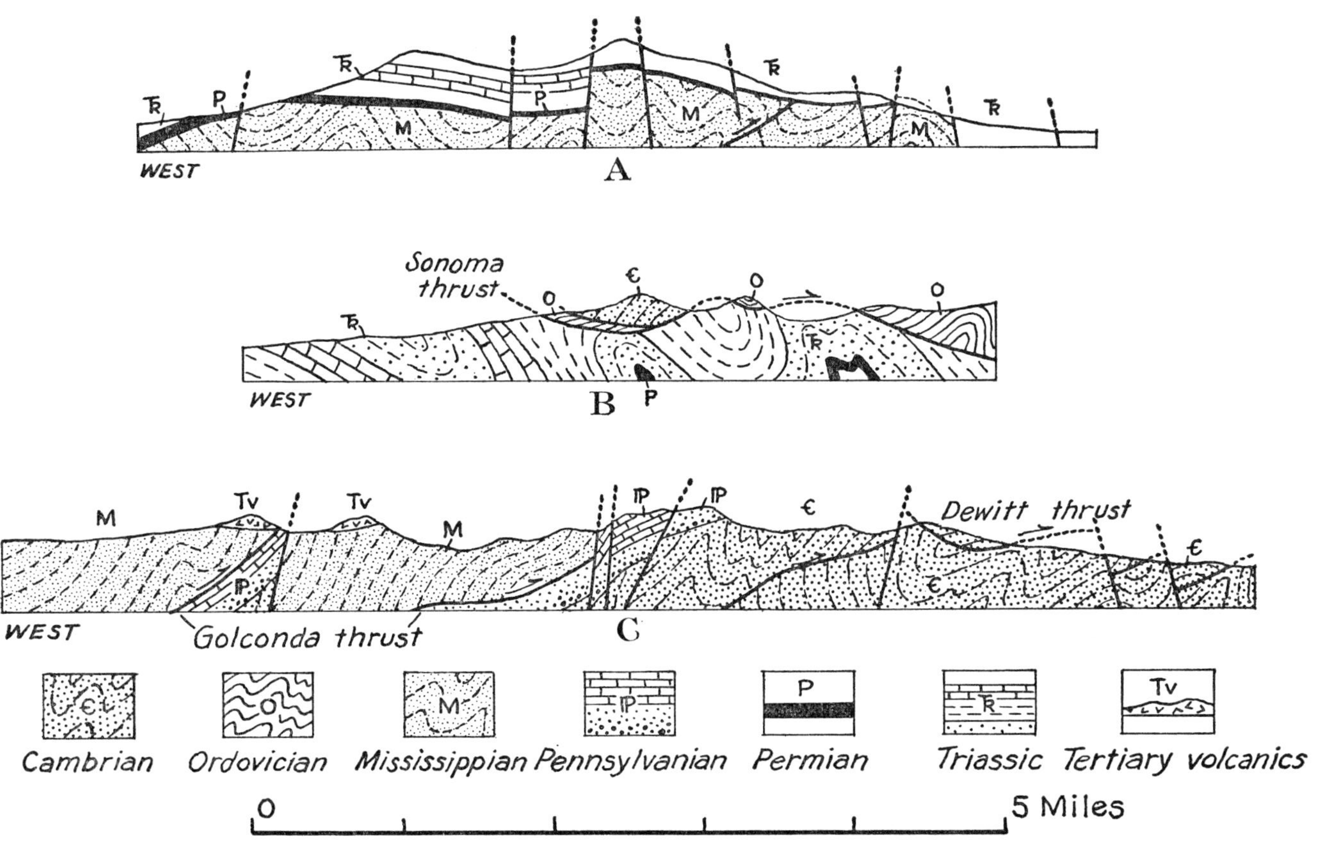

Fig. 84. Sections showing deformed Paleozoic and Mesozoic eugeosynclinal rocks of north-central Nevada. (A) North end of Tobin Range. (B) West side of Sonoma Range. (C) Battle Mountain. The Sonoma and Golconda thrusts are of post-Triassic age; the Dewitt thrust, of pre-Pennsylvanian age. The Triassic rocks of section A are of a different facies from those of section B and are believed to lie on the upper plate of a major thrust. After Ferguson, Muller, and Roberts, 1951, 1952.

marked unconformities were discovered in the sequence, such as one between Permian and Ordovician in the Manhattan district of south-central Nevada. These unconformities and the marked differences between both Paleozoic and Mesozoic rocks to the east and west suggested the existence during those eras of a positive axis along the center of the Great Basin which became known as the *Manhattan geanticline.*

ROCKS AND STRUCTURES OF NORTH-CENTRAL NEVADA. More detailed geologic work done in the last few decades has greatly amplified and modified these earlier concepts.

In north-central Nevada between Eureka and Winnemucca a complex body of Paleozoic eugeosynclinal rocks has been discovered, made up of argillites, cherts, graywackes, and interbedded volcanics. Compared with the miogeosynclinal strata they are poorly fossiliferous, but occasional fossil discoveries from one place to another indicate the presence of many of the Paleozoic systems from the Cambrian upward. Details of the sequence need not trouble us—it is still under investigation, anyway—but one of its characteristic parts is the Vinini formation which contains graptolites of Ordovician age. In contrast to the eugeosynclinal rocks of the Sierra Nevada, those of north-central Nevada are virtually unmetamorphosed.

These eugeosynclinal rocks have been transported eastward across the miogeosynclinal carbonate rocks in broad thrust sheets of relatively small thickness, now so broken by Basin and Range faulting that they are preserved as fragments in successive mountain ranges (Fig. 84). The easternmost of these, the *Roberts Mountain thrust*, carries eugeosynclinal rocks of various older Paleozoic systems over miogesynclinal rocks of about the same ages for distances of more than fifty miles. The *Golconda thrust* farther west similarly carries one sequence of Triassic rocks for miles over another sequence of about the same age.

More remarkable than the prevalence and magnitude of the thrusting is the fact that thrusting of much the same character recurred in the region at widely separated times. The Roberts Mountain thrust and its kindred, which involve early Paleozoic rocks, were emplaced in late Mississippian or early Pennsylvanian time and are overlapped unconformably from the east by Pennsylvanian conglomerates and limestones; unconformities higher in the Pennsylvanian and Permian indicate continuing crustal instability of the region. The Golconda thrust and its kindred, which involve Triassic rocks, formed much later, probably as a phase of the Nevadan orogeny (Fig. 85).

In our discussion of the Appalachian area we alluded to the problem of the relation between the miogeosynclinal and eugeosynclinal rocks—whether miogeosyncline and eugeosyncline were merely two parts of a single trough or were separated through much of their history by a geanticline (Chapter IV, Section 5; Fig. 35).

In north-central Nevada we meet the same problem again, and although the region furnishes much pertinent information it is still unsafe to translate this into any general principles. We learn that in this region the boundary between miogeosyncline and eugeosyncline was unstable; the Manhattan geanticline (if we may still use the term for this feature) was much more complex than first supposed. Parts of the body of eugeosynclinal rocks were carried from time to time

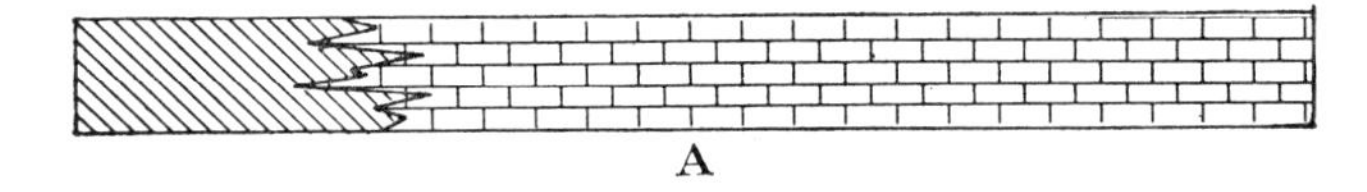

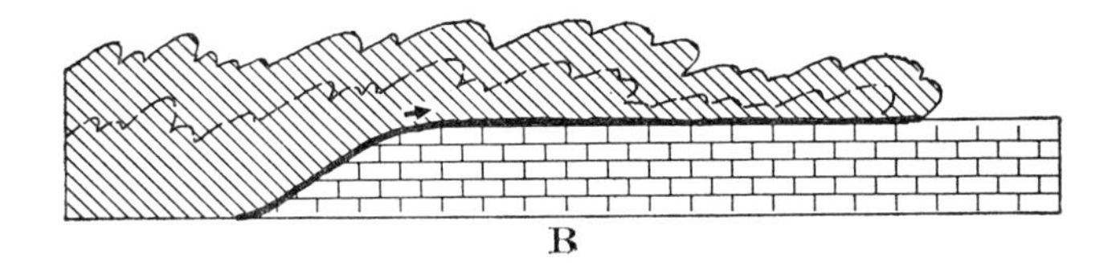

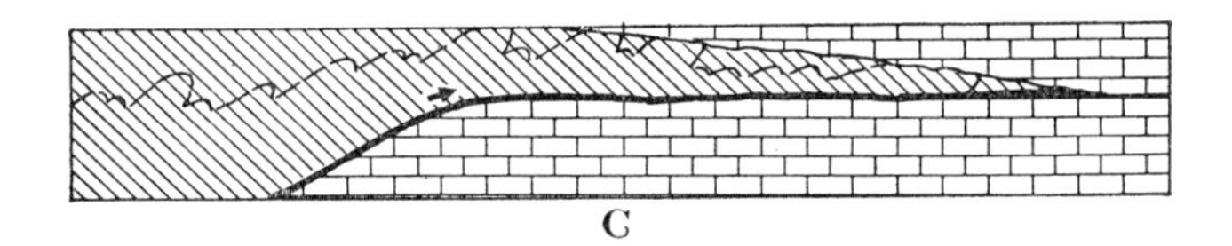

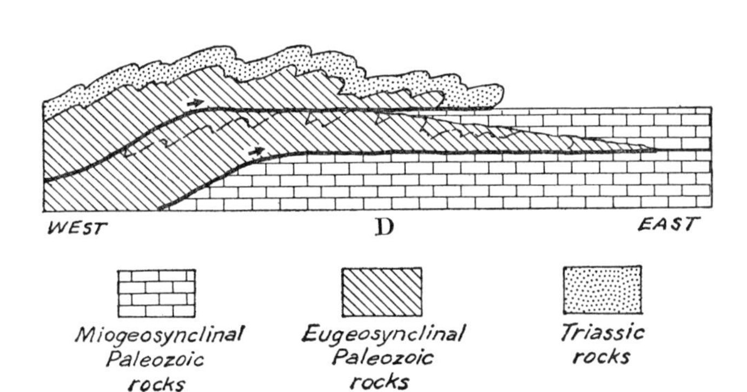

FIG. 85. Diagrammatic sections to illustrate part of the structural history of north-central Nevada. (A) Early to middle Paleozoic time; deposition of contrasting geosynclinal facies. (B) Late Mississippian or early Pennsylvanian time; eastward thrusting of eugeosynclinal over miogeosynclinal deposits along Roberts Mountain thrust and its kindred. (C) Pennsylvanian and Permian time; westward overlap of miogeosynclinal deposits over thrust complex. (D) Mid-Mesozoic time; renewal of eastward thrusting along Golconda thrust and its kindred.

far over the miogeosynclinal rocks by some mechanism which we can as yet only dimly envision. During lulls in the process the miogeosynclinal environment of the area thus invaded reasserted itself, permitting westward advances of carbonate rocks over the eroded surfaces of the thrust sheets (Fig. 85 C).

REFERENCES

2. *Cordilleran miogeosyncline*

Fenton, C. L., and Fenton, M. A., 1937, Belt series of the north; stratigraphy, sedimentation, paleontology: *Geol. Soc. America Bull.*, v. 48, p. 1873-1969.

Lochman-Balk, Christina, 1956, The Cambrian of the Rocky Mountains and southwestern deserts of the United States and adjoining Sonora province, *in* El sistema Cámbrico, su paleogeographía y el problema de su base: *XX Int. Geol. Cong. (Mexico)*, v. 2, p. 529-661.

Nolan, T. B., 1943, The Basin and Range province in Utah, Nevada, and California: *U.S. Geol. Survey Prof. Paper 197 D*, p. 141-196 (esp. p. 145-163).

Nolan, T. B., Merriam, C. W., and Williams, J. S., 1956, The stratigraphic section in the vicinity of Eureka, Nevada: *U.S. Geol. Survey Prof. Paper 276*, 77 p.

Reesor, J. E., 1957, The Proterozoic of the Cordillera in southeastern British Columbia and southwestern Alberta: *Roy. Soc. Canada Spec. Publ. 2*, p. 150-177.

Ross, C. P., 1956, The Belt series in relation to the problems of the base of the Cambrian system, *in* El sistema Cámbrico, su paleogeographía y el problema de su base: *XX Int. Geol. Cong. (Mexico)*, v. 2, p. 683-669.

Warren, P. S., 1951, The Rocky Mountain geosyncline in Canada: *Royal Soc. Canada Trans.*, 3rd ser., v. 45, sec. 4, p. 1-10.

3. *Structure of eastern part of main Cordillera*

Daly, R. A., 1912, Geology of the North American Cordillera at the forty-ninth parallel: *Canada Geol. Survey Mem.* 38, 799 p.

Eardley, A. J., 1944, Geology of the north-central Wasatch Mountains, Utah: *Geol. Soc. America Bull.*, v. 55, p. 819-894.

Hewett, D. F., 1956, Geology and mineral resources of the Ivanpah quadrangle, California and Nevada: *U.S. Geol. Survey Prof. Paper 275*, 172 p.

Hume, G. S., 1957, Fault structures in the foothills and eastern Rocky Mountains of southern Alberta: *Geol. Soc. America Bull.*, v. 68, p. 395-412.

Longwell, C. R., 1950, Tectonic theory viewed from the Basin Ranges: *Geol. Soc. America Bull.*, v. 61, p. 413-414.

Knopf, Adolph, 1957, The Boulder bathylith of Montana: *Am. Jour. Sci.*, v. 255, p. 81-103.

North, F. K., and Henderson, G. G. L., 1954, Summary of the geology of the Southern Rocky Mountains in Canada: *Alberta Soc. Petrol. Geol. 4th Ann. Field Conf. Guidebook* (Banff-Golden-Radium), p. 15-100.

Willis, Bailey, 1902, Stratigraphy and structure, Lewis and Livingston Ranges, Montana: *Geol. Soc. America Bull.*, v. 15, p. 305-352.

4. *Cordilleran eugeosyncline and structures that formed from it*

Cloos, Ernst, 1935, Mother Lode and Sierra Nevada batholith: *Jour. Geology*, v. 45, p. 225-249.

Curtis, G. H., Everenden, J. F., and Lipson, J., 1958, Age determination of some granitic rocks in California by the potassium-argon method: *California Div. Mines Spec. Rept. 54*, 16 p.

Eardley, A. J., 1947, Paleozoic Cordilleran geosyncline and related orogeny: *Jour. Geology*, v. 55, p. 309-342.

Hinds, N. E. A., 1932, Paleozoic eruptive rocks of the southern Klamath Mountains, California: *California Univ. Dept. Geol. Sci. Bull.*, v. 20, p. 375-410.

———, 1935, Mesozoic and Cenozoic eruptive rocks of the southern Klamath Mountains, California: *California Univ. Dept. Geol. Sci. Bull.*, v. 23, p. 313-380.

Jenkins, O. P., ed., 1948, The Mother Lode country; geologic guidebook along Highway 49, Sierran gold belt: *California Div. Mines Bull. 141*, 165 p.

Misch, Peter, 1952, Geology of the northern Cascades of Washington: *The Mountaineer*, v. 45, p. 4-22.

Roberts, R. J., Hotz, P. E., Gilluly, James, and Ferguson, H. G., 1958, Paleozoic rocks of north-central Nevada: *Am. Assoc. Petrol. Geol. Bull.*, v. 42, p. 2813-2857.

Smith, A. R., and Stevenson, J. S., 1955, Deformation and igneous intrusion in southern British Columbia, Canada: *Geol. Soc. America Bull.*, v. 66, p. 811-818.

Taliaferro, N. L., 1942, Geologic history and correlation of the Jurassic of southwestern Oregon and California: *Geol. Soc. America Bull.*, v. 53, p. 71-112.

Woodford, A. O., 1939, Pre-Tertiary diastrophism and plutonism in southern California and Baja California: *6th Pacific Sci. Cong. Proc.*, p. 253-258.

CHAPTER IX

CENOZOIC ROCKS AND STRUCTURES OF THE MAIN PART OF THE CORDILLERA: LATER MODIFICATIONS OF THE FUNDAMENTAL STRUCTURE

1. YOUTHFUL STRUCTURES AND TOPOGRAPHY

The remainder of this book will be devoted primarily to those Tertiary and Quaternary modifications that have so greatly changed the aspect of the western part of the Cordillera after the great deformations of Mesozoic and Paleocene time, creating the modern landscape and the topographic mountains which we see today.

We have already discussed such modifications in the Eastern Ranges and Plateaus (Chapter VII, Section 7), where the low-standing ranges and basins that existed at the close of Laramide orogeny were transformed into the present lofty mountains, plateaus, and plains. This was accomplished in part by folding and faulting later than the main deformation, but to a much larger extent by a great regional upwarp, shared not only by the Rocky Mountains but by the Great Plains on the east and the Colorado Plateau on the west. Differentiation of mountains, plateaus, and plains was thus primarily a result of etching out of the upwarped rocks by erosion.

In that part of the Cordillera west of the Eastern Ranges and Plateaus the later modifications of the fundamental structure were more varied, and crustal activity manifested itself not only regionally but locally—by folding and faulting of the rocks and by construction of volcanic plateaus and mountains. Here a great number of modern topographic features are a direct result of crustal forces and only to a lesser degree the result of erosion and sedimentation.

2. BASIN AND RANGE PROVINCE

DRAINAGE, TOPOGRAPHY, AND STRUCTURE. The first of the younger features which we will discuss will be those of the Basin and Range province, some of which were given passing mention in earlier chapters. Before going further, however, we should clarify our terminology and keep in mind the distinctions between:

(1) The *Great Basin* or region of interior drainage, from which no streams flow to the sea, and which occupies much of Nevada and adjacent parts of Utah, Oregon, and California.

(2) The *Basin and Range province*, which includes not only the Great Basin, but also extensive regions of exterior drainage on the south and southeast in southern Arizona, much of New Mexico, and parts of western Texas.

(3) *Basin and Range topography*, or the characteristic landscape of the Basin and Range province—the peculiar, short, sub-parallel ranges and intervening desert basins whose appearance on the map suggested to Major Dutton "an army of caterpillars crawling northward out of Mexico."

(4) *Basin and Range structure*, or the later deformations of the crust that influenced or produced the Basin and Range topography (Fig. 86). To some extent the topography is a product of the erosion of a deformed terrane, but to a greater extent than in most other regions it appears to be a direct product of the deformation itself.

In some areas all four of these terms are applicable; in others, only one or two. Much of Nevada possesses

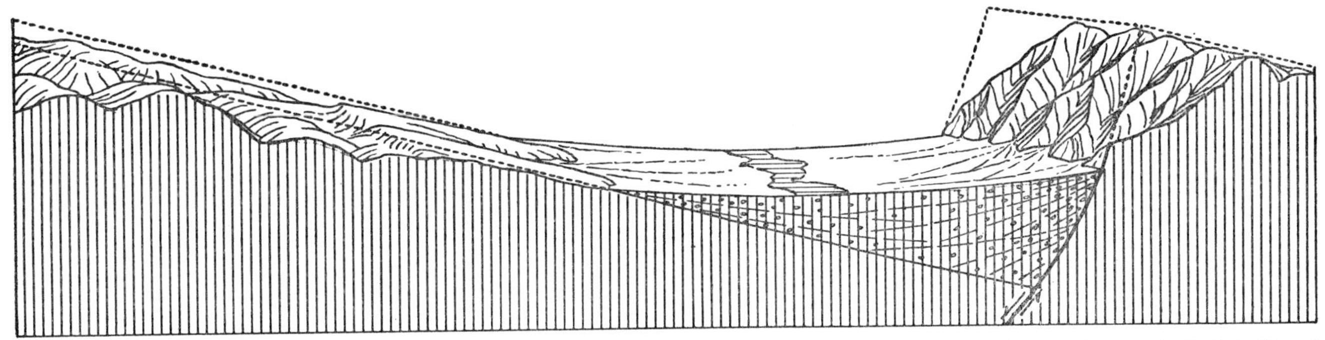

FIG. 86. Block diagram showing typical topography of the Basin and Range province and the block faulting which is believed to have shaped it. After W. M. Davis, 1927.

an interior drainage of Great Basin type as well as topography and structure of Basin and Range type; its interior drainage is due to some extent to its topography and structure, but quite as much to its scanty rainfall and remoteness from the sea. Wide areas to the south and southeast with Basin and Range topography are not part of the Great Basin and are drained to the sea by tributaries of the Colorado River and Rio Grande. Finally, many areas of Basin and Range topography have been so greatly eroded that there is no means of determining the original Basin and Range structure, whatever it may have been.

It is well to bear in mind also that the characteristic topography (and probably the structure, too) of the Basin and Range province has been superposed on a variety of earlier rocks and structures. In the Great Basin it was superposed on eugeosynclinal and miogeosynclinal rocks that were strongly deformed by the Cordilleran orogenies; in New Mexico it was superposed on rocks that were laid down on the continental platform and only lightly deformed later (compare the complex internal structures of the ranges in Fig. 84 with the simple internal structures of Fig. 72). In southeastern Oregon it involves at the surface only Tertiary and later volcanic rocks, and the nature of the rocks and structures beneath them is unknown. Basin and Range topography and structure were thus not a necessary sequel of any particular sort of fundamental Cordilleran structure; the structures on which they were superposed are diverse.

INTERPRETATION OF BASIN AND RANGE TOPOGRAPHY AND STRUCTURE. Origin of Basin and Range topography and its structural implications have long been debated among geologists; the question has sometimes been dignified as the "Basin Range problem."

Many geologists, including Clarence King's men of the Fortieth Parallel Survey, remarked on the strong deformation of the rocks in the ranges—especially in the Great Basin and to a lesser extent elsewhere—and concluded that the topography is merely an effect of erosion working on deformed rocks in an arid climate. It was thus assumed that if folded and faulted rocks such as those in the Valley and Ridge province of the Appalachians were placed in such a climate, erosion would wear down parts of them into desert basins and leave other parts projecting as discontinuous mountain ranges.

To some extent this is so, but not wholly, for in many places folds and faults within the ranges strike out to their edges where they are cut off; their extensions must lie beneath the basins. Yet where the basins have been entrenched by rivers such as the Colorado and Rio Grande, it may be seen that these are thickly filled by Tertiary and Quaternary sediments laid down in troughs much like the present ones. It would appear, therefore, that some other structure has been interposed between the Cordilleran structures exposed in the mountains and the modern topography of ranges and basins.

G. K. Gilbert, whose work for the Powell Survey we have already noted (Chapter VI, Section 3), early suggested a probable explanation: that the ranges are bordered on one or both sides by faults which have raised or rotated them in relatively recent geologic time and depressed the intervening basins. Basin and Range topography would thus have been caused directly by late Tertiary and Quaternary faulting, whose effects are still manifest in the modern landscape.

Somewhat later *William Morris Davis*, the great geomorphologist, followed the ideas suggested by Gilbert and systematized these heterogeneous observations into a sequence (Fig. 87):

(A) The *King formations* (named for Clarence King of the Fortieth Parallel Survey), or the sequence of stratified rocks of the region, primarily of Paleozoic and Mesozoic ages (Fig. 87 A).

(B) The *King folds* (also named for King), or the deformed structures produced by the Cordilleran orogenies (Fig. 87 B).

(C) The *Powell surface* (named for J. W. Powell, whose Survey demonstrated the far-reaching effects of denudation in the Cordillera), or the erosion surface

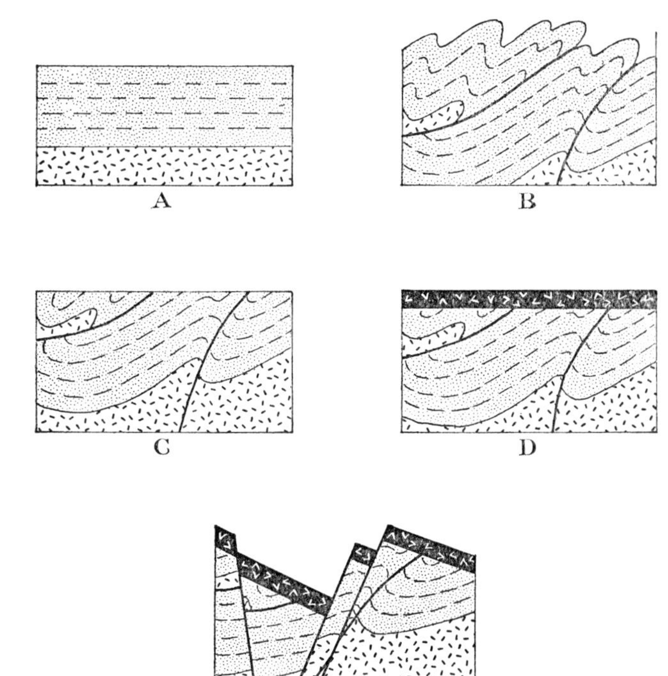

FIG. 87. Diagrams showing the sequence of features which developed in the Basin and Range province, as postulated by W. M. Davis. (A) The King formations. (B) The King folds. (C) The Powell surface. (D) The Louderbacks. (E) The Gilbert fault blocks.

developed on the deformed terrane after the orogeny (Fig. 87 C).

(D) The *Louderbacks* (named for G. D. Louderback of the University of California, who made significant observations in the Great Basin), or the Tertiary and Quaternary lavas which were spread as thin to thick sheets over the eroded surfaces of the earlier rocks (Fig. 87 D).

(E) The *Gilbert fault blocks* (named for G. K. Gilbert), into which all these prior features were broken later in geologic time (Fig. 87 E).

This outline is useful as a statement of the general manner in which successive features of the Basin and Range province were superposed on each other, but one should keep in mind that it is an extreme simplification of diverse features that developed through a long period of time and over a very wide area. The Davis sequence may very well apply in its entirety in a few areas and be only partly represented in others. In most of the ranges the neatly pigeon-holed categories are overlapped and blended.

A similar comment applies to the criticisms of the theories of Gilbert and Davis, of which there have been many. Neither the theory itself nor the alternative explanations can be wholly right or wholly wrong, for many sorts of structures and land forms are found from one part of the Basin and Range province to another. We can accept these as facts, yet we may find it necessary to reject at least some of the large generalizations that have been built upon them.

FAULTS ALONG EDGES OF RANGES. Gilbert conceived of his explanation of Basin and Range structure while engaged in some of the early geologic exploration of the West, when he did not have an opportunity to study closely the bedrock structure of the ranges. He therefore *inferred the structure which produced the ranges very largely from the topographic forms*—a procedure which was novel at the time.

Novelty of the method aroused much skepticism among geologists; the skeptics pointed out that numerous faults could be observed within the ranges which had no direct relation to the topography, and that if great faults actually bordered the ranges they were concealed by alluvial deposits along their bases. Some of the skeptics confidently challenged anyone to find actual exposures of a fault that blocked out a range or that had aided in producing the present topography.

We now know, however, that faulting has gone on for a long period in the Basin and Range province—in fact, from the time of the Cordilleran orogenies of Mesozoic time down to the present. Sequences of faulting vary from one area to another, in places ending well back in the Tertiary, in others continuing until recent time; some of this faulting accords with the classic concept of Basin and Range structure, some does not. Obviously, direct topographic effects of the older faults would have been obliterated long since by erosion, and the effects would become progressively greater the younger the fault (Fig. 88).

In parts of the region, as in southern Arizona and the Mojave Desert of California, blocking out of the ranges must have been an ancient event that was little renewed recently. Here most of the ranges are zig-zag, spiny ridges that have been deeply frayed and embayed by erosion (Fig. 88 C, D, and E). The plains around many of them are erosional surfaces (or pediments) produced by wearing down of the same rocks as those which form the mountains. The plains in some other areas are covered by Quaternary or later Tertiary sediments that overlap the edges of the ranges without an intervening fault at the surface. If any block-faulting shaped ranges such as these, it must have been at a time so remote that the original block form has been destroyed by erosion. As the original form of the range has been destroyed, it may well have been something quite different from a set of fault blocks.

Nevertheless, many other ranges throughout the Basin and Range province are actually block-like in form, do possess straight, steep edges on one or both sides which do cut across structures produced by

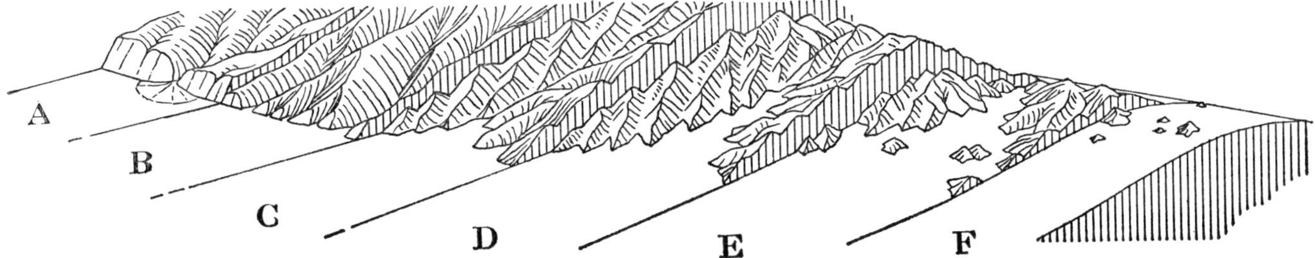

FIG. 88. Block diagram showing stages of erosion of a mountain in the Basin and Range province. An initial fault-block form is assumed in stage A, but succeeding stages might have developed from any initial tectonic form. (A) Fan-free and fan-based. (B) Fan-dented. (C) Fan-bayed. (D) Fan-frayed. (E) Fan-wrapped. (F) Pan-fan. After Davis, 1925.

earlier deformations where such ranges consist of deformed rocks (Fig. 88 A and B).

Despite the skeptics, faults have been proved along the edges of many of the ranges. In places little hills of bedrock project from the desert basins and consist of the same rocks as those that occur high up in the adjacent ranges; they demonstrate that the bedrock beneath has been so depressed that only the highest parts are still visible as "islands" nearly submerged by the basin deposits. Some of these hills lie close to, or even against the lower part of the bedrock of the range scarp, with a contact so abrupt that it can only be the result of faulting rather than of folding or warping.

If a geologist is fortunate to discover such a locality, he may also be able to observe the actual fault contact between the bedrock of the range and the hill in the basin. Moreover, at some other places where no bedrock hills are present, renewal of range uplift has brought the bordering fault into view, with bedrock moved up against the Quaternary and Tertiary deposits that had been shed off the range after the earlier and greater period of block-faulting. In many places, too, the basin deposits themselves are faulted close to the range fronts and are stepped up toward them in little fault scarps; some of these scarps can be related directly to recorded earthquakes.

The faults that have been observed along the edges of the ranges dip toward the basins or downthrown blocks and are therefore *normal faults* (Fig. 85). Their dip is generally much steeper than the slope of the range escarpment behind; this is to be expected, as an exposed fault scarp will be worn back by erosion until it reaches the angle of repose for the material in transit across it. The highest part of the scarp, being first to emerge, will have been eroded longest and worn back farthest; the lower part, which emerged later, will be less eroded; if it has been uplifted recently and rapidly enough its slope may not yet have attained erosional equilibrium.

FAULT SCARPS VERSUS FAULT-LINE SCARPS. Modern critics of the theories of Gilbert and Davis have generally conceded the existence of great faults along the margins of the ranges, but have objected that the range fronts were not direct products of deformation, or *fault scarps*, but were produced by removal of weak beds from the downthrown sides by erosion, hence are *fault-line scarps*.

The critics point out that many of the features that have been interpreted as evidence for direct tectonic control of the scarps could have been produced as well by erosional processes. Also, that many of the plains that slope away from the bases of the scarps, which had been considered depositional surfaces, are actually erosional surfaces or *pediments* cut on the edges of more or less deformed Tertiary deposits. It is even conceivable that the basins were once filled by poorly resistant Tertiary deposits to the levels of the intervening mountain tops, and that they have since been excavated by erosion.

Where rivers such as the Colorado and Rio Grande cross the Basin and Range province, the basin deposits have been planed off by erosion so that their original depositional surfaces have been destroyed. But in nearby basins with interior drainage the surfaces of the deposits have remained depositional; in these, where no streams have flowed out for a long period, it seems unlikely that any extensive excavation and removal of material could have been accomplished. It is true that the earlier deposits of most of the basins have been deformed, especially near the edges of the mountains, but the basins did not become stable afterwards as many are broken by fresh scarplets near or along their edges.

Thus, the fault-line scarp hypothesis of Basin and Range topography and structure does not alter their essential nature; if valid, it moves the time of formation of the gross features farther back into the past. The chief virtue of the explanation is to emphasise that understanding of the Basin Range problem cannot be attained from the present aspect of the region alone, but requires knowledge of how its features evolved through time.

EVOLUTION OF GREAT BASIN AREA. Let us, therefore, examine the evolution of Basin and Range topography and structure since the time of the Cordilleran orogenies. To illustrate this it will be profitable to sample the history of one area where many data are available—the Great Basin between the Colorado Plateau and the Sierra Nevada.

In doing so we will find that deposits were laid down on the lower parts of the surface during a considerable span of Tertiary time, during which the geography evolved as a result of continued crustal movements. Earlier deposits were derived from uplands different from those of the present ranges in form and even in position; as these deposits appear now only in fragments it is not easy to decipher the geography under which they were formed. With further movements the earlier deposits were faulted, tilted, and eroded, and those in which resistant lavas are embedded were raised in places to mountainous heights. Later Tertiary and Quaternary deposits bear a closer relation to modern geography, yet even these are more or less deformed and eroded.

The fragmentary record of earlier Tertiary (Paleocene and Eocene) time suggests that the eastern half of the Great Basin received deposits in lakes, swamps, and floodplains which were probably westward extensions of those in the Rocky Mountain region, and like them, they stood at altitudes only a thousand feet or so above sea level. The western half of the Great Basin and the Sierra Nevada beyond was a low upland. The marine Ione formation deposited along the western edge of the highland in California contains clays of remarkable purity, evidently carried westward by sluggish streams from a worn down and deeply decayed surface of unroofed Nevadan plutonic rocks. In early Tertiary time the topography that had developed on the site of the Great Basin from the orogenically deformed miogeosynclinal and eugeosynclinal rocks had thus become decadent, but the basins and ranges of later time had not yet developed; much of the region apparently possessed an exterior drainage.

During middle Tertiary (Oligocene and Miocene) time the western part of the site of the Great Basin was widely overspread by volcanic rocks which form a succession of lavas, agglomerates, and tuffs. The older of these are now much disturbed and mineralized and have been raised into mountain ranges; they were the hosts of the fabulously rich *Comstock Lode* of the Virginia City district. In parts of the region the volcanics pass into continental sedimentary deposits with more or less admixtures of volcanic detritus. The nature of some of these sediments suggests that they were laid down in basins much like the modern ones, as though a Basin and Range topography had now begun to develop. In some areas there was strong faulting and tilting of the rocks in mid-Miocene time, after which late Miocene and early Pliocene deposits were spread over their eroded surfaces.

Fossil plants occur at many places in the late Miocene and early Pliocene deposits; these provide a good index of the geography, as they are enough like modern forms to have lived in environments of similar altitude and rainfall. In places on the crest of the Sierra Nevada at present altitudes of 9,000 feet are late Miocene deciduous plants that could not have lived higher than at 2,500 feet; in the Great Basin at about the same time were coniferous forests like those that now grow at the margins of woodland and chaparral country. The environments indicated by the different plant assemblages suggest that the crest of the Sierra Nevada stood at altitudes of less than 3,000 feet, with the Great Basin about a thousand feet lower. The ridge of the Sierra Nevada created no more than an ineffective rain shadow over the country to the east, and was evidently crossed in many places by streams draining to the Pacific from headwaters well back in the Great Basin.

Later in the Pliocene the floras of the Great Basin changed to a savanna and grassland assemblage, adapted to less than fifteen inches of rainfall. The Sierra Nevada block was now being uplifted and exerted a climatic influence over the region to the east. Uplift continued into Pleistocene time until the block had been raised 5,000 to 6,000 feet in the north and 7,500 to 9,000 feet in the south. At the same time the Great Basin area was disrupted by extensive block faulting, which disturbed not only the Paleozoic and Mesozoic bedrock, but the Tertiary deposits as well. From Pliocene into Pleistocene time the basin floors were raised from altitudes of less than 2,000 feet to their modern altitudes of 3,000 to more than 5,000 feet.

This record indicates that the western part of the Cordillera—in contrast to the Rocky Mountains where much of the regional uplift had been accomplished by mid-Tertiary time—remained low until late in the Tertiary, after which, in Pliocene and Pleistocene time, it likewise underwent regional uplift as well as major block-faulting.

Absolute uplift and absolute depression. The odd structures that have developed in the Basin and Range region in the near past arouse our curiosity and stimulate our speculation. Before speculating, however, let us explore a question that bears considerably on their origin—What were the *actual* movements of the basins and ranges?

Obvious movements in the region were, of course, the uplift of the ranges and depression of the basins, but these movements were *relative*. Actual movements may have been the same, or they may have been something quite different in relation to a sea-level datum—

for example, subsidence of the whole region with the basins going down the most, or uplift of the whole region with the ranges raised the farthest.

There is, nevertheless, much evidence for an absolute depression of some of the basins and an absolute uplift of some of the ranges. The floor of Death Valley now lies 282 feet below sea level, and recent geophysical work indicates that it is underlain by about 7,000 feet of basin deposits; its rock floor must have been lowered far below its original level—so far that subsequent filling never caught up with its subsidence. On the other hand, mountains at the edge of the Basin and Range province—the Sierra Nevada, the Wasatch Range, and the eastern of the New Mexico ranges—are hinged on more stable crustal blocks and were absolutely uplifted above the remainder of these blocks. Even well within the Basin and Range province some of the mountains project higher than they ever could at an earlier period; the Ruby Range in the Great Basin near Elko, Nevada rises to heights of 11,536 feet, with landscapes as alpine as those in the Sierra Nevada, and nearby ranges stand as high or higher; even if the surroundings of these ranges had risen as a unit the ranges themselves were forced still higher.

Clues as to the regional behavior of at least the Great Basin part of the Basin and Range province are afforded by its Tertiary history which we have outlined. Evidently the Great Basin has not subsided as a whole since the Cordilleran orogenies, but has been raised—the basins from a position a thousand feet or so above sea level in early Tertiary time to more than 5,000 feet today, and the ranges still higher.

In the Great Basin part of the province regional uplift must have been accompanied by broad arching. The explorer Frémont remarked on the fact that the modern lakes of the Great Basin are at its eastern and western edges in Utah and western Nevada—an effect heightened during Pleistocene time when the now-vanished *Lake Bonneville* and *Lake Lahontan* were spread more widely over the same areas. In the central Great Basin of eastern Nevada even the basin floors stand several thousand feet higher than those along the edges, and the crests of the Ruby and other ranges attain the greatest heights in the Great Basin.

Origin of Basin and Range structure. Let us now speculate on possible mechanisms of the formation of Basin and Range structure.

The normal faults that border the ranges, and which are an essential item in their structure, were a product of tension rather than compression, or of extension rather than shortening of the crust. Did this tension pervade the whole region during the time of formation of the Basins and Ranges, or was it merely a local component of some other stress pattern?

We have learned that during the Cordilleran orogenies the Basin and Range province was being deformed by compressional forces. With the close of these orogenies compression relaxed, perhaps to such an extent that the region was placed under tension. Some of the earlier geologists suggested that the Cordilleran orogenies produced a great arch in the Basin and Range area, and that during relaxation the arch collapsed, its parts subsiding into a mosaic of fault blocks. But an arch of the implied breadth and height seems so incredible that no geologist today believes it ever existed. Besides, as we have seen, the Tertiary record indicates that the region as a whole has risen steadily, rather than subsided.

A modern variant of this explanation is that the whole Cordilleran area stood at approximately the same height after the climactic orogenies, but was later modified by subcrustal transfer of material. The Rocky Mountains—and the Great Plains and Colorado Plateau alongside them—were bulged upward during Tertiary time by addition of material at their bases that was derived from beneath the Great Basin; thus the Great Basin either subsided or its uplift lagged behind that of adjoining areas so that its rocks were placed under tension.

This notion has its appealing features but seemingly applies only to the Great Basin segment of the whole Basin and Range province. If the remainder of the province were produced by the same sub-crustal removal of material, where were the complementary upbulges to which the subcrustal material was transferred? Nowhere except in the Great Basin is a great upbulged area so neatly placed alongside an area of Basin and Range structure.

But was the tension implied by the normal faulting in the Basin and Range province so all-pervading? During late Tertiary and Quaternary time considerable parts of the Cordillera were not under tension but compression. In the Coast Ranges of California to the west, accumulated Tertiary sediments were folded, faulted, and thrust during many epochs, with culminations during the middle of the Miocene and early in the Pleistocene. Even in the Basin and Range province itself, middle and late Tertiary sediments of the Mojave Desert section were folded and thrust, apparently at about the same time as the block-faulting elsewhere in the province. These relations suggest that to explain the block-faulting of the province we must seek a more complex interaction of forces than is provided by mere relaxation of the crust after the Cordilleran orogenies.

The tension suggested by the local structures may

be a component of a more pervasive crustal compression—set up, for example, by wrenching and shifting of large blocks of the crust during later phases of the development of the Cordillera. Shifting of such large crustal blocks is suggested by the pattern of the ranges in many parts of the province—over wide areas the ranges trend nearly parallel, yet these differ in trend from those of adjacent areas, the change taking place along well-defined to poorly-defined zones.

North of the *Garlock fault*, which extends eastward into the Mojave Desert, high ranges trend northward; south of the fault are less systematic, lower ranges. The Garlock fault has a known left-lateral strike-slip displacement of many miles and must have influenced profoundly the growth of ranges and basins on either side. A zone about which less is known, the *Walker line*, lies a little northeast of and parallel to the California-Nevada boundary; ranges northeast of it trend northward, those southwest of it, northwestward. Near Las Vegas folds and faults are bent sharply across the Walker line by a right-lateral movement that is probably no younger than the Laramide orogeny, yet deformation associated with modern earthquakes elsewhere on the line is a small-scale replica of the same movement.

The extent and magnitude of such zones in other parts of the Basin and Range province is uncertain as yet, and it remains to be determined whether strike-slip movements along them are sufficient to have produced the crustal tension and normal faulting in the intervening blocks. Nevertheless, such a possibility seems more promising as an explanation of Basin and Range structure than the other theories that have been suggested.

3. NORTHWESTERN VOLCANIC PROVINCE

Volcanic modifications of fundamental structure. We will now consider another sort of modification of fundamental Cordilleran structure—one produced by volcanism, and specifically the volcanism of Tertiary and Quaternary time—which has formed a mighty body of rocks and structures in the northwestern states of Washington and Oregon and contiguous parts of Idaho and California.

We have already discussed the modifications resulting from volcanism in the Eastern Ranges and Plateaus—local volcanic fields of long or short duration in various stages of destruction by erosion, and their associated stocks, dikes, and laccoliths. But these are intermingled with other features of various kinds and ages and are merely an incidental part of the total structure.

In Oregon and Washington, by contrast, volcanic rocks and structures dominate the scene—lava plateaus, volcanic cones, and eruptive ranges produced by upbuilding, and folded or block-faulted ranges produced by deformation of the volcanic materials. In these states no rocks older than Tertiary volcanics and sediments are exposed in an area extending 300 miles parallel with the coast or 400 miles inland; they also cover an even greater area to the south and southeast where older rocks emerge in places (Fig. 89). The volcanic regime was prolonged and extended through most of Tertiary and Quaternary time from one place to another. Nothing comparable to this great volcanic area occurs elsewhere in North America except far to the south in Mexico and Central America (Plate I).

Subdivisions of volcanic province. Broader topographic features of the northwestern province are not unlike those of the southwestern province, in California. Backbone of the province, with its highest peaks, is the Cascade Range which is set well back from the coast like the Sierra Nevada. Along the coast is a lower Coast Range, and between this and the Cascades is a belt of depressions which remind us of the Great Valley of California. East of the Cascade Range is lower country, broadly termed the Columbia Plateaus, in the same position as the Great Basin.

Geologically, however, differences between the topographically analogous features are so great as to lead one to suspect that the resemblances are no more than coincidence. If the features in the two regions were produced by the same forces, these have worked on materials so unlike that they have expressed themselves in a very different manner.

The *Coast Ranges* of Oregon and Washington begin on the south at the edge of the Klamath Mountains and terminate on the north at the latitude of Seattle in the Olympic Peninsula. Most of the Coast Ranges are low; their highest peaks in the *Olympic Mountains* attain only 7,954 feet, although the latter support snow fields and local glaciers because of their northern position. The Coast Ranges consist primarily of broadly folded early to middle Tertiary sediments with interbedded basaltic lavas; in Oregon they form a single anticlinorium, but in Washington their trend is crossed by southeast-trending warps creating a succession of mountain knots and intervening sags.

Between the Coast and Cascade Ranges is a partly discontinuous belt of topographic depressions, forming the *Puget Trough* in Washington and the *Willamette Valley* in Oregon, which have a general synclinal structure.

The *Cascade Range* is a belt of lofty peaks and rugged mountain ridges which originates on the south between the Sierra Nevada and Klamath Mountains and extends northward beyond the International

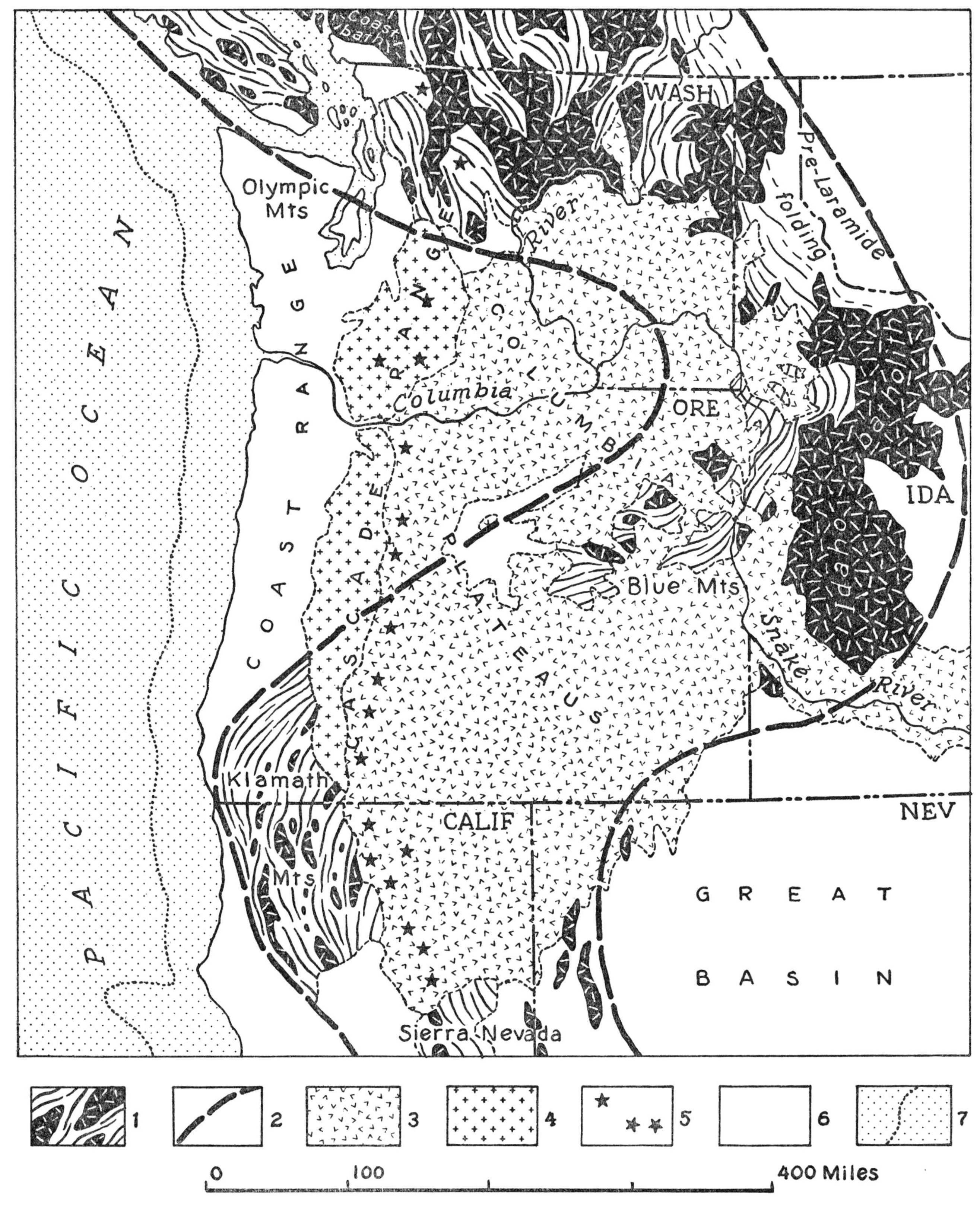

FIG. 89. Map of the northwestern volcanic province of Washington, Oregon, and adjacent states, showing its relation to the arc of the Nevadan orogenic belt. After King, 1958.

Explanation of symbols. 1. Nevadan basement; metamorphic rocks lined, plutonic rocks solid. 2. Inferred margins of Nevadan orogenic belt. 3. Plateau basalts of Miocene and later age. 4. Andesitic volcanics of Cascade Range. 5. Volcanic cones of Cascade Range, mostly of Quaternary age. 6. Other rocks; mainly sedimentary rocks of Mesozoic and Tertiary ages, but including some older Tertiary volcanics. 7. Edge of continental area.

Boundary. Its highest summits are great volcanic cones built up late in geologic time on a foundation of earlier rocks and structures. One of these cones, Lassen Peak, is still active; the others are perhaps merely dormant; the highest, Mount Ranier, attains an altitude of 14,408 feet, or only a little short of the highest summits of the Sierra Nevada and Rocky Mountains.

From central Washington northward the Cascade Range was formed of an emerged basement that was metamorphosed and plutonized during Nevadan and subsequent Mesozoic orogenies, but southward this is buried by an increasing thickness of Eocene to Pliocene rocks, mainly andesitic volcanics (Fig. 90). The volcanic part of the ranges owes its height partly to upwarping, partly to excessive upbuilding.

(Chapter VIII, Section 4), are an emerged segment on the southeast side of the Nevadan orogenic belt between the Klamath Mountains on the southwest and the highlands of Idaho on the northeast. Upwarp of the Tertiary rocks represents renewal of movement along the same trend.

Volcanic rocks continue beyond these uplifts into southeastern Oregon and adjacent states, forming a region which has sometimes been called the *Malheur Plateaus*, composed of more varied rocks and structures than the Columbia Plateau proper. The volcanics here are mainly younger than those farther northwest; to the south they are involved in the block-faulting of the Basin and Range province.

Eastward these younger volcanics fill the great

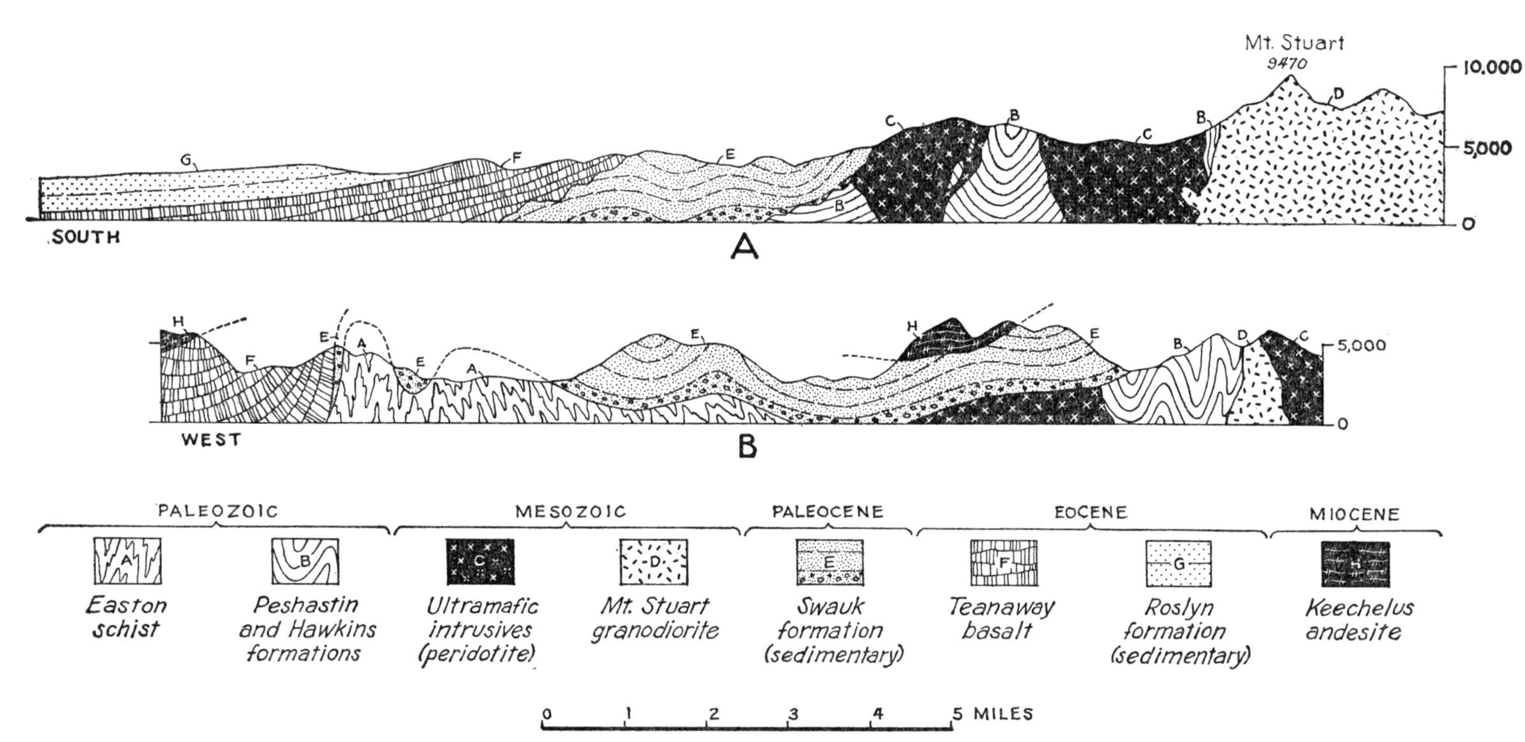

Fig. 90. Sections showing relations between the Tertiary sedimentary and volcanic rocks of the northern Cascade Range and their Mesozoic and Paleozoic basement. Sections are on southeast side of the range in central Washington. After Smith, 1904; and Smith and Calkins, 1906.

East of the Cascade Range are the broad, diversified *Columbia Plateaus.* The *Columbia Plateau proper* is that part in southeastern Washington and northeastern Oregon which is overspread by middle Tertiary basaltic lava, parts of which have been so little deformed as to produce a plateau topography. At the southeast edge of this area in northeastern Oregon the basalts and associated strata have been raised in a chain of folds and uplifts trending west-southwest, in which basement rocks have been revealed by erosion in the highlands of the *Blue Mountains* and nearby areas (Fig. 89). The basement rocks, as we have seen

transverse downwarp of the *Snake River Plain*, extending nearly across the Cordillera, of which mention was made earlier (Chapter VIII, Section 3).

Foundation of the volcanic province. What was the cause of this remarkable localization of volcanic rocks and structures in the northwestern states? To my knowledge there has been little speculation on the question, although much able work has been devoted to learning conditions of formation of the individual rock suites. No doubt the answers are to be found in the basement rocks on which the volcanics lie, but little information on these is available except in "islands" of the older rocks that emerge in uplifts like the Blue Mountains.

The volcanic activity may be related to the great *arc of the Nevadan orogenic belt* which we have inferred extends from northern Washington through central Idaho and the Blue Mountains into the Klamath Mountains (Fig. 89). This arc is reflected in the Tertiary cover—by west-northwest-trending folds that cross the Cascade and Olympic Mountains of northwestern Washington, and by west-southwest-trending folds near the Blue Mountains of northeastern Oregon.

An intriguing question arises—Could the western edge of the Nevadan arc have formed the continental border until late in Mesozoic time, before which its recess was floored by oceanic rather than continental crust? If so, the recess was made continental later by filling with sediments and volcanics, and the coast was straightened by growth of a new set of volcanic structures along the Cascade Range.

THE BORDERLAND OF "CASCADIA." Such an explanation is at variance with the theory that now vanished borderlands once existed off the edges of the present continent. We have already set forth objections to the existence of an "Appalachia" off the southeastern coast of North America (Chapter IV, Section 5); the same objections apply to the supposed borderland of "Cascadia," whose type area was offshore from the present Cascade Range.

Existence of a "Cascadia" is perhaps most strongly indicated along the California coast farther south where strips of crystalline basement strike out to sea; even here other explanations are possible, as we shall find later (Section 4 of this chapter).

But in the type area farther north available information affords little basis for the notion of a "Cascadia." All the earlier rocks are deeply covered by Tertiary sediments and volcanics and no record of them is available; the Tertiary rocks themselves were laid down on a continental margin with only open ocean to the west. Moreover, the continental shelf is very narrow along the Pacific Coast in Washington, Oregon, and California, so that there would have been little room for any earlier offshore lands—unless one accepted some unlikely process of foundering of the lands to oceanic depths, and their subsequent conversion from sialic to simatic crust.

EARLY TERTIARY COASTAL PLAIN OR GEOSYNCLINE. We have allowed ourselves to wander into the fields of speculation. Let us put our feet on the ground once more and consider the known Tertiary and Quaternary history of the volcanic province.

Earlier Tertiary history of the province is revealed mainly in the Coast Ranges, where rocks of these ages are extensively exposed; in the Cascade Range and Columbia Plateaus to the east early Tertiary rocks emerge only in scattered inliers.

The Coast Ranges are principally built of early Tertiary sediments and volcanics, mainly Eocene but extending downward and upward into the Paleocene and Oligocene; they are 10,000 to 15,000 feet thick in most places but are many times thicker in the Olympic Mountains. The sediments are fine to coarse clastics containing volcanic detritus in part, but including components derived from plutonic and other basement rocks farther inland. In the Coast Ranges they are all marine and include thick units with graded bedding, probably deposited by turbidity currents in waters of considerable depth. On the east side of the ranges these pass into littoral and brackish water deposits, and farther inland, into continental, coal-bearing sediments.

Interbedded and interfingered with the sediments are masses of basaltic lava, mostly in the Eocene, partly in the Oligocene. They are generally subordinate to the sediments but in places exceed them in thickness, forming a series of giant lenses whose total volume has been estimated to be about 40,000 cubic miles. A large part of the basalt was erupted beneath the sea, as shown by its pillow structure, but it was locally built up above the water, for some of the flows are amygdaloidal, and weathered zones lie between them. The basalts are of remarkably uniform composition and must have been derived from a simatic layer beneath with little contamination from the sialic crust.

The environment of these early Tertiary sediments and volcanics has been termed variously a coastal plain or a geosyncline. The nature of the sedimentary rocks suggests a comparison with those of about the same age along the Gulf Coast, which we discussed in Chapter V, Section 2—with continental deposits inland and marine deposits offshore, some probably laid down in deep water along or at the foot of the continental slope. The deposits of both regions were seemingly built forward from the continental into the oceanic area. Significant differences are the much larger volume of volcanic material in the northwestern province, and the greater deformation to which its rocks were subjected later; although the early Tertiary deposits of the Gulf Coast and the northwestern province are analogous, the latter have formed on a more mobile part of the crust.

COLUMBIA RIVER BASALT. To continue our story of the volcanic history of the province let us pass to rocks and events in the Columbia Plateaus, leaving those of the Cascade Range for consideration later.

After Eocene and Oligocene time centers of volcanism apparently shifted eastward, or farther into the recess of the Nevadan arc. Here in the Columbia

Plateau proper the *Columbia River basalt* was spread out during Miocene time. The lava flood did not, perhaps, exceed in volume that of the earlier Tertiary basalts farther west, but the latter are only fragmentarily exposed because of subsequent deformation and erosion. Much more of the Columbia River basalt is preserved intact, making it one of the most impressive lava bodies in North America.

The Columbia River basalt covers an area of 100,000 square miles, and has a volume of 35,000 cubic miles. Total thickness is unknown, but more than 5,000 feet is exposed where the lava sequence has been tilted. Individual flows are 100 to 500 feet thick, and must have been very fluid when erupted, as some have been traced 120 miles without reaching their edges. The lava is of strikingly uniform composition; it is all basalt without interbedded flows of other sorts, and all the basalt is *tholeiitic*, a variety free of olivine.

The Columbia River basalt was not erupted from volcanoes, but welled up through fissures. Such fissures, or feeder dikes, are common in northeastern Oregon and southeastern Washington in an area that may have been the center from which the flows originated. Eruption of the lavas must have required both a subsidence of their floor and an upbuilding sufficient to maintain an outward flow. Subsidence tilted the Nevadan basement and earlier Tertiary deposits toward the eruptive center from the north, east, and south; outward flow of the lava blocked the valleys on the tilted Nevadan surface at their lower ends so that lakes were formed in which sediments accumulated. Toward the west near the lower course of the Columbia River the lavas spread seaward across the site of the Cascade Range, and their feather edges interfingered with marine deposits.

In general, the basin form of the Columbia River basalt is still preserved; the basalt lies nearly flat over wide areas or is tilted gently toward the center of the basin. In some places, however, it has been warped or folded, mainly by reactivation of earlier structures.

The Columbia River basalt is of the class termed *plateau or flood basalt*, which implies certain structural and compositional features. Like plateau basalts elsewhere in the world it is of very uniform composition; it welled up from fissures and was not erupted from volcanoes; from the fissures it spread over a wide area. Plateau basalts are supposed to have ascended in areas of crustal tension and fracturing from one of the more basic lower layers of the crust without contamination from the more acidic upper layers through which they passed. Petrogeneticists have debated as to the nature of the basic source layer—whether it was a basaltic or simatic substratum underlying continents and oceans alike, a tholeiitic subdivision of the substratum lying only beneath the continents, or a still deeper peridotitic stratum. Be that as it may, interpretation of the source of the Columbia River basalt would seem to require consideration of its peculiar structural relations—within the recess of the Nevadan arc where truly continental crust may not have developed until after Mesozoic time, and where such crust may have been thinner than normal at the time of the eruptions.

Younger Tertiary lavas to the southeast and east. Most of the surface of the Malheur Plateaus and Snake River Plain southeast of the Blue Mountain axis is formed of lavas younger than the Columbia River basalt, of Pliocene and even Pleistocene age. This suggests a southeastward spread of volcanism with time, but in places earlier volcanic rocks are exposed beneath them, as though volcanism had begun much earlier.

In contrast to the uniform composition of the Columbia River basalt, the lavas to the southeast are diversified; some basalts are tholeiitic like the Columbia River, others are olivine-bearing; many lavas are andesitic, dacitic, or rhyolitic. Some lavas of this southeastern area ascended through fissures, but others were spread from central vents and shield volcanoes.

Contrast between the composition and habit of the lavas northwest and southeast of the Blue Mountain axis suggests that the latter, lying beyond the recess of the Nevadan arc, are in a region where sialic crust was thicker, hence that it could interact to a greater degree with the lavas during their ascent.

Evolution of the Cascade Range. In our story of the volcanic province we have so far by-passed the most prominent part, the Cascade Range, leaving its complex story until the last.

The Cascade Range, like the Coast Ranges and Columbia Plateaus, is built largely of volcanic rocks; their volume amounts to about 25,000 cubic miles. These volcanics contrast with the adjacent basaltic ones, for about seventy-five percent are pyroxene andesite or basaltic andesite. Also, instead of being largely erupted as flows, they include an equal or greater volume of breccias, tuffs, and mudflows formed by explosive volcanism; they contain numerous inclusions of earlier Tertiary sediments, and they were subjected to extensive hydrothermal alteration. The andesites of the Cascade Range were erupted during a long span of Tertiary time, the oldest being Eocene and the youngest Pliocene, but most are probably of Oligocene and Miocene age—of about the same age as parts of the basaltic outpourings in the Coast Ranges and Columbia Plateau.

Evidently some special volcanic condition unlike that on either side existed along the axis of the Cascade Range during Tertiary time. As a result of the growth of the Cascade structure the earlier Tertiary geosynclinal sediments may have been buckled down to such depths that they could contaminate the ascending basaltic magmas with water, silica, and alkalis, and change them into lavas of andesitic composition.

The Cascade Range, partly a belt of uplift, partly a belt of excessive volcanic accumulation, developed in Tertiary time directly across the recess of the Nevadan arc in an area where there is little evidence of prior deformation. Growth of the range is shown by comparing the Eocene and modern vegetation east and west of it; Eocene floras on both sides both grew in tropical lowland forests and were closely related; modern floras on the east side are adapted to much drier climate than those on the west because of the interposition of the Cascade barrier.

In Oregon the Tertiary rocks of the western part of the Cascades are complexly fractured and tilted eastward. In Washington the Cascade axis crosses a succession of west-northwest-trending folds expressed in part by alignments of the drainage. These parallel the grain of the basement rocks on the northwestern side of the Nevadan arc, but run well out into the Columbia Plateau on the east where they deform the Columbia River basalt, and extend westward across the Puget Trough into the Olympic Mountains. The west-northwest folds evidently antedate the north-south Cascade axis, yet their later growth involves rocks formed after the axis had begun to rise.

At some time during the Tertiary the andesites and other rocks of the Cascade Range were invaded by numerous masses of granodiorite and quartz diorite, which form some of the youngest plutonic bodies in the Cordillera. The largest body is the *Snoqualmie batholith* in central Washington, twenty miles in diameter, which invades continental sediments and andesites of Eocene and later age. After these plutonic masses had been unroofed by erosion in mid-Tertiary time they were again partly covered by other andesitic volcanics during a final and mainly Quaternary period of volcanism.

Quaternary events in Cascade Range. Growth of the Cascade Range during Tertiary time was followed by a period of quiescence and deep erosion. Volcanic activity was renewed again later, probably mainly in Quaternary time when there were again andesitic eruptions.

During this later period of activity a great chain of volcanic cones was built on the foundation of older rocks and structures; these now stand as a line of sentinels along the crest of the range (Fig. 89)—in Washington: Glacier Peak, Mount Baker, Mount Ranier, Mount St. Helens, and Mount Adams; in Oregon: Mount Hood, Mount Jefferson, Mount Washington, Three Sisters, "Mount Mazama," and Mount Loughlin; in California: Mount Shasta and Lassen Peak. One of these, "Mount Mazama," has lost its top since the glacial period and is now a caldera filled by the waters of Crater Lake. Another, Lassen Peak, has been active during the present century, and is the only active volcano in the continental United States.

The "Cascadian Revolution." The Cascade Range is of incidental interest as the type area of the alleged "Cascadian Revolution" which, according to many textbooks, neatly ended the Tertiary period and ushered in the Quaternary period. But in the story of the Cascade Range as we have outlined it, what is so revolutionary? As we have seen, the range was long in growth and the end of Tertiary time seems to be a time of quiescence rather than of tectonic and volcanic activity. Even if the concept of a Cascadian Revolution were apt for its type area, the combination of volcanic upbuilding and deformation by which the range was created seems too special to deserve application to a general milestone in geologic history.

History of the Columbia River. The Columbia River, whose sources are in the Northern Rocky Mountains on the east, enters the Pacific in the midst of the volcanic province. Much of its drainage basin is blessed with greater rainfall than the country farther south, and its volume far exceeds that of any other stream on the western slope of the Cordillera. Similar greater rainfall probably prevailed through much of the post-Mesozoic history of the Cordillera, so that an ancestor of the Columbia no doubt existed well back in Tertiary time.

With the onset of the eruptions of Columbia River basalt the course of the river in the coastal plain downstream from the highlands of the Nevadan orogenic belt was obliterated by the flood of lava. The lower ends of the valleys of the river and the valleys of its tributaries in the highlands to the north and northeast were dammed by the basalt and formed a series of lakes. Each lake drained around a spur end to the next lower one on the west, and thus established an exit for the waters along the edge of the volcanic field; with downcutting this became the new course of the Columbia, part of which it still follows. Farther downstream in southern Washington, however, the river is deflected well eastward into the lava country, probably because of outbuilding of andesitic debris from the Cascade Range on the west.

In southern Washington the river also crosses several anticlinal folds in the basalt that plunge southeast from the Cascade Range. Each fold is expressed topographically by a ridge, and in each ridge the river and several of its tributaries have cut deep gorges. The Columbia and its tributaries were probably *antecedent* to the anticlines—they had much their present courses before the folding and maintained them by downcutting as the anticlines were raised. Farther downstream the river cuts a much larger gorge through the complexly upbuilt and upwarped Cascade Range; many geologists believe that the river is antecedent to the growth of the Cascade Range also, but some would ascribe its course through the range to a complex process of superposition.

4. COAST RANGES OF CALIFORNIA

In this final section of the book we will discuss the Coast Ranges of California. These make a large subject, to whose varied facets many geologists have devoted years of labor. Within the scope of this book we can scarcely hope to exhaust the subject, yet perhaps we can set forth the fundamental features of the region and suggest some possible answers to its problems.

TOPOGRAPHIC AND GEOLOGIC PROVINCES. Let us review the topographic and geologic provinces of California (Fig. 91), keeping in mind that they were products of relatively late earth movements that modified the fundamental Cordilleran structures of Mesozoic time.

We have already spoken of some of these provinces:

(A) The *Great Basin* part of the Basin and Range province, with its discontinuous ranges and intervening desert basins, formed by late deformation, mainly block faulting, that was superposed on the deformed eugeosynclinal and miogeosynclinal rocks of the Cordillera. The Great Basin lies mainly in Nevada but its western edge extends over the border into California.

(B) The *Mojave Desert* part of the Basin and Range province, lying southwest of the Great Basin and projecting far westward as a peculiar wedge between other provinces of southern California.

(C) The *Sierra Nevada*, lying west of the Great Basin in central California—a great block of eugeosynclinal and plutonic rocks that was upfaulted on its eastern side and tilted westward toward the Great Valley.

(D) The *Klamath Mountains* of northern California which, like the Sierra Nevada, are made up of eugeosynclinal and plutonic rocks. Here, however, the present mountains were not raised by block faulting but by broad warping.

Let us now continue with the remaining provinces, about which we have said little hitherto.

(E) The *Great Valley*, or San Joaquin and Sacramento Valleys west of the Sierra Nevada in central California. This is a great alluvial-floored depression with the structure of a complex synclinorium, yet it was formed while sedimentation was still in progress so that it is also a sort of geosyncline. Its eastern side is underlain by the down-tilted western edge of the Sierra Nevada block.

(F) The *Coast Ranges proper* between the Great Valley and the Pacific coast—a series of sub-parallel ridges lower than the Sierra Nevada. The Coast Ranges are broken through at mid-length by San Francisco Bay, formed by local subsidence, so that it is convenient to speak of Northern and Southern Coast Ranges. The Coast Ranges include considerable bodies of deformed Tertiary rocks lying on a basement of Mesozoic and older rocks; unlike the Sierra Nevada they are not a single tilted block but a series of fault slices trending generally northwestward at a slight angle to the coast.

(G) The *Transverse Ranges*, trending east and west, against which the southeast-trending Coast Ranges proper terminate on the south. That part to the west—the Santa Ynez Mountains, the Santa Monica Mountains, and the Channel Islands—are largely covered by Tertiary rocks like the Coast Ranges farther north; that part to the east, north of the Los Angeles lowland—the San Gabriel and San Bernardino Mountains—are much more uplifted and expose mainly metamorphic and plutonic rocks.

(H) The *Peninsular Ranges*, south of the Transverse Ranges and Los Angeles lowland and extending into Baja California. Like the Coast Ranges proper farther north they trend south-southeast, but like the San Gabriel and San Bernardino Mountains and the Sierra Nevada they consist mainly of metamorphic and plutonic rocks, and their sedimentary cover is thin and patchy.

FAULTS OF CALIFORNIA. On these physiographic and geologic provinces are imposed the great faults of California (Fig. 91):

Throughout its 400-mile length the *Sierra Nevada* faces the Great Basin on the east in a series of lofty scarps that have been outlined mainly by faults, although the faults are not continuous and are offset en echelon in many places. So far as we know these faults have relatively simple dip-slip displacements, the side on the east being dropped, and the side on the west in the Sierra Nevada being raised and rotated. Movements have occurred on some of these faults within historic time, and have produced scarps in the alluvial deposits of Owens Valley.

More complex and less well understood is the *San Andreas fault*, which extends diagonally southeastward through the Coast Ranges of California, running out into the Pacific north of San Francisco and probably passing under the Gulf of California on the southeast. The true nature of the San Andreas fault

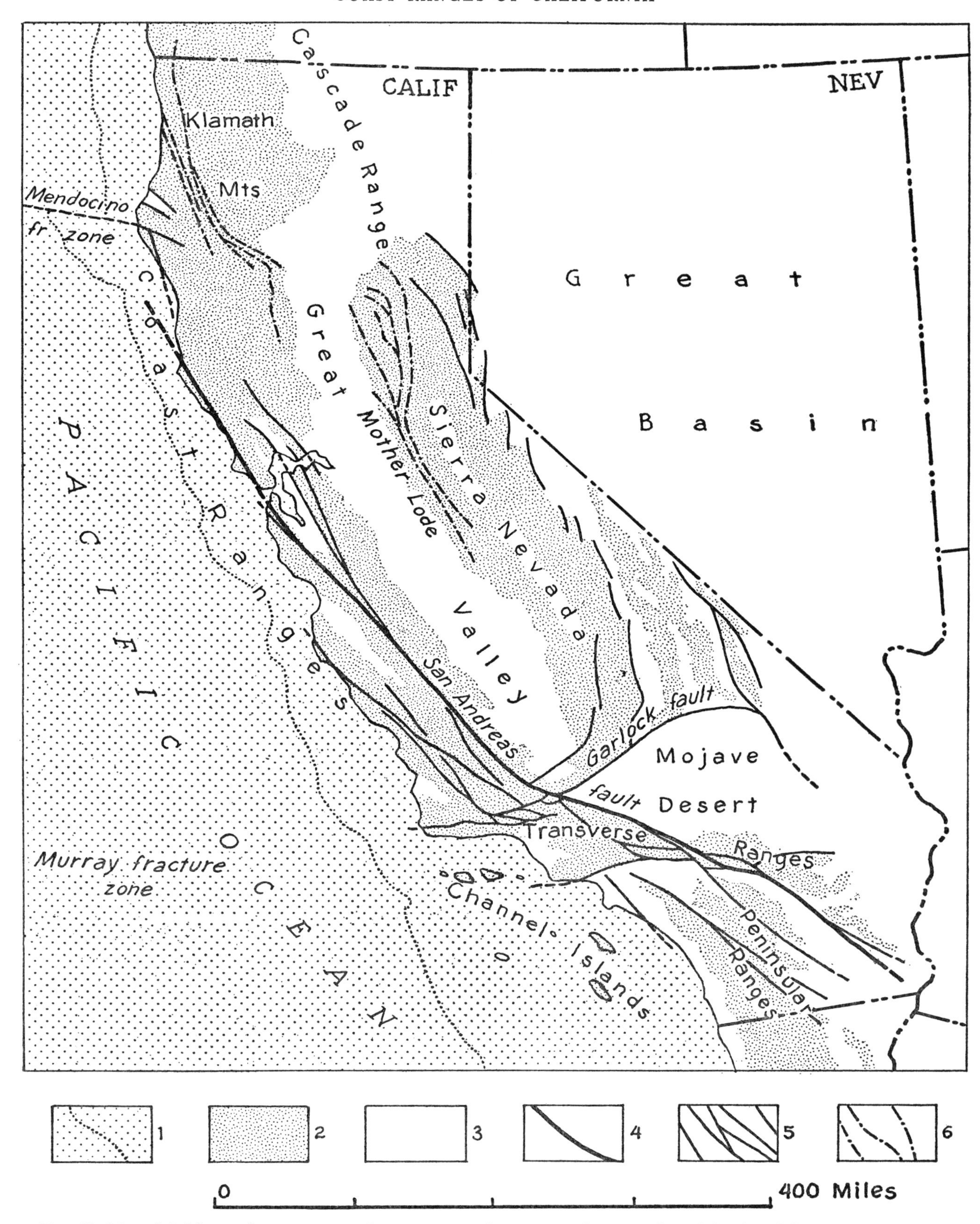

FIG. 91. Map of California showing topographic provinces and the principal faults. Compiled from Tectonic Map of United States, 1944; and other sources.

Explanation of symbols. 1. Ocean, with approximate edge of continental crust indicated by dotted line. 2. Major mountain areas. 3. Lowlands and minor mountain areas. 4. San Andreas fault. 5. Other high-angle faults, active or recently active. 6. Inactive high-angle faults of earlier systems.

was not fully appreciated until the great San Francisco earthquake of April 18, 1906, which made clear that it was a major, through-going fracture, still in process of movement. Unlike the faults of the Sierra Nevada, movement on the San Andreas fault is dominantly *strike-slip*, or sidewise, in a *right-lateral direction;* this is illustrated by the saying "San Francisco and Los Angeles are coming closer together," the two cities lying respectively northeast and southwest of the fault. Nature of the latest strike-slip movements on the San Andreas fault can be demonstrated, among other ways, by offset of stream valleys that cross it; streams flowing northeastward turn southeast along the fault for a quarter of a mile or so before resuming their northeastward course. In places great vertical movements are suggested by rocks and topography on opposite sides of the fault, but this is probably an illusion resulting from sideward shift of high-standing areas against low-standing areas. Total amount of this sideward movement is much debated; we will explore the possibilities later.

The San Andreas fault is only the largest and most spectacular of a host of kindred faults in the Coast, Transverse, and Peninsular Ranges, many of great magnitude in their own right, which branch from it at low angles, or run nearly parallel with it for long distances. Of the many we might mention specifically the *Hayward fault* east of San Francisco Bay and the San Andreas fault, and the *San Jacinto fault* in the Peninsular Ranges west of the San Andreas fault.

Besides these faults another set, mainly in the Transverse Ranges, trends more to the eastward, some showing dip-slip, others strike-slip displacement. The most conspicuous of these, actually north of the Transverse Ranges, is the *Garlock fault*, which branches northeastward from the San Andreas and forms the northern edge of the peculiar wedge of the Mojave Desert. Offset of features along it indicates a *left-lateral strike-slip displacement* (or the opposite of that on the San Andreas). The San Andreas fault shows its only notable deflection in its crossing of the Transverse Ranges, where it bends from its usual southeastward course to the east-southeast for a hundred miles or so.

SCARPS ON PACIFIC OCEAN FLOOR. No discussion of the faults of California, either of the San Andreas system or the Transverse Range system, would be complete without considering the remarkable system of scarps, probably produced by faulting, that recently have been discovered on the floor of the Pacific Ocean on the west (Plate I).

The *Mendocino Escarpment* extends westward from Cape Mendocino in northern California, and has been traced for at least 1,000 miles into the Pacific Ocean basin. Near the shore off Cape Mendocino the scarp faces north, but farther out to sea it is reversed and faces south, in places rising to heights of more than 10,000 feet. This scarp is notable in that it *offsets the edge of the continental shelf* by more than fifty miles in a right-lateral direction (Fig. 91). Moreover, at Shelter Cove a little south of Cape Mendocino the ground was disturbed at the time of the San Francisco earthquake, indicating that movements related to those on the San Andreas fault extended into this area.

The *Murray Escarpment* trends westward from southern California, and has been traced at least 1,900 miles across the Pacific Ocean basin near to or past the Hawaiian Islands. The escarpment faces north and its relief is less than that of the Mendocino Escarpment, being seldom more than 5,000 feet, but it is paralleled closely by other scarps and ridges that are probably parts of the same fracture system. The Murray Escarpment lies on the same trend as the Transverse Ranges; the connection between them is nearly lost on the continental slope, yet the two sets of features are undoubtedly parts of the same system of deformation.

BASEMENT OF THE COAST RANGES. Let us now deal specifically with the Coast Ranges and examine details of their geology.

We have said that the Coast Ranges are characterized particularly by their Tertiary rocks, but while these attain great thickness in parts of the ranges, they do not cover its whole surface. The Tertiary rocks lie on older rocks of various sorts and ages which emerge in places. In some areas the Tertiary lies on Cretaceous strata and these in part on Jurassic strata. Elsewhere the Tertiary or the Cretaceous lies on a true basement of crystalline rocks. The areas in which these crystalline rocks occur have a significant distribution (Fig. 92).

One long strip where crystalline rocks lie at or near the surface extends from Point Reyes and the Farallon Islands near San Francisco southeastward nearly to the Transverse Ranges; it has been named *Salinia* for its characteristic development near the Salinas Valley. The crystalline basement of Salinia consists of highly metamorphosed sedimentary rocks, the Sur series, invaded by plutonic rocks such as the Santa Lucia granodiorite. The crystallines have been assigned to a pre-Franciscan age, as the Franciscan, where present nearby, shows no sign of plutonic alteration (see below for discussion of the Franciscan and its problems). Although suspected to be very old, recent radiometric determinations on the plutonic rocks have yielded ages of 82 to 92 million years, indicating that they are of Cretaceous age like those in the eastern part of the Sierra Nevada.

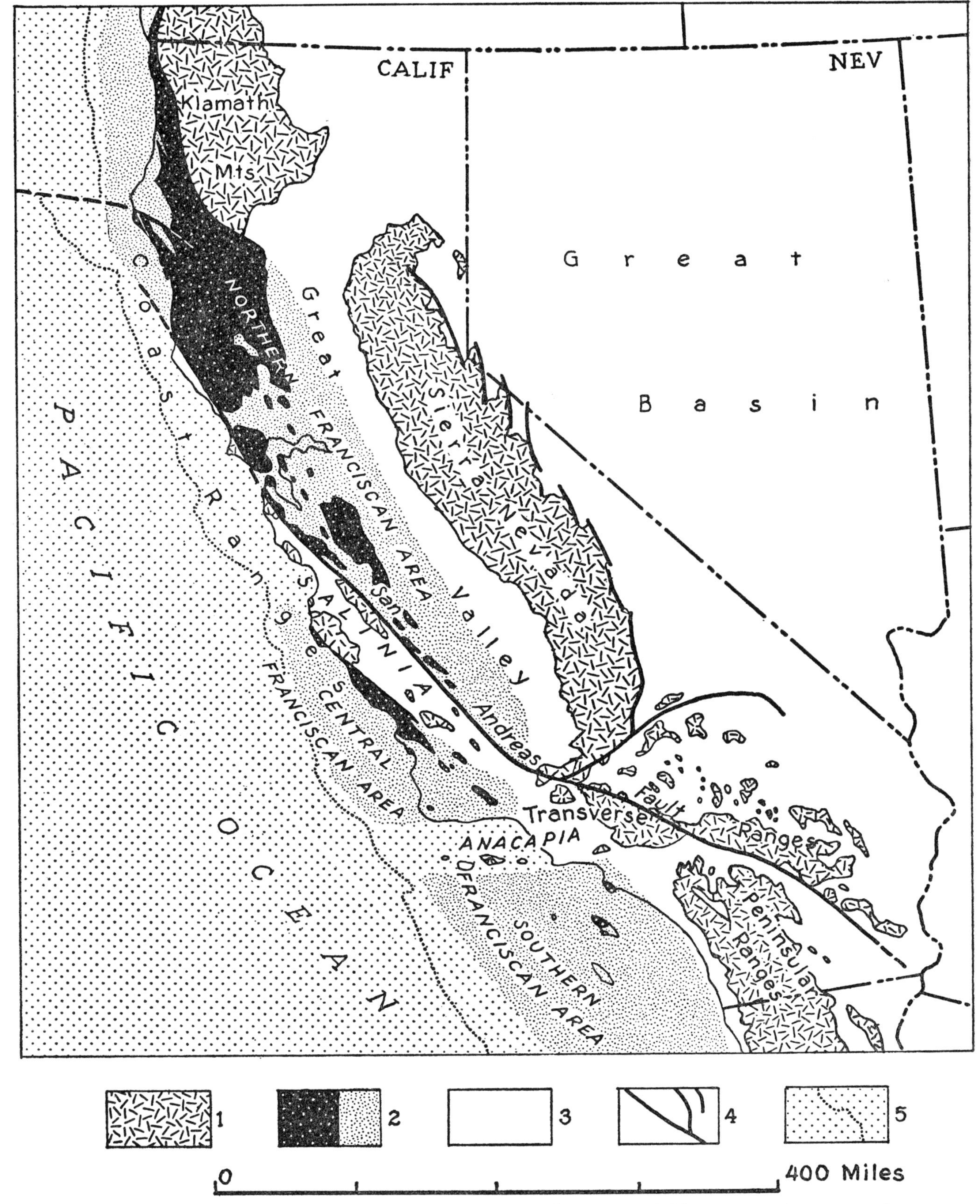

FIG. 92. Map of California showing areas of crystalline basement and the three areas of Franciscan basement. Adapted from Reed, 1933; and Reed and Hollister, 1936; after King, 1958.

Explanation of symbols. 1. Outcrops of crystalline basement (metamorphic and plutonic rocks which were part of the Nevadan orogenic belt). 2. Areas of Franciscan basement (outcrops solid, areas where Franciscan is overlain by younger rocks shaded). 3. Areas where basement is covered by Cretaceous, Tertiary, and Quaternary rocks (also areas of undifferentiated rocks of various ages and structures east of Sierra Nevada). 4. Major faults. 5. Ocean, with edge of continental crust shown by dotted line.

Another strip of crystalline rocks follows the Transverse Ranges from the Channel Islands on the west into the San Bernardino Mountains; from one of the islands it has been called *Anacapia*. This basement is more heterogeneous than that of Salinia and more is known about its age—anorthosite in the San Gabriel Mountains has been dated radiometrically at 900 million years or middle Precambrian; later Paleozoic fossils have been found in meta-sedimentary rocks of the San Bernardino Mountains; the Pelona schist of the San Gabriel Mountains, although called Precambrian, is probably Paleozoic or ever younger; Triassic slate occurs in the Santa Monica Mountains.

Besides the strips of crystalline rocks in Salinia and Anacapia there are broader areas of crystalline rocks in the Peninsular Ranges farther south which closely resemble those in the Sierra Nevada. They consist of metamorphosed Paleozoic and early Mesozoic surficial rocks that were invaded by plutonic rocks of later Mesozoic age. Some of the latter in southern California have been dated radiometrically at 115 million years, or early Cretaceous, but others in Baja California farther south invade middle Cretaceous rocks and are overlain unconformably by the latest Cretaceous.

Franciscan group. Between the areas of crystalline rocks the younger sedimentary rocks of the Coast Ranges lie on a quite different basement, the Franciscan group. Although this can be called a "basement" also, and although it is much deformed and indurated, it is much less altered than the crystalline basement rocks.

The Franciscan group occurs in three separate areas (Fig. 92). The *northern Franciscan area* makes up most of the Coast Ranges north of San Francisco Bay and extends a long distance southeast of the bay, northeast of the San Andreas fault and Salinia. The *central Franciscan area* forms that part of the southern Coast Ranges southwest of Salinia (from which it is separated by the Nacimiento fault) and north of the Transverse Ranges; its Franciscan rocks are exposed along the Pacific coast for many miles southward from Point Sur. The *southern Franciscan area* lies mostly at sea off southern California, south of the Transverse Ranges and west of the Peninsular Ranges. Its rocks emerge in the Palos Verdes Hills, Catalina Island, and various islands and promontories along the coast of Baja California; otherwise its extent is suggested mainly by occurrence of its characteristic debris in younger sedimentary rocks along the coast.

Rocks of the Franciscan group are an old-looking, much deformed, thoroughly indurated series of graywackes, shales, bedded cherts, limestone lenses, and interbedded basaltic lavas cut by many ultramafic serpentine intrusions; near the latter they have been altered to glaucophane schist. This assemblage is typically eugeosynclinal.

A curious feature of the Franciscan is that its base is nowhere visible. Geologists have long supposed that it should lie unconformably on the metamorphic and plutonic rocks of the crystalline basement, yet wherever these are juxtaposed the contact is a fault. There is thus no evidence of the sort of crustal material on which the Franciscan rocks were laid; it might have been something quite different from the adjacent crystalline basement.

The age of the Franciscan has long been a puzzle as it contains very few fossils. In most places it is overlain unconformably by Tertiary rocks, and in a few by Cretaceous rocks (although these are generally latest Cretaceous). In some places it is reported to be overlain by the latest Jurassic Knoxville formation. Commonly it has been believed to be older than the latest Jurassic, yet because of its lack of severe metamorphism, it is thought to be younger than the Nevadan orogeny (Fig. 83 A).

But these interpretations have been thrown into doubt by recent fossil discoveries. Ammonoids and pelecypods from the sandstones, and Foraminifera from the limestones at widely separated localities have proved to be as young as early Late Cretaceous in age (Albian and Cenomanian), although at a few localities fossils as old as Late Jurassic have been discovered in similar rocks (Fig. 83 B). This is especially puzzling because along the eastern edge of the Coast Ranges a conformable fossiliferous sequence nearly eight miles thick extends from Late Cretaceous into Late Jurassic (Chico, Horsetown, Paskenta, and Knoxville formations). Most of this sequence is thin-bedded sandstone and shale and lacks any beds of Franciscan lithology. How could the Franciscan "basement" farther west be as young or younger than these strata?

A possible explanation may be that the Franciscan is a peculiar facies—eugeosynclinal, if you will—that formed at different times from one place to another, into which various later Mesozoic formations passed westward toward the edge of the Pacific. At one time the Franciscan was thought to be a shallow water or even a continental deposit. Could not the Franciscan have been a deposit laid down in quite deep water instead, on or at the foot of the continental slope of the time, its sediments being transported thence from nearer the shore by turbidity currents? Equivalent shelf sediments might at first have been the Late Jurassic Knoxville formation; later on, as the shelf built westward, various units in the Cretaceous. During the Cenozoic folding and fault-slicing of the Coast Ranges, these continental shelf and slope deposits were forced up and made part of the continent.

Theories on origin of the Coast Ranges. To one like myself who has followed from afar for many years the development of ideas on Coast Range geology, one of its intriguing features has been the diversity of interpretations which have been offered to explain it. Truly the difficulties of the area have been a challenge to every geologist who has worked there, to which each has responded to the best of his ability—yet not always with harmonious results.

The pattern was set by the giants of the past: *Andrew C. Lawson* of the University of California at Berkeley and *Bailey Willis* of Stanford University. Although both men lived into their nineties and nearly to our own day, most of their work and conflicts on the Coast Ranges occurred many years before. Of greater interest now are the interpretations of their successors of a decade or two ago, and especially those of three geologists.

Bruce L. Clark, who was a professor of paleontology at the University of California at Berkeley, did outstanding work on the Tertiary stratigraphy of the Coast Ranges, during which he also evolved a theory of Coast Range evolution and structure. He supposed that the ranges were a mosaic of fault blocks, each with a more or less independent history during Cenozoic and earlier times, rising or sinking erratically so that at one time a block might be uplifted and eroded, at another submerged and covered by sediments. Folds which occur within the fault blocks were believed to be an incidental result of their slicing and compression.

Ralph D. Reed, who was chief geologist in Califorian for The Texas Company, opposed the notions of Clark and had another picture of Coast Range structure. Folding was believed to be the normal, dominant feature; faulting, incidental and in places merely an effect of the folding. He emphasised the contrasting nature of the basement in different parts of the ranges, and suggested that the three areas of Franciscan rocks were formed in independent geosynclinal basins separated by axes of crystalline basement in Salinia and Anacapia; this framework of the basement much influenced the nature of the folding and faulting in the overlying Cretaceous and Tertiary sediments.

Nicholas L. Taliaferro has been professor of geology at the University of California at Berkeley for many years, but most of the results of his work on the Coast Ranges were published somewhat later than those of Clark and Reed. Of interest is his conclusion that Salinia, Anacapia, and the Franciscan areas were not ancient positive and negative elements, but originated no farther back than Late Cretaceous time. Franciscan deposits were supposed to have been laid on the crystalline basement entirely across the Coast Ranges, but during a Late Cretaceous orogeny the basement was raised along certain belts from which the Franciscan was removed by erosion.

Viewing Clark's and Reed's theories in retrospect one can find many virtues in both; very possibly the final answer lies somewhere between the extremes they presented. To a large extent Taliaferro's interpretations reconcile these opposing views, but they may require further modification as suggested below.

Displacement on San Andreas fault. Strangely, all three geologists placed little emphasis on the San Andreas fault and the proved strike-slip movements that have taken place on it and its kindred. The great lateral shifts which some geologists suspected along them were thought to be incredible and without proof, the really important displacements being better explained by vertical uplift or depression of the adjacent blocks.

In the meantime a number of other geologists have been at work on the San Andreas fault and the rocks adjoining it. Among them *Levi F. Noble* has studied for the U.S. Geological Survey the segment between Soledad Pass and Cajon Pass on the north side of the San Gabriel Mountains in southern California (Fig. 93). Here scars are still visible that were produced during the Fort Tejon earthquake of 1857, a disturbance which equalled in magnitude the one of 1906 farther north. Stream valleys that cross the fault are shifted by as much as half a mile in a right-lateral direction, and Pleistocene deposits are shifted in the same direction by five or six miles (Fig. 94 A and B). Later Tertiary formations are still more discordant; one must go twenty or thirty miles eastward to find a counterpart of the rocks on the south side (Fig. 94 C). The older Tertiary formations show even greater anomalies but are too fragmentarily exposed to yield decisive evidence. In the segment of the San Andreas fault studied by Noble displacement of the various Cenozoic formations have been progressively greater the older the rocks, suggesting that the fault has been moving steadily, or perhaps spasmodically, at least since early Tertiary time.

Total displacement on the San Andreas fault should be indicated by offset of the basement rocks, which have shared all its later movements. Matching of basement rocks on the two sides is as yet inconclusive, but geologists working in southern California have found remarkable resemblances between the basement of the San Gabriel Mountains southwest of it and those of the Orocopia Mountains northeast of it, which lie near the Salton Sea 150 miles to the southeast (Fig. 94 E). The anorthosites and the Pelona schist of the San Ga-

briel Mountains appear to be matched by units in the Orocopia Mountains.

It is thus possible that earlier syntheses of Coast Range history will require modification, as the blocks into which the area is divided may have shifted laterally with respect to each other through time, by amounts great enough to have produced significant changes in the geography.

A PROPOSED HISTORY OF THE COAST RANGES. With this background we can propose still another interpretation of the history of the Coast Ranges—though knowing full well that any such proposal will create anomalies as yet unexplainable.

Let us begin with the Nevadan orogeny of mid-Mesozoic time. This, we will recall, deformed, metamorphosed, and plutonized the Cordilleran eugeosynclinal rocks of earlier Mesozoic and Paleozoic time along a belt extending from the Klamath Mountains through the Sierra Nevada to the Peninsular Ranges.

There is little evidence that Nevadan orogeny affected the site of the Coast Ranges on the west, although this is commonly assumed. It is equally likely that the zone of greatest deformation lay farther east and that the Coast Range area was less disturbed. The Nevadan deformed belt may have been the western edge of the continent at the time, being bordered beyond by a continental shelf and slope where the Franciscan, Knoxville, Paskenta, and later deposits accumulated (Fig. 95 A). These may in part have formed while this orogeny was in progress, instead of being younger than the Nevadan orogeny.

The great orogeny in the western area, according to Taliaferro, took place late in Cretaceous time but before its close, when the previous Late Jurassic and

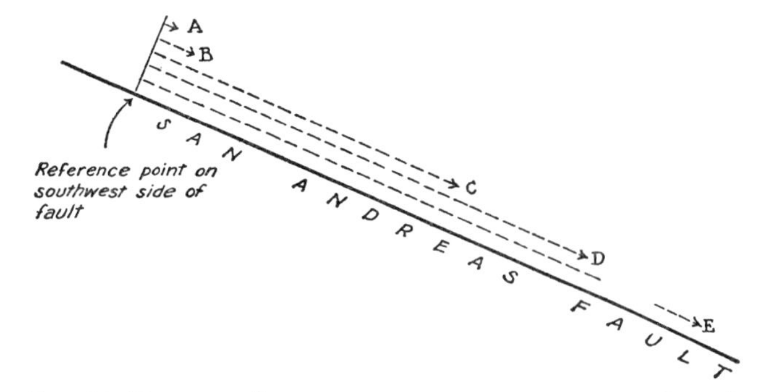

FIG. 94. Diagram illustrating progressive movements through time on the San Andreas fault, as observed between Soledad Pass and Cajon Pass, southern California. (A) Offset of modern streams, ½ mile. (B) Of Pleistocene deposits, 5 to 6 miles. (C) Of Miocene and Pliocene formations, 20 to 30 miles. (D) Of early Tertiary formations, amount undetermined but greater than C. (E) Of basement rocks, amount undetermined but may be as great as 150 miles. Data from Noble, 1954.

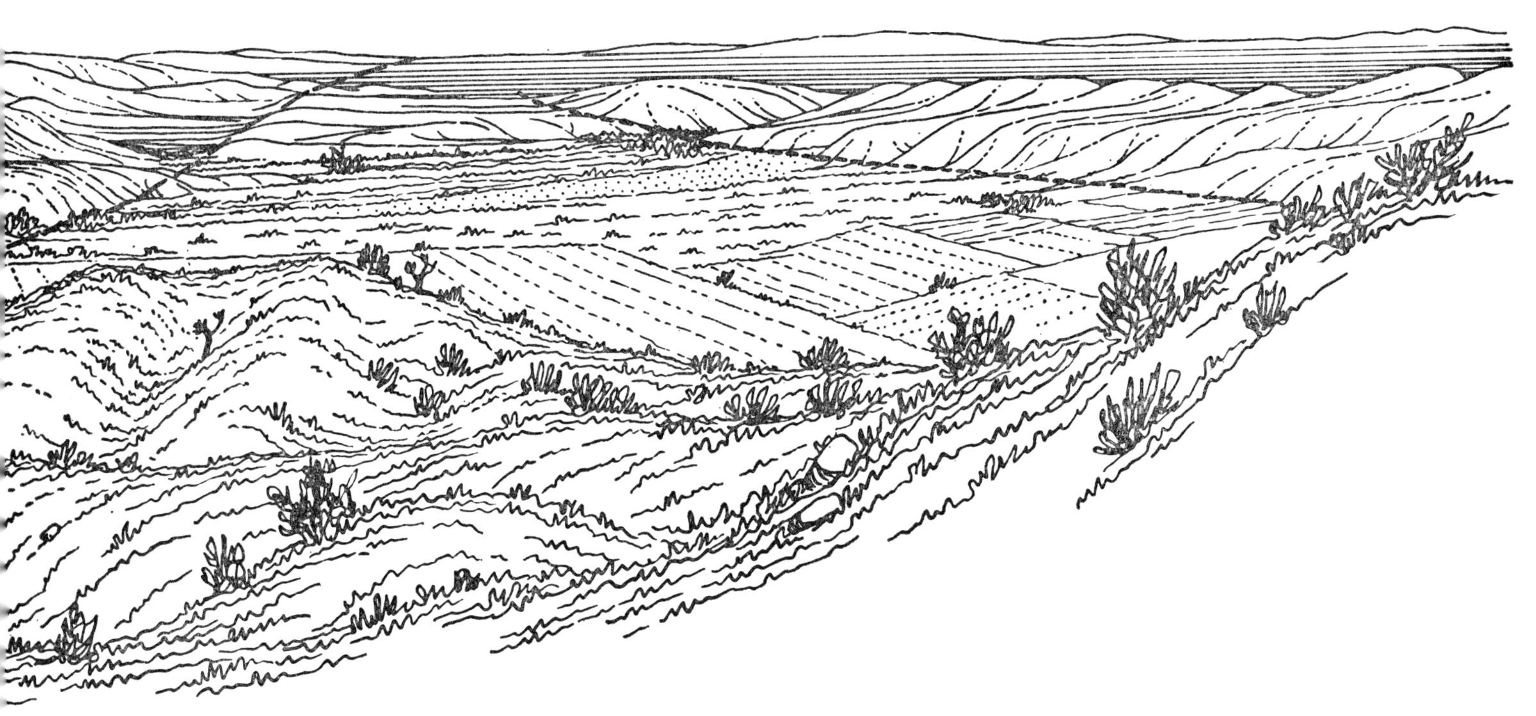

FIG. 93. View of San Andreas fault zone at Valyermo on north side of San Gabriel Mountains, southern California. View is west-northwest, with San Gabriel Mountains on left and Mojave Desert on right. Long dashes indicate the trace of the main San Andreas fault. Short dashes indicate the traces of lesser faults at the north and south edges of the fault zone, which are about 3 miles apart. After Noble, 1954 (drawn by P. B. King).

Early Cretaceous geosyncline was fragmented into blocks, the uplifted parts of which were vigorously eroded, so that the later Cretaceous and the Tertiary deposits were laid down in some places over the earlier geosynclinal rocks, and in other places over crystalline rocks. But if we should superpose on this interpretation the concept of a progressive lateral shift of the crust along the San Andreas and later faults, it would appear that the first great shift occurred at least as early as Late Cretaceous time.

By such an assumption we have, of course, greatly complicated any attempt to reconstruct the geography of the Coast Ranges during Mesozoic time, as lateral shifting during and since then would have carried various blocks far out of their original relations to each other, some northwest or southeast on the San Andreas trend, others east or west on the Transverse Range trend. Like Humpty-Dumpty, all the King's horses and all the King's men cannot put the Coast Ranges back together again.

As a start, though, let us assume that the principal strike-slip movement was on the San Andreas fault. If so, the crystalline rocks of the Transverse and Peninsular Ranges southwest of the San Andreas fault would have lain several hundred miles farther south and a lesser distance east of their present positions.

And so also—and here is the real shocker—would the crystalline rocks of Salinia. Salinia may not have been a geanticline after all, raised and stripped of a Franciscan cover as a result of a late Cretaceous orogeny; instead it may have been a sliver that was peeled off the western margin of the Nevadan orogenic belt and shifted to the northwest (Fig. 95 B). The Cretaceous age of the plutonics of Salinia and absence of any intrusive relations between them and the nearly contemporaneous Franciscan rocks can now be viewed in a different light; they may have been far distant from the Franciscan at the time when they were intruded.

Under this interpretation the northern and central Franciscan areas were neither separate geosynclinal basins nor parts of one basin which was later split by a geanticlinal uplift. Instead the two areas were segments of the continental shelf and slope west of the Nevadan orogenic belt that orginally lay north and south of each other; by movements on the San Andreas fault they were later shifted alongside.

Having set forth this general concept, we can introduce a few of the many possible complications:

(1) Kindred faults of the San Andreas system may simply have moved by lesser amounts than the San Andreas itself, but at least some of them could have

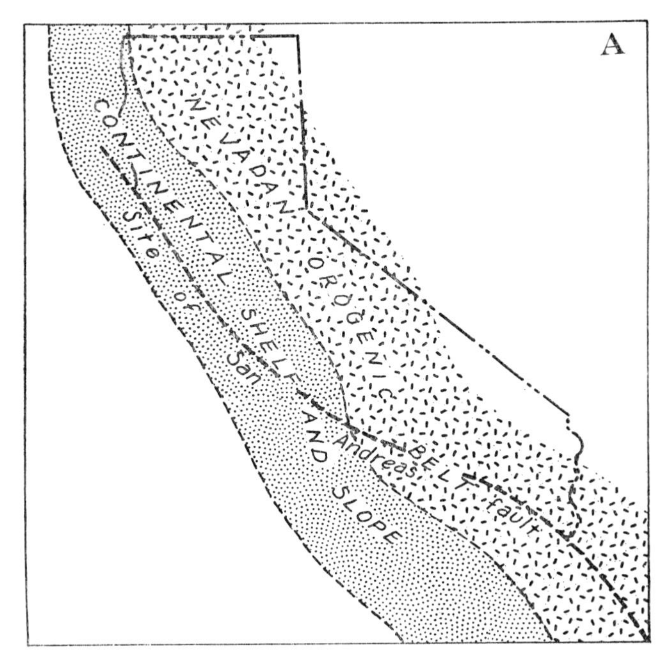

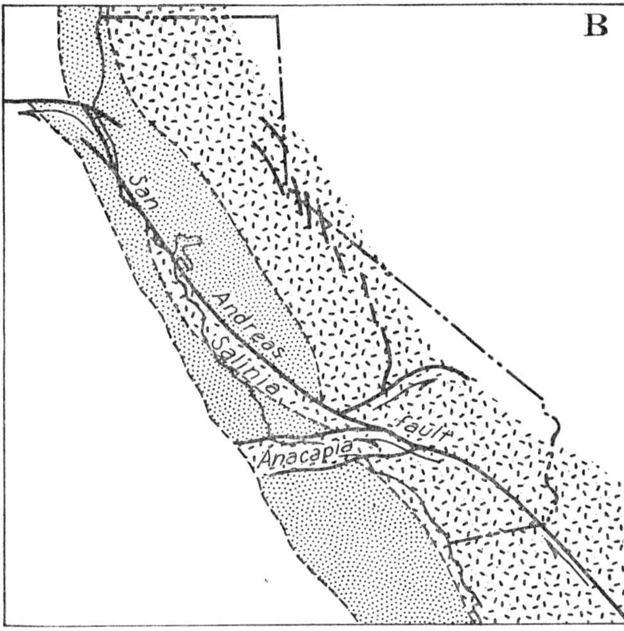

Fig. 95. Sketch maps to illustrate an hypothesis for the structural evolution of California. (A) Conditions in Late Jurassic and Early Cretaceous time during deposition of Franciscan and related units on the continental shelf and slope west of the Nevadan orogenic belt. (B) Present conditions after great lateral shift on San Andreas fault and westward deflections along Mendocino fracture zone (in north) and in Anacapia (in south).

absorbed the main movement at times when the San Andreas itself was inactive. Something of the sort seems to be in progress now in southern California where the San Jacinto fault west of the Salton Sea appears to be more active than the southern segment of the San Andreas to the east. A similar relation at an earlier period is suggested on the San Gabriel fault northwest of Los Angeles on which many miles of strike-slip movement occurred before late Tertiary time, but none later.

Absorption of the main movement by one fault or another through time might account for conflicting evidence as to the amount of displacement on the main San Andreas fault in different parts of its course. Total shift of the whole area is likely to be greater than the indicated displacement on any one fault.

(2) If great lateral shift on the San Andreas fault is assumed, a puzzling feature is its failure to offset the coastline where it passes out to sea at Mussel Rock south of San Francisco. Still more puzzling is the lack of offset of the edge of the continental shelf farther northwest along the apparent prolongation of the San Andreas fault.

East of the prolongation of the San Andreas fault, however, at Shelter Cove south of Cape Mendocino, significant ground movements occurred at the time of the San Francisco earthquake of 1906; west of the cape the continental shelf is offset by more than fifty miles on the east-west-trending Mendocino fracture zone. In this northern area, shift of the rocks on the oceanward side of the San Andreas fault may thus have been absorbed by kindred faults east of the main trend, and then have been deflected westward onto the Mendocino fracture zone (Fig. 95 B).

(3) The westward deflection of the rocks on the oceanward side of the San Andreas system onto the Mendocino fracture zone in the north may suggest an explanation for the complex interaction between the San Andreas system and the other east-west system farther south—that of the Transverse Ranges and the Murray fracture zone. Neither the San Andreas nor the Transverse Range systems is clearly younger or older than the other; movements on the two may have alternated, or even may have been simultaneous.

At times the northwestward shift of the Coast Range area along the San Andreas trend may have been deflected westward along the Transverse Range trend in the same manner as it appears to have been on the Mendocino trend (Fig. 94 B). If such a deflection were great enough, the crystalline axis of Anacapia might have been derived in the same manner as suggested for Salinia—from a sliver off the Nevadan orogenic belt farther south.

Tertiary sedimentation and deformation. The framework of the Coast Ranges was thus established by Nevadan orogeny, by deposition of Jurassic and Cretaceous continental shelf and slope deposits west of the orogenic belt, and by disordering of the pattern by progressive lateral shift of the crustal blocks, beginning at least as early as the time of the late Cretaceous orogeny. These primarily Mesozoic events leave much room for doubt and speculation; the record becomes plainer in the Tertiary, although no less complex.

During Tertiary and Quaternary time compression of the rocks of the Coast Ranges was exerted at frequent intervals. Some areas show a record of almost continuous deformation for long periods, and some of the sedimentary sequences contain so many unconformities that it is difficult to generalize them into any orogenic climaxes. Nevertheless, one receives a rather blurred impression of two principal climaxes: one near the middle of the Miocene, the other early in the Pleistocene. During the latter the Pliocene and early Pleistocene deposits in the basins of southern California were steeply upended, and their eroded edges were covered by later Pleistocene deposits which were little deformed afterwards. To this early Pleistocene orogeny the German geologist Hans Stille has given the name *Pasadenan orogeny.*

Tertiary sedimentation and deformation were strongly influenced by the nature of the basement on which the deposits were laid; deposits laid over the relatively rigid crystalline belts were much less deformed than those laid over the weaker rocks of the Franciscan group. In places Tertiary and Quaternary sedimentation took place in more or less separate basins between higher standing crustal blocks, formed mainly of crystalline basement. Two of these, the *Los Angeles* and *Ventura basins* of southern California, were rapidly deepened in late Tertiary time, then filled by enormous thicknesses of Pliocene and Pleistocene deposits laid down in progressively shallower water; later on these were deformed by the early Pleistocene Pasadenan orogeny.

The mid-Miocene deformation shaped the ridges and troughs of the Coast Ranges into about their present configuration but left wide tracts of country still submerged; the mid-Pleistocene deformation brought about a widespread emergence. During times of greatest submergence the Coast Range area probably resembled the present offshore region of southern California—with shallow shelves and banks interspersed with deeper troughs, and with linear islands like the present Channel Islands. During times of greatest emergence the Coast Range area probably resembled the present topography around San Francisco Bay—with mountain ridges and intervening troughs and valleys, in which continental sediments were being deposited and whose lowest parts, like the present bay, were covered by shallow, ramifying seas.

REFERENCES

2. *Basin and Range province*

Axelrod, D. I., 1957, Late Tertiary floras and the Sierra Nevada uplift: *Geol. Soc. America Bull.*, v. 68, p. 19-45.

Davis, W. M., 1925, The Basin Range problem: *Natl. Acad. Sci. Proc.*, v. 11, p. 387-392.

Gilbert, G. K., 1928, Studies of Basin-Range structure: *U.S. Geol. Survey Prof. Paper 153*, 92 p.

Hubbs, C. L., and Miller, R. R., 1948, The Great Basin with emphasis on glacial and post-glacial times, II, The zoological evidence; correlation between fish distribution and hydrographic history in the desert basins of western United States: *Utah Univ. Bull.*, v. 38, p. 17-166.

Louderback, G. D., 1904, Basin Range structure of the Humboldt region: *Geol. Soc. America Bull.*, v. 15, p. 289-346.

Nolan, T. B., 1943, The Basin and Range province in Utah, Nevada, and California: *U.S. Geol. Survey Prof. Paper 197 D*, p. 141-196 (esp. p. 178-186).

Van Houten, F. B., 1956, Reconnaissance of Cenozoic sedimentary rocks of Nevada: *Am. Assoc. Petrol. Geol. Bull.*, v. 40, p. 2801-2825.

3. *Northwestern volcanic province*

Hodge, E. T., 1938, Geology of the lower Columbia River: *Geol. Soc. America Bull.*, v. 49, p. 831-930.

Snavely, P. D., Jr., and Baldwin, E. R., 1948, Siletz River volcanic series, northwestern Oregon: *Am. Assoc. Petrol. Geol. Bull.*, v. 32, p. 805-812.

Waters, A. C., 1955, Volcanic rocks and the tectonic cycle: *Geol. Soc. America Spec. Paper 62*, p. 663-684.

Williams, Howel, 1940, The geology of Crater Lake National Park, Oregon, with a reconnaissance of the Cascade Range southward to Mount Shasta: *Carnegie Inst. Washington Publ. 540*, 162 p.

4. *Coast Ranges of California*

Clark, B. L., 1930, Tectonics of the Coast Ranges of middle California: *Geol. Soc. America Bull.*, v. 41, p. 747-828.

Hill, M. L., and Dibblee, T. W., Jr., 1953, San Andreas, Garlock, and Big Pine faults, California: *Geol. Soc. America Bull.*, v. 64, p. 443-458.

Jahns, R. H., 1954, Investigations and problems of southern California geology, *in* Geology of southern California: *California Div. Mines Bull. 170*, chap. 1, p. 5-29.

Menard, H. W., 1955, Deformation of the northeastern Pacific Ocean basin and the west coast of North America: *Geol. Soc. America Bull.*, v. 68, p. 1149-1198.

Noble, L. F., 1954, The San Andreas fault zone from Soledad Pass to Cajon Pass, California, *in* Geology of southern California: *California Div. Mines Bull. 170*, chap. 4, p. 37-48.

Reed, R. D., and Hollister, J. S., 1936, Structural evolution of southern California: *Am. Assoc. Petrol. Geol. Bull.*, v. 20, p. 1533-1704.

Stille, Hans, 1936, Present tectonic state of the earth: *Am. Assoc. Petrol. Geol. Bull.*, v. 20, p. 849-880.

Taliaferro, N. L., 1943, Geologic history and structure of the central Coast Ranges of California: *California Div. Mines Bull. 118*, p. 119-163.

EPILOGUE

With our examination of the Coast Ranges of California we have completed our arm-chair journey across North America. Our journey has been more circuitous than an actual traveler would care for—we commenced in the far north in the Canadian Shield, passed into the Interior Lowlands and across the Appalachian Mountains to the Atlantic Ocean, then proceeded southward across the Gulf Coastal Plain to the West Indies; finally we arrived at the front of the Cordillera and crossed this westward from the Great Plains to the Pacific Coast.

Our journey has also taken us through vast reaches of geologic time—from the earliest beginnings that can be discovered in the Precambrian rocks of the nucleus of the continent, through the mountains that were built up later in successive belts around the edges of the original nucleus. In the end, we arrived amidst the Cenozoic rocks and structures that formed, and are still forming, along the Pacific coast.

During our journey we enlisted the aid of many skills—the strictly geological, by which the little deformed and strongly deformed rocks have been analysed; the geophysical, by which the deeper layers of the crust have been probed; the geochemical, by which the plutonic and volcanic activity of the earth has been explained, and which has furnished us, as well, with a new means of dating the rocks and measuring geologic time.

We have finally, like the Forty-niners, made it through to the Golden Gate and have come out on the farther side of the continent. In our dash to the coast we have, like the pioneers, had to discard many things that might seem precious—in our case, areas or geologic features of much interest in themselves which did not seem to illustrate fully our general theme. Despite this loss, it is to be hoped that the essentials have been saved—an exposition of the evolution of North America, both in its grand design and in some of its infinite variations.

FIG. 96. The Pacific Coast of California. Steeply tilted early Tertiary strata at San Pedro Point, San Mateo County, south of San Francisco. After Lawson, 1915.

SOURCES OF ILLUSTRATIONS

End papers. Inside front cover: King, P. B., and Stupka, Arthur, 1950, The Great Smoky Mountains—their geology and natural history: *Sci. Monthly*, v. 71, fig. 5, p. 38 (by P. B. King).

Inside back cover: Noble, L. F., 1954. The San Andreas fault zone from Soledad Pass to Cajon Pass, California, *in* The geology of southern California: *California Div. Mines Bull. 170*, chap. IV, fig. 8, p. 45 (by P. B. King).

Plate 1. Compiled from tectonic maps of United States and Canada, geologic maps of Alaska, Mexico, and Central America, and other sources. Base and projection copied from map of American Geographical Society, 1942 (used with permission of the Society).

Figure 1. Wilson, J. T., 1954, The development and structure of the crust, in *The Earth as a Planet*: Univ. Chicago Press, fig. 2, p. 148; Ewing, Maurice, and Press, Frank, 1955, Geophysical contrasts between continents and ocean basins: *Geol. Soc. America Spec. Paper 62*, fig. 1, p. 5. (Published with permission of J. Tuzo Wilson).

Figure 2. Original drawing.

Figure 3. Original drawing.

Figure 4. Original drawing.

Figure 5. Original drawing.

Figure 6. Original drawing.

Figure 7. Geological Association of Canada, 1950, *Tectonic Map of Canada*: Geol. Soc. America; Gill, J. E., 1949, Natural divisions of the Canadian Shield: *Roy. Soc. Canada Trans.*, ser. 3, v. 43, sec. 4, p. 61-69; Wilson, J. T., 1954, The development and structure of the crust, in *The Earth as a Planet*: Univ. Chicago Press, fig. 20, p. 196.

Figure 8. Leith, C. K., Lund, R. J., and Leith, Andrew, 1935, Precambrian rocks of the Lake Superior region: *U.S. Geol. Survey Prof. Paper 184*, pl. 1.

Figure 9. A—Lee, Wallace, 1956, Stratigraphy and structural development of Salina basin area: *Kansas Geol. Survey Bull. 121*, pl. 4. B—Bridge, Josiah, 1930, Geology of the Eminence and Cardareva quadrangles: *Missouri Bur. Geol. Mines*, v. 24, 2nd ser., geologic map, section B B′.

Figure 10. Kay, Marshall, 1951, North American geosynclines: *Geol. Soc. America Mem. 48*, pl. 12, p. 32.

Figure 11. A—Richardson, G. B., 1909, Description of the El Paso district: *U.S. Geol. Survey Geol. Atlas*, folio no. 166, fig. 8, p. 8. B—Darton, N. H., 1928, "Red beds" and associated formations of New Mexico: *U.S. Geol. Survey Bull. 794*, fig. 23, p. 99.

Figure 12. Original drawing.

Figure 13. A—Cumings, E. R., and Shrock, R. R., 1928, Niagaran coral reefs of Indiana and adjacent states and their stratigraphic relations: *Geol. Soc. America Bull.*, v. 39, p. 579-619. B—King, P. B., 1948, Geology of southern Guadalupe Mountains, Texas: *U.S. Geol. Survey Prof. Paper 215*, pl. 7.

Figure 14. Original drawing.

Figure 15. Wasson, Theron, and Wasson, I. B., 1927, Cabin Creek field, West Virginia: *Am. Assoc. Petrol. Geol. Bull.*, v. 11, fig. 2, p. 710; Lafferty, R. C., 1941, Central basin of Appalachian geosyncline: *Am. Assoc. Petrol. Geol. Bull.*, v. 25, figs. 4 and 5, p. 786-789.

Figure 16. A—Cohee, G. V., 1948, Cambrian and Ordovician rocks in Michigan basin and adjacent areas: *Am. Assoc. Petrol. Geol. Bull.*, v. 32, fig. 8, p. 1440. B—*op.cit.*, fig. 9, p. 1444. C—Cohee, G. V., 1948, Thickness and lithology of Upper Ordovician and Lower and Middle Silurian rocks in the Michigan basin: *U.S. Geol. Survey, Oil and Gas Inves.*, Chart 33, sheet 2, fig. 10. D—Landes, K. K., 1945, The Salina and Bass Island rocks of the Michigan basin: *U.S. Geol. Survey, Oil and Gas Inves.*, Map 40, fig. 1. E—Cohee, G. V., 1947, Lithology and thickness of the Traverse group in the Michigan basin: *U.S. Geol. Survey Oil and Gas Inves.*, Chart 28, fig. 1. F—Stose, G. W., 1946, Geologic map of North America: *Geol. Soc. America.*

Figure 17. A—Cohee, G. V., 1948, Cambrian and Ordovician rocks in Michigan basin and adjacent areas: *Am. Assoc Petrol. Geol. Bull.*, v. 32, fig. 2, p. 1428-1429. B—*op.cit.* fig. 6, p. 1434-1435. C—Cohee, G. V., 1948, Thickness and lithology of Upper Ordovician and Lower and Middle Silurian rocks in the Michigan basin: *U.S. Geol. Survey Oil and Gas Inves.*, Chart 33, sheet 2, fig. 7. D—*op.cit.*, fig. 9. E—Landes, K. K., The Salina and Bass Island rocks of the Michigan basin: *U.S. Geol. Survey Oil and Gas Inves.*, Map 40, fig. 3. F—Cohee, G. V., 1947, Lithology and thickness of the Traverse group in the Michigan basin: *U.S. Geol. Survey Oil and Gas Inves.*, Chart 28, fig. 2.

Figure 18. Tester, A. C., 1937, Geologic map of Iowa: *Iowa Geol. Survey*; Stose, G. W., and Ljungstedt, O. A., 1933, Geologic map of United States: *U.S. Geol. Survey*; Ballard, Norval, 1942, Regional geology of Dakota basin: *Am. Assoc. Petrol. Geol. Bull.*, v. 26, fig. 9, p. 1579; Wanless, H. R., 1955, Pennsylvanian rocks of Eastern Interior Basin: *Am. Assoc. Petrol. Geol. Bull.*, v. 39, fig. 2, p. 1760-1761.

Figure 19. Original drawing.

Figure 20. King, P. B., 1942, Permian of west Texas and southeastern New Mexico: *Am. Assoc. Petrol. Geol. Bull.*, v. 26, fig. 18, p. 665. (Published with permission of American Association of Petroleum Geologists).

Figure 21. West Texas Geological Society, 1942, Resume of geology of the south Permian basin, west Texas: *Geol. Soc. America Bull.*, v. 53, pl. 2, p. 560.

Figure 22. King, P. B., 1948, Geology of southern Guadalupe Mountains, Texas: *U.S. Geol. Survey Prof. Paper 215*, pl. 17, sections E E′ and K K′.

Figure 23. A—Keith, Arthur, 1901, Description of the Maynardville quadrangle (Tennessee): *U.S. Geol. Survey Geol. Atlas*, folio no. 74. B—Darton, N. H., 1899, Description of the Monterey quadrangle (Virginia and West Virginia): *U.S. Geol. Survey Geol. Atlas*, folio no. 61.

Figure 24. Butts, Charles, 1927, Fensters in the Cumberland Mountain overthrust block in southwestern Virginia: *Virginia Geol. Survey Bull. 28*, pl. 2; p. 4-5; Rich, J. L., 1934, Mechanics of low-angle overthrust faulting as illustrated by Cumberland thrust block, Virginia, Kentucky, and Tennessee: *Am. Assoc. Petrol. Geol. Bull.*, v. 18, figs. 4 and 5, p. 1589; Miller, R. L., and Fuller, J. O., 1947, Geologic and structure contour maps of Rose Hill oil field, Lee County, Virginia: *U.S. Geol. Survey Oil and Gas Inves.*, map 76.

Figure 25. Rodgers, John, 1953, The folds and faults of the Appalachian Valley and Ridge province: *Kentucky Geol. Survey Spec. Publ. 1*, figs. 2 and 3, p. 160-161, 164-165; Stose, G. W., and Ljungstedt, O. A., 1931, Geologic map of Pennsylvania: *Pennsylvania Geol. Survey*; Wilson, C. W., Jr., 1956, Pennsylvanian geology of the Cumberland Plateau: *Tennessee Div. Geol. folio*: pl. 7-8.

Figure 26. Billings, M. P., 1955, Geologic map of New Hampshire: *New Hampshire Planning and Development Commission*; White, W. S., and Jahns, R. H., 1950, Structure of central and east-central Vermont: *Jour. Geology*, v. 58, fig. 2, p. 182.

Figure 27. Keith, Arthur, 1907, Description of the Pisgah quadrangle (North Carolina and South Carolina): *U.S. Geol. Survey Geol. Atlas*, folio no. 147; Keith, Arthur, Geologic map of Cowee quadrangle, unpublished.

Figure 28. A—Barrell, Joseph, 1915, Central Connecticut in geologic past: *Connecticut Geol. Survey Bull. 23*. B—Longwell, C. R., 1933, The Triassic belt of Massachusetts and

Connecticut, *in* Eastern New York and western New England: *16th Int. Geol. Cong. Guidebook 1*, fig. 23, p. 102.

Figure 29. Stose, A. J., and Stose, G. W., 1944, Geology of the Hanover-York district, Pennsylvania: *U.S. Geol. Survey Prof. Paper 204*, pl. 16, p. 57.

Figure 30. A—Billings, M. P., Thompson, J. B., Jr., and Rodgers, John, 1952, Geologic map of east-central New York, southern Vermont, and southern New Hampshire: *Geol. Soc. America, Guidebook for field trips in New England (Boston meeting)*. B—King, P. B., 1955, Geologic map of parts of North and South Carolina and adjacent states: *Geol. Soc. America, Guides to southeastern Geology (New Orleans meeting)*.

Figure 31. A—Ewing, Maurice, and others, 1950, Geophysical investigations in the emerged and submerged Atlantic Coastal Plain, part V, Woods Hole, New York, and Cape May sections: *Geol. Soc. America Bull.*, v. 61, fig. 1, p. 884. B—Ewing, Maurice, and others, 1937, Geophysical investigations in the emerged and submerged Atlantic Coastal Plain, part 1, Methods and results: *Geol. Soc. America Bull.*, v. 48, fig. 32, p. 790.

Figure 32. Original drawing.

Figure 33. Moore, R. C., 1941, Stratigraphy: *Geol. Soc. America 50th Anniversary Volume*, fig. 7, p. 200 (based on work of G. H. Chadwick and G. A. Cooper).

Figure 34. Original drawing.

Figure 35. Kay, Marshall, 1951, North American geosynclines: *Geol. Soc. America Mem. 48*, pl. 9, p. 26.

Figure 36. Original drawing.

Figure 37. King, P. B., 1937, Geology of the Marathon region, Texas: *U.S. Geol. Survey Prof. Paper 187*, pl. 15, p. 118.

Figure 38. King, P. B., 1937, Geology of the Marathon region, Texas: *U.S. Geol. Survey Prof. Paper 187*, pl. 1 B, p. 6.

Figure 39. A and B—Hendricks, T. A., and others, 1947, Geology of western part of Ouachita Mountains of Oklahoma: *U.S. Geol. Survey Oil and Gas Inves.*, map 66, sheet 3, sections A A′ and C C′. C—Purdue, A. H., and Miser, H. D., 1923, Description of Hot Springs district, Arkansas: *U.S. Geol. Survey Geol. Atlas*, folio no. 215, fig. 6, p. 6.

Figure 40. Miser, H. D., 1954, Geologic map of Oklahoma: *U.S. Geol. Survey*; Hendricks, T. A., and others, 1947, Geology of western part of Ouachita Mountains of Oklahoma: *U.S. Geol. Survey Oil and Gas Inves.*, map 66; Ham, W. E., and McKinley, Myron, 1955, Geologic map and sections of the Arbuckle Mountains, Oklahoma, *in* Field conference on geology of the Arbuckle Mountains region: *Oklahoma Geol. Survey Guidebook 3*.

Figure 41. A—Tomlinson, C. W., 1929, The Pennsylvanian system in the Ardmore basin: *Oklahoma Geol. Survey Bull. 46*, section A A′, pl. 17. B—Dott, R. H., 1934, Overthrusting in Arbuckle Mountains, Oklahoma: *Am. Assoc. Petrol. Geol. Bull.*, v. 18, section A A′, fig. 11, p. 601; Ham, W. E., and McKinley, Myron, 1955, Geologic map and sections of the Arbuckle Mountains, Oklahoma; *in* Field conference on geology of the Arbuckle Mountains region: *Oklahoma Geol. Survey Guidebook 3*, section B B′. C—Dott, *op.cit.*, section B B′, fig. 11, p. 601; Ham and McKinley, *op.cit.*, section D D′.

Figure 42. A—Swesnik, R. M., and Green, T. A., 1950, Geology of the Eola area, Garvin County, Oklahoma: *Am. Assoc. Petrol. Geol. Bull.*, v. 34, fig. 7, p. 2188. B—Selk, E. L., 1951, Types of oil and gas traps in southern Oklahoma: *Am. Assoc. Petrol. Geol. Bull.*, v. 35, fig. 16, p. 602. (Published with permission of American Association of Petroleum Geologists).

Figure 43. Hiestand, T. C., 1935, Regional investigations, Oklahoma and Kansas: *Am. Assoc. Petrol. Geol. Bull.*, v. 19, fig. 9, p. 963; Tulsa Geological Society, 1941, Possible future oil provinces of northern mid-continent region: *Am. Assoc. Petrol. Geol. Bull.*, v. 25, fig. 8, p. 151; Mallory, W. W., 1948, Pennsylvanian stratigraphy and structure, Velma pool, Stephens County, Oklahoma: *Am. Assoc. Petrol. Geol. Bull.*, v. 32, fig. 2, p. 1954-1955.

Figure 44. Fisk, H. N., and McFarlan, E., Jr., 1955, Late Quaternary deltaic deposits of the Mississippi River: *Geol. Soc. America Spec. Paper 62*, fig. 7, p. 288.

Figure 45. Original drawing.

Figure 46. Imlay, R. W., 1940, Lower Cretaceous and Jurassic formations of southern Arkansas and their oil and gas possibilities: *Arkansas Geol. Survey Inf. Circular 12*, section E E′, p. 3; Hazzard, R. T., Spooner, W. C., and Blanpied, B. W., 1947, Notes on the stratigraphy of formations which underlie the Smackover limestone in south Arkansas, northeast Texas, and north Louisiana, *in* Reference report on certain oil and gas fields of north Louisiana, south Arkansas, Mississippi, and Alabama: *Shreveport Geol. Soc.*, v. 2, p. 483-503.

Figure 47. Carsey, J. B., 1950, Geology of Gulf Coastal area and continental shelf: *Am. Assoc. Petrol. Geol. Bull.*, v. 34, fig. 5, p. 367. (Published with permission of American Association of Petroleum Geologists).

Figure 48. Lowman, S. W., 1949, Sedimentary facies in Gulf Coast: *Am. Assoc. Petrol. Geol. Bull.*, v. 33, fig. 23, p. 1972. (Published with permission of American Association of Petroleum Geologists).

Figure 49. A 1 and A 2—Fisk, H. N., 1944, Geological investigation of the alluvial valley of the lower Mississippi River: *Mississippi River Commission*, pl. 33, p. 64. B—Compiled from various sources, including Thompson, W. C., 1937, Geologic sections in Texas and adjoining states: *Am. Assoc. Petrol. Geol. Bull.*, v. 21, pl. B, p. 1084; Storm, L. W., 1945, Résumé of facts and opinions on sedimentation in Gulf Coast region of Texas and Louisiana: *Am. Assoc. Petrol. Geol. Bull.*, v. 29, fig. 8, p. 1329; Weaver, Paul, 1950, Variations in history of continental shelves: *Am. Assoc. Petrol. Geol. Bull.*, v. 34, fig. 3, p. 356.

Figure 50. Original drawing.

Figure 51. Raitt, R. W., Fisher, R. I., and Mason, R. G., 1955, Tonga Trench: *Geol. Soc. America Spec. Paper 62*, fig. 9, p. 253.

Figure 52. Senn, Alfred, 1940, Paleogene of Barbados and its bearing on history and structure of the Antillean-Caribbean region: *Am. Assoc. Petrol. Geol. Bull.*, v. 24, fig. 3, p. 1596. (Published with permission of American Association of Petroleum Geologists).

Figure 53. Original drawing.

Figure 54. Burbank, W. S., and Lovering, T. S., 1933, Relation of stratigraphy, structure, and igneous activity to ore deposits of Colorado and southern Wyoming; *in* Ore deposits of the western states (Lindgren volume): *Am. Inst. Min. Met. Eng.*, fig. 11, p. 274; Committee on Tectonics, National Research Council, 1944, Tectonic map of United States: *Am. Assoc. Petrol. Geol.*; also state geologic maps.

Figure 55. McKee, E. D., 1931, *Ancient landscapes of the Grand Canyon region*: fig. on p. 8. (Published with permission of E. D. McKee).

Figure 56. Darton, N. H., 1925, A résumé of Arizona geology: *Arizona Bur. Mines Bull. 119*, fig. 13, p. 182.

Figure 57. Burbank, W. S. and Lovering, T. S. 1933, Relation of stratigraphy, structure, and igneous activity to ore deposits of Colorado and southern Wyoming, *in* Ore deposits of the western states (Lindgren volume): *Am. Inst. Min. Met. Eng.*, figs. 13 and 14 p. 280 and 282; Read, C. B., and Wood, G. H., Jr., 1947, Distribution and correlation of Pennsylvanian rocks in late Paleozoic sedimentary basins of northern New Mexico: *Jour. Geology*, v. 55, fig. 2, p. 226; Brill, K. C., Jr., 1952, Stratigraphy in the Permo-Pennsylvanian zeugogeosyncline of Colorado and northern New Mexico: *Geol. Soc. America Bull.*, v. 63, pl. 3, p. 818.

Figure 58. Original drawing.

Figure 59. Darton, N. H., and others, 1924, Geologic map of state of Arizona: *Arizona Bur. Mines*; Andrews, D. A., and Hunt, C. B., 1948, Geologic map of eastern and southeastern Utah: *U.S. Geol. Survey Oil and Gas Inves.*, map 70.

Figure 60. Gregory, H. E., 1950, Geology and geography of the Zion Park region, Utah and Arizona: *U.S. Geol. Survey Prof. Paper 220*, pl. 5.

Figure 61. Original drawing.

Figure 62. King P. B., 1955, Orogeny and epeirogeny through time; *Geol. Soc. America Spec. Paper 62*, fig. 4, p. 731; based on unpublished compilation of Erling Dorf, and Cobban, W. A., Reeside, J. B., Jr., 1952, Correlation of the Cretaceous formations of the western interior of the United States: *Geol. Soc. America Bull.*, v. 63, p. 1011-1044.

Figure 63. Original drawing.

Figure 64. Powell, J. W., 1876, Report on the geology of the eastern portion of the Uinta Mountains and a region of country adjacent thereto: *U.S. Geol. Geograph. Survey of the Territories, 2nd Div.*, pl. 4 of atlas.

Figure 65. Original drawing.

Figure 66. Stose, G. W., 1946, Geologic map of North America: *Geol. Soc. America.* King, P. B., 1958, Evolution of modern surface features of western North America, *in* Zoogeography: *Am. Assoc. Advancement of Science Mem.*, fig. 7. (Published with permission of American Association for the Advancement of Science).

Figure 67. Weed, W. H., 1899, Description of the Little Belt Mountains quadrangle: *U.S. Geol. Survey Geol. Atlas*, folio no. 56, structure section sheet; Ross, C. P., Andrews, D. A., and Witkind, I. J., 1955, Geologic map of Montana: *U.S. Geol. Survey.*

Figure 68. Original drawing.

Figure 69. Burbank, W. S., and Lovering, T. S., 1933, Relation of stratigraphy, structure, and igneous activity to ore deposits of Colorado and Southern Wyoming: *Am. Inst. Min. Met. Eng.*, Ore deposits of the western states (Lindgren volume), fig. 15, p. 284; Committee on Tectonics, National Research Council, 1944, Tectonic map of United States; *Am. Assoc. Petrol. Geol.*

Figure 70. A—Lovering, T. S., 1935, Geology and ore deposits of the Montezuma quadrangle, Colorado: *U.S. Geol. Survey Prof. Paper 178*, pl. 4, section B B′; Van Tuyl, F. M., and McLaren, R. L., 1933, Geologic map of Golden area, *in* Colorado: *16th Int. Geol. Cong. Guidebook 19*, pl. 15, p. 138. B—Burbank, W. S., and Goddard, E. N., 1937, Thrusting in Huerfano Park, Colorado, and related problems of orogeny in the Sangre de Cristo Mountains: *Geol. Soc. America Bull.*, v. 48, fig. 4, p. 968.

Figure 71. Burbank, W. S., and Lovering, T. S., 1933, Relation of stratigraphy, structure, and igneous activity to ore deposits of southern Wyoming: *Am. Inst. Min. Met. Eng.*, Ore deposits of western states (Lindgren volume), fig. 16, p. 289; Committee on Tectonics, National Research Council, 1944, Tectonic map of United States: *Am. Assoc. Petrol. Geol.*

Figure 72. A—Wood, G. H., and Northrop, S. A., 1946, Geology of Nacimiento Mountains, San Pedro Mountains, and adjacent plateaus of Sandoval and Rio Arriba Counties, New Mexico: *U.S. Geol. Survey Oil and Gas Inves.*, map 57, section E E′. B—Read, C. B., and others, 1944, Geologic map and stratigraphic sections of Permian and Pennsylvanian rocks of parts of San Miguel, Santa Fe, Sandoval, Bernalillo, and Valencia Counties, north-central New Mexico: *U.S. Geol. Survey Oil and Gas Inves.*, map 21, section A A′. C—Kelley, V. C., and Wood, G. H., 1946, Lucero uplift, Valencia, Socorro, and Bernalillo Counties, New Mexico: *U.S. Geol. Survey Oil and Gas Inves.*, map 47, section D D′. D—Denny, C. S., 1940, Tertiary geology of the San Acacio area, New Mexico: *Jour. Geology*, v. 48, fig. 3, section A A′, p. 82. E—Wilpolt, R. H., and Wanek, A. A., 1951, Geology of the region from Socorro and San Antonio east to Chupadera Mesa, Socorro County, New Mexico: *U.S. Geol. Survey Oil and Gas Inves.*, map 121, sheet 2, section F F′. F—King, P. B., 1948, Geology of the southern Guadalupe Mountains, Texas: *U.S. Geol. Survey Prof. Paper 215*, pl. 3, section B B′, and fig. 18, p. 122.

Figure 73. Original drawing.

Figure 74. Powell, J. W., 1876, Report on the geology of the eastern portion of the Uinta Mountains and a region of country adjacent thereto: *U.S. Geol. Geograph. Survey of the Territories, 2nd Div.*, fig. 3, p. 15.

Figure 75. From many sources including: Okulitch, V. J., 1949, Geology of parts of the Selkirk Mountains in the vicinity of the main line of the Canadian Pacific Railway, British Columbia: *Canada Geol. Survey Bull. 14*, fig. 4, p. 18; Okulitch, V. J., 1956, The Lower Cambrian of western Canada and Alaska, *in* El sistema Cámbrico, su paleogeographía ye el problema de su base: *20th Int. Geol. Cong. (Mexico)*, fig. 4, p. 270.

Figure 76. Warren, P. S., 1927, Banff area, Alberta: *Canada Geol. Survey Mem. 153*, fig. 1, p. 4.

Figure 77. Based on Longwell, C. R., 1950, Tectonic theory viewed from the Basin Ranges: *Geol. Soc. America Bull.*, v. 61, pl. 1, p. 413; and other sources.

Figure 78. A—North, F. K., and Henderson, G. G. L., 1954, Summary of the geology of the southern Rocky Mountains of Canada: *Alberta Soc. Petrol. Geol., 4th Ann. Field Conf. Guidebook (Banff-Golden-Radium)*, section A A′; B—*op.cit.*, section B B′; C—Clark, L. M., 1954, Cross-section through the Clarke Range of the Rocky Mountains of southern Alberta and southern British Columbia: *Alberta Soc. Petrol. Geol., 4th Ann. Field Conf. Guidebook (Banff-Golden-Radium)*, section E E′, p. 106. D—Daly, R. A., 1912, Geology of the North American Cordillera at the forty-ninth parallel: *Canada Geol. Survey Mem. 38*, sheets 1-3; Link, T. A., 1935, Types of foothills structures of Alberta, Canada: *Am. Assoc. Petrol. Geol. Bull.*, v. 19, fig. 2, p. 1431, fig. 32, p. 1465. (Section D published with permission of American Association of Petroleum Geologists).

Figure 79. Gallup, W. B., 1951, Geology of the Turner Valley oil and gas field, Alberta, Canada: *Am. Assoc. Petrol. Geol. Bull.*, v. 35, fig. 5, p. 806; Hume, G. S., 1957, Fault structures in the foothills and eastern Rocky Mountains of southern Alberta: *Geol. Soc. America Bull.*, v. 68, fig. 2, p. 398. (Published with permission of American Association of Petroleum Geologists).

Figure 80. Willis, Bailey, 1902, Stratigraphy and structure, Lewis and Livingston Ranges, Montana: *Geol. Soc. America Bull.*, v. 13, fig. 5, p. 334.

Figure 81. Original drawing.

Figure 82. A—Lindgren, Waldemer, 1911, The Tertiary gravels of the Sierra Nevada, California: *U.S. Geol. Survey Prof. Paper 73*, section A B, pl. 9; Lindgren, Waldemer, and Turner, H. W., 1895, Description of the Smartsville sheet: *U.S. Geol. Survey Geol. Atlas*, folio 18, section B B′; Lindgren, Waldemer, 1900, Description of the Colfax quadrangle: *U.S. Geol. Survey Geol. Atlas*, folio 66, section A A′; Lindgren, Waldemer, 1897, Description of the Truckee quadrangle: *U.S. Geol. Survey Geol. Atlas*, folio 39, section A A′. B—Lindgren, *op.cit.*, 1911, section C D, pl. 9; Lindgren, Waldemer, 1894, Description of the Sacramento sheet: *U.S. Geol. Survey Geol. Atlas*, folio 5, section A A′; Lindgren, Waldemer, and Turner, H. W., 1894, Description of the Placerville sheet: *U.S. Geol. Survey Geol. Atlas*, folio 3, section A A′; Lindgren, Waldemer, 1896, Description of the Pyramid Peak quadrangle: *U.S. Geol. Survey Geol. Atlas*, folio 31, section A A′. C—Turner, H. W., and Ransome, F. L., 1897, Description of the Sonora quadrangle: *U.S. Geol. Survey Geol. Atlas*, folio 41, section B B′; Cloos, Ernst, 1936, Die Sierra-Nevada-Pluton in Californien: *Neues Jahrb.*, Beilage-Band 76, Heft 3, Abt. B, Taf. 18; Calkins, F. C., 1930, The granitic rocks of the Yosemite region, *in* Geologic history of Yosemite Valley: *U.S. Geol. Survey Prof. Paper 160*, pl. 51; Putnam, W. C., 1949, Quaternary geology of the June Lake district, California: *Geol. Soc. America Bull.*, v. 60, pl. 1, p. 1281.

Figure 83. A—Anderson, F. M., 1938, Lower Cretaceous deposits in California and Oregon: *Geol. Soc. America Spec.*

Paper 16, table 1, p. 38; Taliaferro, N. L., 1942, Geologic history and correlation of the Jurassic of southeastern Oregon and California: *Geol. Soc. America Bull.*, v. 53, p. 71-112. B—McKee, E. D., and others, 1956, Paleotectonic maps of the Jurassic system: *U.S. Geol. Survey Misc. Geol. Inves.*, map 1-175, table 2, p. 5; Irwin, W. P., 1957, Franciscan group in Coast Ranges and its equivalents in Sacramento Valley, California: *Am. Assoc. Petrol. Geol. Bull.*, v. 41, p. 2284-2297.

Figure 84. A—Ferguson, H. G., Roberts, R. J., and Muller, S. W., 1952, Geology of the Golconda quadrangle, Nevada: *U.S. Geol. Survey Geol. Quadrangle May 15*, section A A′. B—Ferguson, H. G., Muller, S. W., and Roberts, R. J., 1951, Geology of the Winnemucca quadrangle, Nevada: *U.S. Geol. Survey Geol. Quadrangle Map 11*, section D D′; Muller, S. W., 1949, Sedimentary facies and geologic structures in the Basin and Range province: *Geol. Soc. America Mem. 39*, fig. 3, section A, p. 52. C—Roberts, R. J., 1951, Geology of the Antler Peak quadrangle, Nevada: *U.S. Geol. Survey Geol. Quadrangle Map 10*, section C C′.

Figure 85. Original drawing.

Figure 86. Davis, W. M., 1927, lecture sketch.

Figure 87. Original drawing.

Figure 88. Davis, W. M., 1925, The Basin Range problem: *Natl. Acad. Sci. Proc.*, v. 11, fig. 2, p 392.

Figure 89. King, P. B., 1958, Evolution of modern surface features of western North America, *in* Zoogeography: *Am. Assoc. Advancement of Science Mem.*, fig. 9. (Published with permission of American Association for the Advancement of Science).

Figure 90. A—Smith, G. O., 1904, Description of the Mount Stuart quadrangle: *U.S. Geol. Survey Geol. Atlas*, folio 106, section A A′. B—Smith, G. O., and Calkins, F. C., 1906, Description of the Snoqualmie quadrangle; *U.S. Geol. Survey Geol. Atlas*, folio 134, section B B′.

Figure 91. Committee on Tectonics, National Research Council, 1944, Tectonic map of United States: *Am. Assoc. Petrol. Geol.*; U.S. Geological Survey, 1955, State of California (1:1,000,000, with shaded relief).

Figure 92. Reed, R. D., 1933, Geology of California: *Am. Assoc. Petrol. Geol.*, fig. 16, p. 73; Reed, R. D., and Hollister, J. S., 1936, Structural evolution of southern California: *Am. Assoc. Petrol. Geol. Bull.*, v. 20, fig. 2, p. 1549. King, P. B., 1958, Evolution of modern surface features of western North America, *in* Zoogeography: *Am. Assoc. Advancement of Science Mem.*, fig. 10. (Published with permission of American Association for the Advancement of Science).

Figure 93. Noble, L. F. 1954, The San Andreas fault zone from Soledad Pass to Cajon Pass, California, *in* The geology of southern California: *California Div. Mines Bull. 170*, chap. 4, fig. 5, p. 42 (drawn by P. B. King).

Figure 94. Data from Noble, L. F., 1954, *op.cit.*, p. 37-48.

Figure 95. Original drawing.

End piece and Jacket. Lawson, A. C., 1915, Description of the San Francisco district: *U.S. Geol. Survey Geol. Atlas*, folio 193, pl. 1.

GENERAL INDEX

INDEX OF AUTHORS CITED

THE LIBRARY OF CONGRESS HAS CATALOGUED THIS BOOK AS FOLLOWS:

King, Philip Burke, 1903—The evolution of North America. Princeton, Princeton University Press, 1959. 189 p. illus. 29 cm. Includes bibliography. I. Geology—North America. I. Title. QE71.K54 (557) 59-5598 ‡ Library of Congress.

PBK